REA's Test Prep Books Are The Best!
(a sample of the hundreds of letters REA receives each year)

" A guaranteed "5" on the AP Biology. The book's strength is its six full-length exams with multiple-choice questions. "
AP Biology Student, Stanford University, Stanford, CA

" REA's AP Biology test prep is a good reference source with good tests. It has an excellent review section and the tests are complete. "
AP Biology Student, Detroit, MI

" Your book was such a better value and was so much more complete than anything your competition has produced — and I have them all! "
Teacher, Virginia Beach, VA

" Compared to the other books that my fellow students had, your book was the most useful in helping me get a great score. "
Student, North Hollywood, CA

" Your book was responsible for my success on the exam, which helped me get into the college of my choice... I will look for REA the next time I need help. "
Student, Chesterfield, MO

" Just a short note to say thanks for the great support your book gave me in helping me pass the test... I'm on my way to a B.S. degree because of you! "
Student, Orlando, FL

(more on next page)

(continued from front page)

" I just wanted to thank you for helping me get a great score on the AP U.S. History exam... Thank you for making great test preps! "
Student, Los Angeles, CA

" Your *Fundamentals of Engineering Exam* book was the absolute best preparation I could have had for the exam, and it is one of the major reasons I did so well and passed the FE on my first try. "
Student, Sweetwater, TN

" I used your book to prepare for the test and found that the advice and the sample tests were highly relevant... Without using any other material, I earned very high scores and will be going to the graduate school of my choice. "
Student, New Orleans, LA

" What I found in your book was a wealth of information sufficient to shore up my basic skills in math and verbal... The practice tests were challenging and the answer explanations most helpful. It certainly is the *Best Test Prep for the GRE!* "
Student, Pullman, WA

" I really appreciate the help from your excellent book. Please keep up the great work. "
Student, Albuquerque, NM

" I am writing to thank you for your test preparation... your book helped me immeasurably and I have nothing but praise for your *GRE* preparation."
Student, Benton Harbor, MI

REA: THE TEST PREP AP TEACHERS RECOMMEND

AP ENVIRONMENTAL SCIENCE

2nd Edition

TestWare® Edition

Kevin R. Reel
Head of School
The Colorado Springs School
Colorado Springs, Colorado

Research & Education Association
Visit our website at: www.rea.com

Planet Friendly Publishing
✓ Made in the United States
✓ Printed on Recycled Paper
  Text: 10%   Cover: 10%
Learn more: www.greenedition.org

GREEN EDITION

At REA we're committed to producing books in an Earth-friendly manner and to helping our customers make greener choices.

Manufacturing books in the United States ensures compliance with strict environmental laws and eliminates the need for international freight shipping, a major contributor to global air pollution.

And printing on recycled paper helps minimize our consumption of trees, water and fossil fuels. This book was printed on paper made with **10% post-consumer waste**. According to Environmental Defense's Paper Calculator, by using this innovative paper instead of conventional papers, we achieved the following environmental benefits:

Trees Saved: 5 • Air Emissions Eliminated: 1,895 pounds
Water Saved: 965 gallons • Solid Waste Eliminated: 116 pounds

For more information on our environmental practices, please visit us online at **www.rea.com/green**

Research & Education Association
61 Ethel Road West
Piscataway, New Jersey 08854
E-mail: info@rea.com

AP ENVIRONMENTAL SCIENCE
with TestWare®

Copyright © 2010 by Research & Education Association, Inc. Prior editions copyright © 2006, 2005 by Research & Education Association, Inc. All rights reserved. No part of this book may be reproduced in any form without permission of the publisher.

Printed in the United States of America

Library of Congress Control Number 2009937489

ISBN-13: 978-0-7386-0788-7
ISBN-10: 0-7386-0788-6

Windows® is a registered trademark of Microsoft Corporation.

REA® and TestWare® are registered trademarks of Research & Education Association, Inc.

ABOUT OUR AUTHOR

Kevin R. Reel has been in science education for over 25 years. He has written numerous articles for academic journals and has contributed to many textbooks in chemistry, health, and environmental science. He earned his B.A. and M.S. from Stanford University, and is currently the Head of School at the Colorado Springs School.

ABOUT RESEARCH & EDUCATION ASSOCIATION

Founded in 1959, Research & Education Association is dedicated to publishing the finest and most effective educational materials—including software, study guides, and test preps—for students in middle school, high school, college, graduate school, and beyond. Today, REA's wide-ranging catalog is a leading resource for teachers, students, and professionals. We invite you to visit us at *www.rea.com* to find out how "REA is making the world smarter."

ACKNOWLEDGMENTS

In addition to our author, we would like to thank Larry B. Kling, Vice President, Editorial, for supervising development; Pam Weston, Vice President, Publishing, for setting the quality standards for production integrity and managing the publication to completion; John Cording, Vice President, Technology, for coordinating the design and development of REA's TestWare®; Diane Goldschmidt, Senior Editor, for preflight editorial review; Heena Patel, Technology Project Manager, for design contributions and software testing efforts; Weymouth Design for designing our cover; Christine Saul, Senior Graphic Artist, for her graphic arts contributions; and Rachel DiMatteo, for typesetting the revisions.

We also would like to extend our gratitude for permission to reproduce artwork found in this book: "The Nitrogen Cycle" (pg. 20), and "The Sulfur Cycle" (pg. 21), by permission of Kenneth J. Edwards, Jr., Vice President, Aiken-Murray Corp.; "Solar Energy Flow" (pg. 31), adapted from a graphic by Dr. Charles Ophardt, Professor of Chemistry, Elmhurst College, Elmhurst, Ill. "The Rock Cycle" (pg. 38), by permission of Dylan Prentiss, Department of Geography, University of California, Santa Barbara.

CONTENTS

About Our Author v
Independent Study Schedule xi

Chapter 1

**EXCELLING ON THE AP ENVIRONMENTAL
SCIENCE EXAM** 1
    About This Book and TestWare® 3
    About the Advanced Placement Program 3
    About the AP Environmental Science Exam 4
    AP Environmental Science Exam Content 4
    About Our AP Course Review 6
    Scoring REA's Practice Exams 6
    Scoring the Official Exams 7
    Studying for Your AP Exam 8
    Test-Taking Tips 9
    The Day of the Exam 10

Chapter 2

MATTER AND ENERGY IN THE ENVIRONMENT 11
    Organization of Matter 13
    Conservation of Matter 17
    Conservation of Energy 22
    Case Summaries 24

Chapter 3

EARTH SYSTEMS AND GLOBAL CHANGES 27
    Properties That Drive Global Processes 29
    Earth 34
    Atmosphere 42
    Case Summaries 49

Chapter 4
THE BIOSPHERE 53
Productivity and Food Chains 55
Population Interactions 56
Traits of Biological Communities 60
Biomes 61
Biodiversity 68
Case Summaries 74

Chapter 5
POPULATION DYNAMICS 77
Fundamentals of Population Biology 79
Human Populations 87
Case Summary 94

Chapter 6
HUMAN HEALTH 97
Morbidity and Mortality 99
Environmental Health Hazards 99
Identifying Health Risks 103
Factors That Affect Toxicity 105
Case Summaries 112

Chapter 7
ATMOSPHERE RESOURCES 117
Composition of the Atmosphere 119
Air Pollution: Sources and Solutions 120
Effects of Air Pollution 129
Case Summaries 133

Chapter 8
WATER RESOURCES 135
Water Compartments and the Water Cycle 137
Trends in Water Use 139
Water Pollution 143
Wastewater Treatment 147
Case Summaries 150

Chapter 9
LAND RESOURCES 153
    Principles of Land Use.. 155
    Management of Specific Land Uses..................................... 158
    Case Summaries.. 182

Chapter 10
ENERGY RESOURCES 185
    Understanding Energy... 187
    Non-Renewable Energy Sources.. 192
    Renewable Energy Sources.. 198
    Patterns of Human Energy Use.. 202
    Case Summaries.. 204

Chapter 11
ENVIRONMENT AND SOCIETY 209
    Economic Forces... 211
    Environmental Ethics... 218
    Nonmarket Values 219
    Environmental History, Laws, and Regulations 222
    Environmental Laws 226
    Environmental Policy and Management............................... 230
    Case Summaries 235

PRACTICE EXAMS

PRACTICE EXAM 1 237
    Answer Key 266
    Detailed Explanations of Answers 267

PRACTICE EXAM 2 285
    Answer Key 312
    Detailed Explanations of Answers 313

PRACTICE EXAM 3 329
    Answer Key 357
    Detailed Explanations of Answers 358

PRACTICE EXAM 4 373
    Answer Key 397
    Detailed Explanations of Answers 398

ANSWER SHEETS 415

GLOSSARY 429

INDEX 451

Installing REA'S TestWare® 468

AP ENVIRONMENTAL SCIENCE
Independent Study Schedule

The following study schedule allows for thorough preparation for the AP Environmental Science examination. Although it is designed for six weeks, it can be condensed into a three-week course by collapsing each two-week period into a single week. Be sure to set aside at least two hours each day to study. Bear in mind that the more time you spend studying, the more prepared and relaxed you will feel on the day of the exam

Week	Activity
1	Read and study Chapter 1, which will introduce you to the AP Environmental Science Examination.
2 & 3	Carefully read and study the Environmental Science Review included in Chapters 2-11 of this book.
4	Take AP Environmental Science Practice Exams 1 and 2 on CD-ROM. After scoring your exam, carefully review all incorrect answer explanations. If there are any types of questions or particular subjects that seem difficult to you, review those subjects by referring back to the appropriate section of the Environmental Science Review. If possible, read some environmental science-related articles from science magazines (available at your local library or bookstore). This will help familiarize you with current events in science and technology related to environmental science, as well as terminology and style used by scientists (including those who write questions for the AP Environmental Science Exam).
5	Take Practice Exam 3, and after scoring your exam, carefully review all incorrect answer explanations. If there are any types of questions or particular subjects that seem difficult to you, review those subjects by referring back to the appropriate section of the Environmental Science Review.
6	Take Practice Exam 4. After scoring it, review all incorrect answer explanations. Study any areas in which you consider yourself to be weak by using the Environmental Science Review and any other reliable sources you have on hand. Review our practice tests again to be sure you understand the problems that you originally answered incorrectly.

Excelling on the AP Environmental Science Exam

Chapter 1

Excelling on the AP Environmental Science Exam

About This Book and TestWare®

This book, along with our companion TestWare® software, provides an accurate and complete representation of the Advanced Placement Examination in Environmental Science. REA's practice exams are based on the format of the most recently administered AP Environmental Science exam and each includes every type of question that you can expect to encounter on the real test. Following each of our practice tests is an answer key, complete with detailed explanations designed to clarify the material for you. By using the subject reviews, completing both practice tests, and studying the explanations that follow, you will put yourself in the best possible position to do well on the actual test.

Four practice exams are included in this book and two of the practice exams are also included on the enclosed TestWare® CD-ROM. The software provides timed conditions and instant, accurate scoring, which makes it all the easier to pinpoint your strengths and weaknesses. Use them, along with the detailed explanations of answers, to help determine your strengths and weaknesses, and to prepare you to score well on exam day.

About the Advanced Placement Program

The Advanced Placement program is designed to provide high school students with the opportunity to pursue college-level studies. The program consists of two components: an AP course and an AP exam.

Students are expected to gain college-level skills and acquire college-level knowledge of environmental science through the AP course. Upon completion of the course, students take the AP exam. Test results are used to grant course credit and/or determine placement level in the subject when entering college.

AP exams are administered every May. Additional information can be requested from the following source:

AP Services
Educational Testing Service
P.O. Box 6671

AP Environmental Science

Princeton, New Jersey 08541-6671
Phone: (609) 771-7300 or (888) 225-5427
Fax: (609) 530-0482
E-mail: apexams@ets.org
Website: *www.collegeboard.com*

About the AP Environmental Science Exam

The AP Environmental Science exam is three hours long. Each section in each exam is completed separately. You will have 90 minutes to answer 100 multiple-choice questions, which are worth 60% of your final grade. Each correct answer is worth one point, and each incorrect answer takes away 1/4 point.

The free response section is 90 minutes long and has four questions. The first question will ask you to make conclusions based on a set of data, the second question will be document-based, and the third and fourth questions are synthesis and evaluation questions. All four questions are weighted equally.

AP Environmental Science Exam Content

I. Interdependence of Earth's Systems: Fundamental Principles and Concepts (25%)
   A. The Flow of Energy
      1. forms and quality of energy
      2. energy units and measurements
      3. sources and sinks, conversions
   B. The Cycling of Matter
      1. water
      2. carbon
      3. major nutrients
         a. nitrogen
         b. phosphorus
      4. differences between cycling of major and trace elements
   C. The Solid Earth
      1. Earth history and the geologic time scale
      2. Earth dynamics: plate tectonics, volcanism, the rock cycle, soil formation
   D. The Atmosphere
      1. atmospheric history: origin, evolution, composition, and structure
      2. atmospheric dynamics: weather, climate
   E. The Biosphere
      1. organisms: adaptations to their environments
      2. populations and communities: exponential growth, carrying capacity
      3. ecosystems and change: biomass, energy transfer, succession
      4. evolution of life: natural selection, extinction
II. Human Population Dynamics (10%)

A. History and Global Distribution
   1. numbers
   2. demographics, such as birth and death rates
   3. patterns of resource utilization
B. Carrying Capacity— Local, Regional, Global
C. Cultural and Economic Influences

III. Renewable and Nonrenewable Resources: Distribution, Ownership, Use, Degradation (15%)
A. Water
   1. fresh: agricultural, industrial, domestic
   2. oceans: fisheries, industrial
B. Minerals
C. Soils
   1. soil types
   2. erosion and conservation
D. Biological
   1. natural areas
   2. genetic diversity
   3. food and other agricultural products
E. Energy
   1. conventional sources
   2. alternative sources
F. Land
   1. residential and commercial
   2. agricultural and forestry
   3. recreational and wilderness

IV. Environmental Quality (20–25%)
A. Air/Water/Soil
   1. major pollutants
      a. types, such as SO_2, NO_2, and pesticides
      b. thermal pollution
      c. measurement and units of measure such as ppm, pH, pg/L
      d. point and nonpoint sources (domestic, industrial, agricultural)
   2. effects of pollutants on:
      a. aquatic systems
      b. vegetation
      c. natural features, buildings and structures
      d. wildlife
   3. pollution reduction, remediation, and control
C. Impact on Human Health
   1. agents: chemical and biological
   2. effects: acute and chronic, dose-response relationships
   3. relative risks: evaluation and response

V. Global Changes and their Consequences (15–20%)
   A. First-order Effects (changes)
      1. atmosphere: CO_2, CH_4, stratospheric O_3
      2. oceans: surface temperatures, currents
      3. biota: habitat destruction, introduced exotics, overharvesting
   B. Higher-order Interactions (consequences)
      1. atmosphere: global warming, increasing ultraviolet radiation
      2. oceans: increasing sea level, long-term climate change, impact on El Niño
      3. biota: loss of biodiversity
VI. Environment and Society: Trade-Offs and Decision Making (10%)
   A. Economic Forces
      1. cost-benefit analysis
      2. marginal costs
      3. ownership and externalized costs
   B. Cultural and Aesthetic Considerations
   C. Environmental Ethics
   D. Environmental Laws and Regulations (International, National, and Regional)
   E. Issues and options (conservation, preservation, restoration, remediation, sustainability, mitigation)

About Our AP Course Review

As mentioned earlier, this review is designed to prepare you for success on the AP Exam. Therefore, an entire year's work has been distilled into the leanest preparation manual possible to ensure victory on exam day. This text is aimed at students serious about improving their likelihood of success through hard work and attention to the key elements to be tested. This text will also help a student prepare for daily classroom success as well. Students have a variety of learning styles that are not always met by classroom teachers, so this text will serve well as a supplement to daily classroom learning. Components of this review have been field-tested in the classroom by students of varying capacity, and all can attest to their improved performance both in the classroom, as well as on the exam.

Scoring REA's Practice Exams

Scoring the Multiple-Choice Sections

For each multiple-choice section, use this formula to calculate your raw score:

$$\underset{\substack{\text{Number Correct}\\\text{(out of 100)}}}{\underline{}} - (\underset{\text{Number Wrong}}{\underline{}} \times 1/4) = \underset{\substack{\text{Multiple-Choice}\\\text{Score (weighted)}}}{\underline{}}$$

Scoring the Free-Response Sections

For the free-response section, use this formula to calculate your raw score:

Question 1 _____ × $1\frac{1}{2}$ = _____
            (out of 10)   (weighted)

Question 2 _____ × $1\frac{1}{2}$ = _____
            (out of 10)   (weighted)

Question 3 _____ × $1\frac{1}{2}$ = _____
            (out of 10)   (weighted)

Question 4 _____ × $1\frac{1}{2}$ = _____
            (out of 10)   (weighted)

_____ + _____ + _____ + _____ = _____
Question 1   Question 2   Question 3   Question 4   Free-Response
(weighted)   (weighted)   (weighted)   (weighted)   Score (weighted)

The Composite Score

To obtain your composite score for each exam, use the following method.

_____ + _____ = _____
Multiple-Choice   Free-Response   Composite Score
Weighted Score   Weighted Score

Use the following chart to approximate your AP score for each exam:

AP Grade Conversion Chart

*Final Score Range	AP Grade
120 – 160	5
94 – 119	4
68 – 93	3
38 – 67	2
0 – 37	1

* Candidates' scores are weighted by a formula determined in advance each year by the Development Committee.

Scoring the Official Exams

Weighted Multiple-Choice + Weighted Free-Response = Total Composite Score

The College Board creates a formula (which changes slightly every year) to convert raw scores into composite scores grouped into broad AP grade categories. The weights for the multiple-choice sections are determined by the Chief Reader, who uses a process called equating. This process compares the current year's exam performance on selected multiple-choice questions to that of a previous year, establishing a level of achievement for the current year's group and a degree of difficulty for the current exam. This data is combined with historical trends and the reader's professional evaluation to determine the weights and tables.

The AP free-response is graded by teacher volunteers, grouped at scoring tables, and led by a chief faculty consultant. The consultant sets the grading scale that translates the raw score into the composite score. Past grading illustrations are available to teachers from the College Board, and may be ordered using the contact information given on page 3. These actual examples of student responses and a grade analysis can be of great assistance to both the student and the teacher as a learning or review tool.

Composite Score	AP Grade	Percentage of students at this grade (2007)
70 – 90	5 (extremely well qualified)	10.8
54 – 69	4 (well qualified)	23.1
45 – 53	3 (qualified)	17.9
29 – 44	2 (possibly qualified)	17.5
0 – 28	1 (no recommendation)	30.7

Total number of students: 52,416
Mean Grade: 2.66
Source: www.collegeboard.com

Studying for your AP Exam

To best utilize your study time, follow our Study Schedule, which you will find in the front of this book. The schedule is based on a six-week program, but if necessary can be condensed to three weeks by collapsing each two-week period into one week.

Use previous tests and quizzes to provide a study guide outline. Focus particular attention on questions that you got wrong. Do not repeat the same mistake.

Study with another student or in a group. Take turns asking each other questions. You will be amazed at how much you learn by playing the teacher!

Draw all the key models that you are required to be fluent in. Practice the possible movements that may result from change in the variables present in the model. For example, draw the various population curves and age distributions.

Create a study sheet with all of the key formulae. Use the text or past exams to test your ability to solve questions on the population growth rate, for example. Continue until you are error-free.

Prepare for the test over several days. Don't cram. Communicate with your teacher regarding areas that you don't understand. Give your teacher adequate time to prepare an answer.

Whenever possible, take as many practice tests as possible. Review errors with your teacher or other students. Again, past tests and quizzes are invaluable as a learning tool.

Test-taking Tips

This test has time limits. Do not dwell on any one question. For the multiple-choice sections, do not spend more than one minute on a question. Come back to it later if time permits.

Be calm. If you have prepared properly, you are competent in the subject area and the test will prove it.

Immediately write key formulae and models on the test cover. This may save you from simple errors as time constraints pressure you. During the free-response sections, use part of the reading period for this.

Answer questions of a lower degree of difficulty first. The first time through, for example, complete definitions of biomes or different kinds of biodiversity. Circle the questions that require calculations or involve complicated reasoning, like calculating energy use over time.

Use the process of elimination when you are unsure of an answer. If you are unable to narrow a multiple-choice question to two possible right answers, pass on the question. You will be penalized 1/4 point for each wrong answer, but you will not lose any points for leaving it blank. However, do not overuse this method; leaving 15 or 20 questions blank is not advisable.

Assume no knowledge on the part of the test grader. For the free-response questions, writing as though the reader is unfamiliar with the subject will force you to answer more thoroughly.

Devote an equal amount of time to each of the four free-response questions. All the free-response questions are weighted equally and are expected to take approximately an equal amount of time to answer.

Know the directions and format for each section of the exam. Familiarizing yourself with the directions and format of the different test sections will not only save you time, but will also ensure that you alleviate some uncertainty and its corresponding anxiety. For example, in the instructions to the free-response sections is a statement urging you to "emphasize the line of reasoning that generated your results; it is not enough to list the results of your analysis." This type of warning is very significant and might make a big difference in the way you approach this section.

Be prepared to draw and label any graphs necessary to a complete answering of free-response questions. Drawing the models will help you to discover the correct answer. Most of the written answers are enhanced by reference to a model. Be neat, and clearly label all elements. If possible, link models together (for instance, various age distribution charts).

Understand and answer the free-response questions employing the given structure (outline format) of the question. For the ten-point structure, the first part may be a simple reaction or explanation of the concept introduced in the question. The second part may require you to extrapolate from those facts and estimate results. The third part may require you to propose a possible solution to any problems that may come up.

THE DAY OF THE EXAM

Before the Exam

On the day of the test, you should wake up early (preferably after a good night's rest) and have a good breakfast. Make sure to dress comfortably so that you are not distracted by being too hot or too cold while taking the test. Also, plan to arrive at the test center early. This will allow you to collect your thoughts and relax before the test and will spare you the anxiety that comes with being late.

Before you leave for the test center, make sure you have your admission form, social security number, and another form of identification, which must contain a recent photograph, your name, and signature (i.e., driver's license, student identification card, or current alien registration card). You will not be allowed to take the test if you do not have proper identification. Also make sure to bring your school code, as well as several sharpened No. 2 pencils with erasers for the multiple-choice questions and black or blue pens for the free-response questions.

You may wear a watch, but only one without a beeper or an alarm. No dictionaries, textbooks, notebooks, compasses, correction fluid, highlighters, rulers, computers, cell phones, beepers, PDAs, scratch paper, listening and recording devices, briefcases, of packages will be permitted, and drinking, smoking, and eating are prohibited while taking the test.

During the Exam

Once you enter the test center, follow all of the rules and instructions given by the test supervisor. If you do not, you risk being dismissed from the test and having your scores canceled.

After the Exam

When taking the exam, you may immediately register to have your score sent to the college of your choice, or you may wait and later request to have your AP score reported to the college of your choice.

Matter & Energy in the Environment

Chapter 2

Matter and Energy in the Environment

Organization of Matter

The environment is composed of either matter or energy. **Matter** takes up space and has mass. Matter has many levels of organization based on size and whether or not it is considered living matter. At the smallest levels of organization—atoms and compounds—matter is not alive. At progressively larger levels of organization—cells, individuals of a species, populations, and communities—matter is contained in living organisms. At the largest scale of organization on the planet—ecosystems—matter is both living and nonliving. Other levels of organization extend beyond the planet—solar systems, galaxies, etc.—but they also extend beyond this course.

Nonliving Matter: Atoms, Ions, and Compounds

Atoms

Atoms are the smallest unit of elements and are listed on the periodic table. The nucleus of an atom is composed of neutral neutrons and positively charged protons. Negatively charged electrons exist outside the nucleus. All atoms are isotopes of some other atom that has the same number of protons but a different number of neutrons.

Ions

Ions are atoms or combinations of atoms that demonstrate an unbalanced electrical charge. This happens when the total number of electrons does not equal the total number of protons. If there are more electrons than protons, it is a negatively charged ion, or anion. If there are more protons than electrons, it is a positively charged ion, or cation. Monoatomic ions involve only one atom; polyatomic ions involve more than one atom. Below is a chart of the common monoatomic and polyatomic ions typically encountered in AP Environmental Science.

Table 2.1 Common Monoatomic and Polyatomic Ions

	Ion Symbol	Name of Ion
Monoatomic Ions	Cl^-	Chloride
	O^{2-}	Oxide
	S^{2-}	Sulfide
	Ca^{2+}	Calcium
	Pb^{2+}	Lead
	Hg^{2+}	Mercury
Polyatomic Ions	NO_2^-	Nitrite
	NO_3^-	Nitrate
	SO_4^{2-}	Sulfate
	CO_3^{2-}	Carbonate
	PO_4^{3-}	Phosphate

Compounds

Compounds are combinations of atoms held together by ionic or covalent bonds. Ionic compounds are formed when a negatively charged nonmetal combines with an ionic attraction to a positively charged metal, for example, calcium chloride, $CaCl_2$. Molecules are combinations of nonmetal atoms held together by covalent bonds. Organic molecules are carbon-based molecules that also contain hydrogen, and sometimes also contain oxygen, nitrogen, or sulfur.

Polar Molecules

Molecules are compounds of nonmetals covalently bound together. A polar molecule contains a separation of electrical charge within the molecule that helps it to be attracted to ions or other polar molecules. The water molecule is considered polar because the oxygen atom draws the electrons more closely to it than does the hydrogen. This causes a slight negative charge near the oxygen and a slight positive charge near the hydrogen. The polarity of water is strong enough to break apart many ionic compounds, which causes them to dissolve, or go into solution.

Acids and Bases

A single proton, or hydrogen ion (H^+), can be removed from water when some substances are dissolved in it. If a substance creates a hydrogen ion when put in water, it is called an **acid**. A **base** is a substance that creates a hydroxide ion (OH^-) when put in water. The reaction of an acid and a base produces the water molecule in the following neutralization reaction:

$$H^+ + OH^- \rightarrow H_2O$$
$$\text{Acid} \quad \text{Base} \quad \text{Water}$$

The concentration of protons or hydroxide ions in solution is depicted by **pH**. The pH value is the negative exponent of the molar concentration of hydrogen ions, as shown in the following chart. Any pH below 7 is acidic, any above 7 is basic. A pH of 7.0 is neutral.

Table 2.2 Hydrogen Ion Concentration and pH

Molar Concentration of H^+	pH
10^{-2}	2 (acidic)
10^{-5}	5 (acidic)
10^{-7}	7 (neutral)
10^{-10}	10 (basic)
10^{-14}	14 (basic)

States of Matter

Matter exists on earth in one of three states—solid, liquid, or gas. Matter needs to absorb energy and increase thermal motion of the individual atoms in order to pass from solid to liquid, or from liquid to gas. Changes of states may be depicted with an arrow, or "yield" sign, similar to chemical reactions.

$$H_2O\ (s) \quad \xrightarrow{heat} \quad H_2O\ (l) \quad \xrightarrow{heat} \quad H_2O\ (g)$$
$$\text{solid ice} \qquad\qquad \text{liquid water} \qquad\qquad \text{water vapor}$$

Living Matter: From Cells to Communities

Cells

The **cell** is the fundamental unit of living material. Cell functions follow from specific cell structures. The **nucleus** of the cell contains the blueprints of cell function encoded by the sequence of base pairs in the DNA of **chromosomes**. The DNA code is translated into an RNA code. Then it moves outside the nucleus where it directs the **protein synthesis**. Every protein in a living organism corresponds to a specific **gene**, or sequence of base pairs, on a particular chromosome.

Cell growth is accomplished through one of two processes. Sperm and egg cells undergo **meiosis**, which takes a cell with two copies of each gene (for humans, 46 chromosomes in all) and creates four cells, each with one copy of each gene (23 chromosomes in all for humans). One of these resulting cells (the sperm, for example) will combine during fertilization with another cell with a single copy of genetic material (an egg) to form a complete cell with two sets of each gene. After fertilization, cell division is accomplished through **mitosis**, which duplicates a single cell with two copies of each gene to form a second cell with identical genetic material. In a cancerous cell, mitosis is uncontrolled and rampant cell growth creates a tumor or prolific cells. (See Chapter Six to better understand how cancer begins.)

Cell processes require energy. Specific structures within cells perform the necessary chemical reactions to convert either solar or chemical energy into a form of energy that can be used by the cell. In plants, **chloroplasts** convert solar energy, carbon dioxide, and water into glucose and oxygen gas in the following reaction. This process is called **photosynthesis**.

$$6\,CO_2 + 6\,H_2O \rightarrow C_6H_{12}O_6 + 6\,O_2$$

In all cells, **mitochondria** convert glucose and oxygen into carbon dioxide and water in a process that also manufactures adenosine triphosphate (ATP). The ATP produced in this process, which is called **respiration**, is used as a currency for energy use throughout the cell.

$$C_6H_{12}O_6 + 6\,O_2 \rightarrow 6\,CO_2 + 6\,H_2O$$

Tissues and Organ Systems

Different types of cells can be found in different areas of the body; the structure of each cell is related to its function. Groups of cells that come together for like function are called **tissues**. **Organ systems** are groups of tissues that come together to perform a specific function. For example, a nerve cell is a long slender cell that allows for the propagation of chemical and electrical signals. Several nerve cells are bundled together to form neural tissue, and the accumulation of neural tissue that makes up the "command and control" function of our body is called the **nervous system**.

Species

A **species** is a group of organisms with similar enough genetic makeup (and consequently similar tissue and organ structure) to be able to reproduce and produce fertile offspring. Living organisms are classified into increasingly specific groups (kingdom, phylum, class, order, family, genus, species). The two most specific groupings (genus, species) are used to describe the scientific name of each species. The genus name is capitalized and sometimes abbreviated, while the species name is not; both are italicized (e.g., *Homo sapiens*, or *H. sapiens*).

Population

A **population** is an interbreeding group of organisms of the same species that lives in the same general area at the same time. (Population dynamics and growth are discussed in detail in Chapter Five.)

Community

A **community** is a group of interdependent populations whose niches overlap in some way—usually by geographical location. (The interactions between different populations within a community are discussed in more detail in Chapter Five.)

Ecosystems

Ecosystems are composed of biological communities and their physical surroundings. Ecosystems include abiotic factors (nonliving) such as pH, temperature, sunlight, moisture, and nutrients; as well as biotic factors (living) such as predators, prey, and wastes. Ecosystems can be as large as a continent (Antarctica, for example) or as small as the inside of a human mouth. Major earth ecosystems are sometimes called **biomes**, and can be separated into either terrestrial ecosystems or aquatic ecosystems. Aquatic ecosystems can be further divided into marine (salt water) ecosystems and freshwater ecosystems. (The major ecosystems are surveyed more thoroughly in Chapter Four.)

Conservation of Matter

Chemical Reactions

Changes in the arrangement of atoms and molecules are considered chemical reactions. Chemical reactions are the objective manifestation of the **Law of Conservation of Mass**, which states that matter cannot be created or destroyed, it can only change form. (Note: chemical changes contrast with changes in the identities of elements, which are nuclear reactions and will be outlined later.) In a chemical reaction, reactants are turned into products. For example, the combustion of coal forms carbon dioxide.

$$C\,(s) \;+\; O_2\,(g) \qquad CO_2\,(g)$$
$$\text{Reactants} \qquad\qquad \text{Product}$$

There are different types of chemical reactions. The most common types are acid-base reactions, oxidation-reduction reactions, and precipitation reactions. Most **acid-base reactions** involve the transfer of a proton, such as the addition of a proton to the hydroxide ion to form water:

$$H^+ \;+\; OH^- \qquad H_2O$$
$$\text{Acid} \quad\;\; \text{Base} \qquad \text{Water}$$

Instead of moving protons, **oxidation-reduction reactions** move electrons. In this course, oxidation-reduction reactions are most apparent when either the hydrogen atom or the oxygen atom is added to an element. For example, carbon is oxidized when it undergoes combustion in this reaction:

$$C\,(s) \;+\; O_2\,(g) \qquad CO_2\,(g)$$

Precipitation reactions involve the combination of two ions in the aqueous phase to form an insoluble solid. A common example in the marine environment is the use of soluble calcium ions and carbonate ions to build a home for snails and diatoms in this reaction:

$$Ca^{2+} (aq) + CO_3^{2-} (aq) \rightarrow CaCO_3 (s)$$

The insoluble calcium carbonate from the dead organisms can eventually be compressed into limestone, and then eventually into marble.

Once a chemical reaction is complete, an **equilibrium** is established between reactants and products. Sometimes the equilibrium favors the predominant formation of products; sometimes the equilibrium favors only a minimal formation of products. The equilibrium is designated by a two-sided arrow ($\leftrightarrow$). For example, the following is an important equilibrium in human physiology and in the marine environment. Carbon dioxide in the presence of water will form some carbonic acid, which will in turn form some bicarbonate ions and hydrogen ions. The reaction favors the reactants (CO_2 and H_2O) at equilibrium, but a small amount of carbonic acid (H_2CO_3), hydrogen ions (H^+), and bicarbonate ions (HCO_3^-) also exist. All five of these chemical ions and compounds are "in equilibrium" with each other.

$$CO_2 + H_2O \rightarrow H_2CO_3 \rightarrow H^+ + HCO_3^-$$

Nutrient Cycles: Matter Moving Through the Environment

While a chemical reaction demonstrates the conservation of matter in a single chemical reaction, combinations of chemical and physical changes demonstrate the conservation of matter as matter moves through entire ecosystems.

The Carbon Cycle

1. Photosynthesis captures carbon dioxide to form glucose, which animals eat and convert to other biomolecules. These molecules can be combusted to form carbon dioxide, or compressed into fossil fuels and then combusted to form carbon dioxide.

2. Respiration produces carbon dioxide.

3. Carbon dioxide is captured and dissolved by the ocean, in equilibrium with carbonate ion.

4. Carbonate ion is used by microorganisms, which die and are compacted by sediments to form limestone, dolomite, and marble. These minerals, if part of a subduction zone, can become molten magma and be vented as carbon dioxide.

5. Microorganisms build biological molecules from carbon dioxide. When they die, organic material is compressed to form peat, coal, or even crude oil.

6. Crude oil and coal are burned to produce carbon dioxide in the atmosphere.

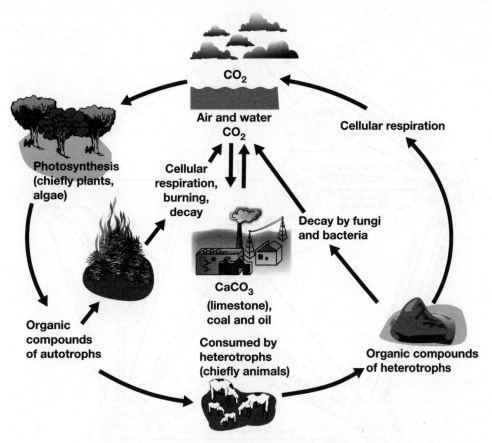

Figure 2.1 The Carbon Cycle.

The Nitrogen Cycle

1. Nitrogen fixation is the process by which atmospheric nitrogen gas is converted into ammonia (NH_3) or ammonium (NH_4^+) by bacterial metabolism. Also, atmospheric nitrogen can be oxidized by combustion to form nitrogen dioxide (NO_2), and later combined with water to form acid rain (HNO_3 or HNO_2).

2. Nitrification is the process by which ammonia is oxidized to form nitrite ions (NO_2^-), and then nitrite ions are oxidized to form nitrate ions (NO_3^-).

3. Nitrate ions are also formed by lightning-induced oxidation of nitrogen gas.

4. Nitrate ions are used as nutrients for plants to form proteins and nucleic acids.

5. Plant proteins are eaten by animals.

6. Animals metabolize proteins and form ammonia, or die and decompose to form ammonia, in a process called ammonification.

AP Environmental Science

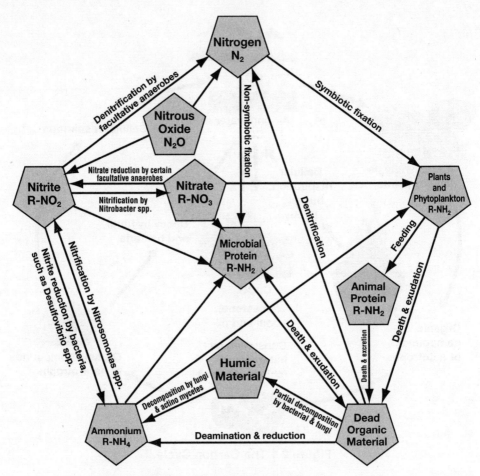

Figure 2.2 The Nitrogen Cycle.

The Phosphate Cycle

Phosphorous remains constantly in the form of the phosphate ion (PO_4^{3-}) during the phosphate cycle, whereas the carbon, nitrogen, and sulfur cycles all involve chemical reactions that oxidize or reduce the element in the cycle. There is no gaseous phase of the phosphate ion in the phosphate cycle.

1. Phosphates are used in animals and plants.

2. Decaying organic matter puts phosphates in soil, which can later be mined and used as fertilizer, or becomes part of ocean bottom sediments and rocks in a geological process.

3. Animals and plants on the land or in the sea pick up phosphates and use them again.

The Sulfur Cycle

1. Sulfur is used by living organisms to construct proteins.

2. Living cells die, decompose, and are compressed into fossil fuels, which produce sulfur dioxide (SO_2) when combusted.

3. Sulfur dioxide combines with water vapor to form sulfuric and sulfurous acid (H_2SO_4 and H_2SO_3, respectively), or acid rain, which deposits onto soil and water and distributes sulfate and sulfite ions (SO_4^{2-} and SO_3^{2-}, respectively) globally.

4. Sulfate ions are taken in by plants as nutrients to produce proteins.

5. Sulfate ions are also reduced to sulfide ions (S^{2-}) and hydrogen sulfide (H_2S) by microorganisms.

6. Volcanic eruptions also produce sulfide ions.

7. Sulfide ions form insoluble precipitates with most metal ions and reside throughout the earth's crust.

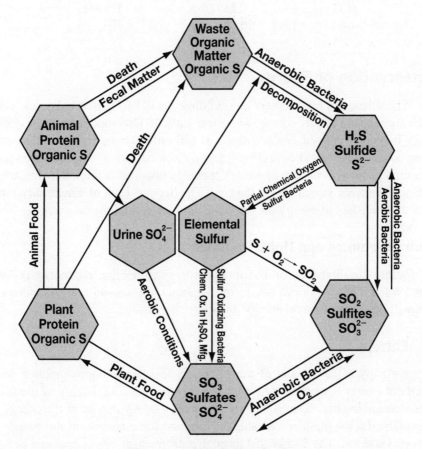

Figure 2.3 The Sulfur Cycle.

8. When some types of marine plankton are warmed, they produce dimethyl sulfide (DMS) that is oxidized to sulfur dioxide (SO_2), and then to sulfate ions. These sulfate ions regulate global climate when they increase the reflectivity of the earth and cool the earth by acting as condensation nuclei, which causes water vapor to condense.

9. Other microorganisms can further reduce sulfide ions to inorganic elemental sulfur.

The Water Cycle

Unlike the nutrient cycles, the hydrologic—or water cycle—is characterized by changes in physical state, rather than chemical reactions. When heated, solid ice melts into liquid water; when heated further, liquid water evaporates into water vapor. The water cycle is addressed further in Chapter Eight, but it is also mentioned here because it is an important method of cycling matter and energy around the earth. The water cycle is powered by solar energy, which changes water from one form to another, and by gravity, which transports water downhill to the sea.

$$H_2O\ (s) \quad \xrightarrow{heat} \quad H_2O\ (l) \quad \xrightarrow{heat} \quad H_2O\ (g)$$
$$\text{solid ice} \qquad \text{liquid water} \qquad \text{water vapor}$$

Conservation of Energy

The science of heat centers around three Laws of Thermodynamics, two of which are relevant to this course. The **First Law of Thermodynamics** states that energy is neither created nor destroyed, it can only change forms. For example, human societies convert chemical energy into electrical energy in a coal-fired power generation plant. The amount of electrical energy obtained is limited by the amount of chemical energy stored within the coal. The **Second Law of Thermodynamics** states that transfers of energy decrease the amount of total useful energy.

Thermodynamics and Heat Transfers

While the First Law of Thermodynamics assures that no energy is lost or gained when it changes from one form to another, the second law guarantees that there will be a loss of *useful* energy during such a change.

Entropy

Entropy is the amount of useful energy lost per amount of matter. In any transfer of energy—whether it be a plant absorbing the solar energy of the sun, or a rabbit absorbing the chemical energy of a plant—some amount of the energy will be available for the plant or the rabbit to use, and some amount of the energy will be wasted and lost. The energy lost through entropy disperses as heat and becomes more scattered, or disordered. In this way, entropy is considered a measurement of disorder; as heat disperses, entropy—or disorder of energy—increases. The more that energy becomes unavailable, the more disordered it is. Every chemical and physical change carries with it a measurable change in entropy.

Convection and Conduction

Heat energy may be transferred from one place to another through convection or conduction. Convection involves the movement of warmed mass from one place to another. Conduction involves one warm mass bumping up against another mass, and thereby imparting some of its energy to it by the collision. Convection and conduction will be covered in more detail in the next chapter.

Efficiency

For any transfer of energy, the amount of energy that is useful from one step to the next is considered the **efficiency** of the transfer, and is usually expressed as a percent. For example, a coal-fire power plant derives an amount of electrical energy that is 38% of the total amount of chemical energy stored in coal. Therefore, the process of obtaining electrical energy from coal is 38% efficient. The rest of the energy has been lost to heat and accounts for the necessary increase in entropy. A chicken is 33% efficient in converting chemical energy stored as grain into chemical energy stored as egg protein. Beef cattle are only about 7% efficient. Human efficiency in metabolizing glucose is about 40%; the rest of the energy is lost as heat. Energy efficiency will be covered in more detail in Chapter Ten.

Energy Movement through Ecosystems

Ecosystems are characterized by how energy is obtained, converted into chemical energy, and transferred from one organism to another. **Productivity** is the amount of biomass that is produced by a community. **Primary productivity** is the amount of biomass produced by photosynthetic organisms. **Secondary productivity** is the amount of biomass produced by organisms that eat photosynthetic organisms.

Food chains are sequences of organisms that begin with a primary producer and trace the movement of biomass through a series of predator/prey relationships. **Food webs** are interconnecting series of food chains. Each step along a food chain is a **trophic level**; in most cases, living organisms are able to convert about 10% of ingested biomass into biomass that is available for the next trophic level. Chapter Four will cover the relationships surrounding productivity, food chains, and food webs in much greater detail.

There is a different amount of volume associated with each tropic level because energy is lost to heat, metabolism, undigested food, or unconsumed food at each level of the pyramid. For example, a rabbit might not eat absolutely all of the lettuce in its surroundings; some might be consumed by decomposers, or become part of geological sediments. Additionally, the lettuce the rabbit does eat is not all digested. Of the amount digested, many of those nutrients help with the rabbit's metabolism and are expelled as unusable heat. Of the total biomass available to the rabbit, only a small portion (usually about 10%) actually becomes part of the rabbit's biomass. Likewise, only a small portion of the energy available from primary productivity becomes energy available from secondary productivity. In this way, a food pyramid is a biological example of the Second Law of Thermodynamics: entropy increases.

Case Summaries

Water Treatment and the Nitrogen Cycle

Principles mentioned in this case:

- **Nitrogen cycle**
- **Natural cycles in industrial processes**

Human metabolism of proteins produces ammonia-rich urine, which is a major component that needs to be eliminated by waste water treatment plants. Once sewage has been filtered and settled, it is sprayed over oxygen-saturated templates that help bacteria grow. Those bacteria that devour the available nutrients tend to grow and reproduce, so the bacteria that are able to consume ammonia as part of the nitrogen cycle are naturally encouraged to grow. As long as there is enough oxygen available, the bacteria will convert the ammonia first to the nitrite ion, and then to the nitrate ion. The resulting nitrate ion is an important nutrient; most sewage treatment plants will use plants or chemical means to remove the nitrate ions before the water is considered fully treated.

Composting Wastes: Combining the Carbon and Nitrogen Cycles

Principles mentioned in this case:

- **Nitrogen and carbon cycles**
- **Recycling and waste reduction**

Any time carbon or nitrogen wastes accumulate, naturally occurring bacteria that metabolize the carbon and nitrogen will grow and decompose the material. Eliminating garden and household wastes that contain carbon or nitrogen can be accomplished by constructing a compost pile that fosters the growth of the right kind of bacteria.

The ideal compost pile contains about a 30:1 ratio of carbon to nitrogen. Too much carbon will cause the compost to degrade too slowly; too much nitrogen will produce an unappealing ammonia odor. Materials that contain carbon include woody stalks, leaves, and paper. Materials that are rich in nitrogen include grass clippings and manure. The bacteria in the compost will also need oxygen and water, and a moderate amount of heat. However, once the bacteria begin to work, the compost will create its own heat.

In general, aerobic bacteria add oxygen to carbon or nitrogen; anaerobic bacteria will add hydrogen. If the compost gets enough oxygen, carbon is converted to carbon dioxide and nitrogen is converted to nitrate nutrients. If the compost does not get enough oxygen, the carbon is converted to methane and nitrogen is converted to ammonia—both of which are unpleasant smelling. The trick to having a good-smelling compost heap that adequately decomposes both carbon and nitrogen without foul odors is to be sure that it gets sufficient oxygen by regularly turning the material.

Energy Efficiency of Lights

Principle mentioned in this case:

- **Efficient use of energy**

Of the electricity consumed in the United States, nearly half is used for lighting. Some types of lighting are more efficient in converting electricity into light. For example, a 15-watt fluorescent light bulb puts out as much light as a traditional 60-watt incandescent bulb, and the fluorescent bulb lasts about ten times longer. If everyone used the new fluorescent bulbs, we could cut this sector of energy usage by 75%.

Earth Systems & Global Changes

Chapter 3

Earth Systems and Global Changes

Properties that Drive Global Processes

Global movement of matter is driven by the properties of water and energy. The energy that drives movement of material around the Earth comes from one of three different sources: photons radiated from the sun, the gravitational potential energy of the moon, or the nuclear reactions at the center of the Earth. Heat energy from those sources can be transferred in one of the following three ways.

- **Radiation**: energy is carried by a photon from one place to another.

- **Conduction**: energy is transferred from one particle to another through a collision between the two particles. When this happens, some heat is produced.

- **Convection**: energy is moved by energy-containing particles from one place to another.

Convection is the primary mechanism of energy transfer in the atmosphere, oceans, and within the solid Earth. The sun heats the air and water to initiate convective currents, and the center of the Earth heats the mantle to initiate convective currents of molten earth material.

Convection in the Mantle, Oceans, and Atmosphere

Convection

Convection is the circulation of material that occurs when the density of the material is decreased upon warming, or increased upon cooling. When the density of the material changes, the material moves from one place to another. In Figure 3.1, the material at point A rises because it is heated, while the material at point B falls because it cools. The principle of convection drives the movement of solid, liquid, and gaseous material throughout the Earth.

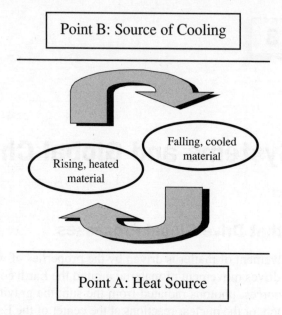

Figure 3.1 A heat source and a heat sink combine to move Earth materials in a convection cycle.

Convection in the Earth's Mantle

Energy from the Earth's core heats the mantle material, which then rises and eventually comes in contact with the cooler crust, or lithospheric plates. The cooled mantle material then descends to be reheated by the core. It may happen that a set of convection cycles develops; the cooled molten mantel does not descend all the way to the core, but descends until it is heated by (and, in turn, cools) a secondary convection cycle.

In some places, the ascending molten magma actually punches through the lithospheric plate and erupts. This is known as a "hotspot" and can be found in the volcanism of Hawaii, Yellowstone, Iceland, and the Galapagos Islands.

Convection in the Oceans

The process of convection is the most dominant force driving ocean currents. However, in the ocean, density differences are not only caused by differences in temperature, but also differences in salinity. A difference—or gradient—that involves both temperature and salinity is called a **thermohaline** gradient. The convection cycle created by differences in temperature and salinity is called **thermohaline circulation**.

Convection Cells in the Atmosphere

1. Differential Solar Heating of the Atmosphere

a. The Fate of Energy from the Sun

As the radiant energy comes into the atmosphere, the high-energy, short wavelength UV light tends to be absorbed by ozone in the stratosphere; visible light penetrates through the atmosphere; and low-energy, long wavelength infrared light tends to be absorbed by carbon dioxide and water in the troposphere. Once the radiation hits Earth, some solar energy is reflected—particularly from surfaces such as water or snow. Dark surfaces, such as soil of forests, absorb much more light. Eventually, all the energy absorbed by the Earth is re-radiated back into space as infrared radiation—some of which is easily trapped by clouds and warms the Earth. (This principle will be discussed more in the section on the Greenhouse Effect.) The energy striking the surface of the Earth warms air, decreases its density, and causes it to rise.

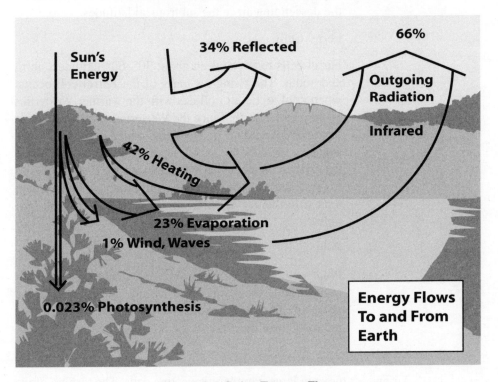

Figure 3.2 Solar Energy Flow.

b. More Solar Energy Strikes the Equator

The sun's rays strike the equatorial latitudes more directly than the polar latitudes. Therefore, the equatorial latitudes receive more solar energy than higher latitudes. Convection cells in the atmosphere are initiated by the warm equatorial air that rises to a higher altitude—creating underneath it a low pressure area. The air then moves north or south at a high altitude until it confronts cool air. Then the cooler combined air mass becomes more dense and sinks—creating underneath it a high pressure area. These high pressure zones tend to produce hot, dry air at about 30° north and 30° south latitude. For this reason, many of the world's deserts tend to exist at those latitudes.

    c.   Major Global Convection Cells

        i.   Hadley Cells

Hadley cells represent the air masses that are heated at the equator, rise, and spread to the north or south. When the high-altitude air confronts cooler air from higher latitudes, it cools and descends in the subtropical zones (about 25°–40° north and south). Tropical rainforests tend to occur under the rising air, and deserts tend to exist under the descending air in the subtropical latitudes.

        ii.   Ferrel Cells

Ferrel cells exist between about 30°–60° latitudes, north and south. The rising air mass of the Ferrel cell occurs when cool arctic air collides with the warmer Westerlies as they move north. Since the Westerlies ultimately come from the descending Hadley cell—which originated in the tropics—then the areas where this collision takes place experience weather that is influenced by both polar and tropical air masses. Seasonal variations of temperature tend to be larger at polar latitudes than at tropical latitudes.

        iii.   Polar Cells

Polar cells are cold, very dry, and dense. As polar air moves south, the warmer mid-latitude air climbs over the polar air and works north—itself becoming a part of the polar cell as it descends at the pole and then works south.

2.   Coriolis Effect

Air masses move in large, circular convection cells. The turning of the Earth underneath these convection cells creates circular movements of air called the **Coriolis Effect**. The typical surface winds that are created by the Coriolis Effect are called **trade winds**. The Coriolis Effect moves trade winds in a clockwise direction in the northern hemisphere and a counterclockwise direction in the southern hemisphere.

3.   Jet Streams and the Circumpolar Vortex

In the upper regions of the troposphere—just under the stratosphere—air can move from one atmospheric convection cycle to another along a sinusoidal path around the globe. These are massive rivers of air called **jet streams.** In the United States, there are generally two different jet streams: a **subtropical jet stream** that flows through the southern portion of the United States, and a **northern jet stream** that brings cold arctic air down from the north. The northern jet stream is part of a movement of cold air that moves from west to east around the northern portion of the globe; this massive movement of air is called the **circumpolar vortex**. As the cold lobes of the circumpolar vortex move across the northern continents, they collide with warmer air and dictate the weather. Temperature differences cause water to precipitate or evaporate, and pressure differences cause air to move as wind from areas of high pressure to areas of low pressure.

The Properties of Water

In addition to energy driving convection currents in the litho-sphere, oceans, and atmosphere, the weather and currents are also affected by the unique properties of water.

Properties that Contribute to Climate

1. Specific Heat

   Specific heat is a physical constant that represents the amount of heat needed to raise the temperature of a specific mass of a material. Water has a specific heat of $4.18 \text{ J/g} \times °C$, whereas sand and most metals have a specific heat that is less than $1.0 \text{ J/g} \times °C$. This means that water requires more than four times as much heat to change its temperature than those other materials. As a result, coastal areas near oceans or inland areas near bodies of water have more moderate temperature fluctuations than those areas that are far from water. This means that global temperatures are more moderate than that of other planets.

2. Energy of Vaporization

   The amount of heat needed to vaporize water, or energy of vaporization, is higher than other liquids. This makes it an ideal substance for organisms to excrete in order to cool themselves. A film of water evaporating from the skin of a hot animal will use considerable heat. On a global scale, as water precipitates in a weather front or due to a change in pressure, the heat that went into the evaporation of water is returned. For this reason, water vapor is a form of heat storage; and the heat of vaporization is sometimes called the latent heat of vaporization because the water holds latent—or stored—heat energy. For example, water that evaporates in the Southern Atlantic during the summer months can carry considerable energy in the form of a hurricane as it moves toward the southern coast of the United States.

Properties that Contribute to Currents

Ocean currents are driven by air currents sweeping across the surfaces of the oceans, and by water moving from one place to another as it changes density. The density of water is determined mostly from its temperature and salinity. The currents that are created by the differential temperature and salinity gradients are called **thermohaline currents**.

A unique aspect of water is responsible for upwelling in lakes during the spring. Water is most dense at 4°C, just above its freezing point. When the surface of the lake freezes, it remains at 0°C or colder, insulating the water underneath. When spring comes, the surface warms and thaws. As the surface water reaches 4°C, it reaches its maximum density and sinks, causing the nutrient-rich water in the bottom of the lake to come up to the surface. This process is called **upwelling**.

Properties that Contribute to Life

Water is sometimes called "the universal solvent," although that title is a misnomer, as it doesn't really dissolve everything. However, there are good reasons why water can dissolve many substances, and the solvent nature of water allows it to be an important carrier of nutrients through the bodies of living organisms and the environment.

Indirectly, the ability of ionic compounds to dissolve in water also contributes to the convection of water in the oceans because thermohaline gradients—which depend on temperature and the amount of dissolved solids in the water—drive ocean currents.

The same aspect of water that makes it a good solvent—its polarity—also allows it to easily undergo a process called **capillary action**. Capillary action occurs when a liquid seems to crawl up a tube of very narrow diameter. Capillary action moves easily through small spaces, which promotes its mobility in both living creatures and the environment.

Earth

Plate Tectonics

The Theory of Continental Drift

In 1915, Alfred Wegener proposed that all the continents were originally a single continent that began to break apart about 180 million years ago. Wegener's theory was supported by the fit between the coastlines and fossil evidence showing the same population on previously adjacent but now land areas thousands of miles apart. However, Wegener's theory implied that the continents simply plowed over the top of the ocean floor, and he died not being able to identify the force that pushed the continents. It was not until the 1960s that scientists found that the mid-

Atlantic rift was actually a point where molten lithospheric material was being pushed to the surface of the crust—pushing apart the two plates on either side of the rift. At this point, the theory of continental drift yielded to the more specific theory of plate tectonics, that describes the surface of the Earth as a set of dynamic plates that move and collide.

Rate of Plate Movement

The rate of plate movement can be measured by examining iron ore samples. The magnetic domains in the ore orient themselves toward the north pole, so at those rare times when Earth's polarity flips, so does the orientation of newly cooled iron ore. Knowing the length of time between these changes in magnetic polarity, scientists can then measure the rate of plate movement. Different plates seem to move at different rates. The Arctic Plate moves the slowest (about 2.5 centimeters per year) and the Pacific Plate moves the fastest (about 15 centimeters per year, measured near Easter Island). Using this method, scientists can surmise how one plate will move relative to another by studying plate boundaries.

Plate Boundaries

1. Divergent Boundaries

   Divergent boundaries occur where magma is pushed up from the asthenospheric mantle and new crust is produced. This upward movement of molten material pushes the two plates apart. This is also called a **constructive boundary** (it constructs new crust) or **sea floor spreading** if the boundary is at the bottom of an ocean. Divergent boundaries also cause **rift valleys** when they exist on land. Examples of divergent boundaries include the mid-Atlantic rift, which runs through Iceland and forms the boundary between the North American and Eurasian plates, and the boundary between the African and Arabian plates that forms the Red Sea.

2. Convergent Boundaries

   If new crust is formed at one place, it stands to reason that some plate must at some point dive back down into the asthenosphere. Such a place is called a **convergent boundary** (where two plates converge) or a **destructive boundary** (where a crustal plate is destroyed). The type of collision that takes place depends on whether the plate involved is an oceanic plate or continental plate.

   a. Oceanic-Continental Convergence

   Continental plates are less dense than oceanic plates, so when they collide, the oceanic plate dives underneath the continental plate. This causes a **subduction zone** that creates a deep trench at the point of collision. The friction of one plate moving against another turns the rock into magma, which then rises to the surface and forms a volcano on the continental plate. An example is the Nazca Plate subducting under the South American Plate—forming the volcanic Andes Mountains.

    b. Oceanic-Oceanic Convergence

When two oceanic plates converge, one is subducted under the other and a **trench** is formed alongside an area of volcanic activity. A striking example is the Marianas Trench where the Pacific Plate dives underneath the Philippine Plate. The Marianas Trench is the deepest trench in any of the oceans—about 33,000 feet. The volcanic activity that results from convergence of oceanic plates typically forms seismically active island arcs that mirror the plate boundary, such as the Aleutians in Alaska.

    c. Continental-Continental Convergence

When two continental plates converge, neither is subducted because both plates are light and resist being pushed into the asthenosphere. Instead, the two plates buckle and push each other upward. A dramatic example is the collision of the Indian and Asian Plates that resulted in the Himalayas—the highest mountain range in the world. Sometimes, one plate will slide underneath the other and raise up a high altitude plateau just beyond the plate boundary. The Tibetan plateau is a good example of this.

3. Transform Boundaries

A boundary between two plates that are sliding past one another is called a **transform boundary**. The seismically active San Andreas fault in California is an example of a transform boundary between the Pacific Plate—which is moving to the northwest—and the North American Plate.

Earthquakes

Plate boundaries are seismically active and prone to earthquakes. An earthquake occurs when a large amount of earthen material previously held by friction quickly adjusts to a new position, thereby releasing a large amount of kinetic energy. The spot where the adjustment takes place is called the **focus**; the spot on the surface of the Earth directly above the focus is called the **epicenter**. When the adjustment takes place, energy radiates out from the focus in the form of waves.

1. Waves

The waves radiating out of the focus creates waves inside the Earth like concentric spheres, called *body waves*. There are two types of body waves, *P waves* from compression and *S waves* from lateral motion. P waves are fast-moving and move through any material. S waves are slower and only travel through solids. When the body waves reach the surface of the Earth, they cause sinusoidal surface waves.

2. The Richter Scale

The Richter Scale is a logarithmic scale that measures earthquake intensity. For every increase in one unit on the Richter Scale, there is a ten-fold increase in ground displacement and a 30-fold increase in energy. Smaller earthquakes at the limit of perception have a Richter value of about 3.0. A magnitude 5.0 earthquake will make it difficult to stand up, items will fall off shelves, and some landslides may take place from liquefaction of the ground. At magnitude 7.0, houses are shaken off foundations and there may be large cracks in the ground. The largest earthquake measured 9.5 (the 1960 earthquake in Chile).

The Rock Cycle

The rock cycle refers to the cycling of rock between three types of material: igneous, sedimentary, and metamorphic rock.

Igneous Rocks

Igneous rocks are formed from magma that has solidified. **Magma** is the molten material that comes from Earth's interior during volcanic eruptions, and is called lava when it is above the Earth's surface. Different igneous rocks include basalt, rhyolite, and granite.

Sedimentary Rocks

After igneous rock has formed and cooled, weathering processes break apart the rock into particles and deposits them in layers which, over time, compress and harden the layers underneath to form sedimentary rocks. Examples of sedimentary rocks include shale, limestone, and sandstone.

1. Mechanical Weathering

   Mechanical weathering takes place when a force is applied against the rock force to break it up. Glaciers, ocean waves, and erosion from wind or water are examples of mechanical weathering.

2. Chemical Weathering

   Chemical weathering takes place when a chemical reaction weakens and breaks apart rock surfaces. The oxygen in the air can react with various minerals and be oxidized. Acid can dissolve many minerals. Both oxidation and reaction with acid weaken materials so that they undergo mechanical weathering more easily.

Sedimentary rock can be further compressed to form metamorphic rock, or melted and subsequently cooled to form igneous rock.

Metamorphic Rocks

Both sedimentary and igneous rocks can change into metamorphic rocks as a result of being exposed to heat or pressure. Metamorphic rocks are usually formed when layers of sedimentary or igneous rocks become further compressed by tectonic plate movement. For example, compression of limestone produces marble; compression of shale produces slate; and compression of sandstone produces quartz. Metamorphic rocks can be melted and subsequently cooled to form igneous rock.

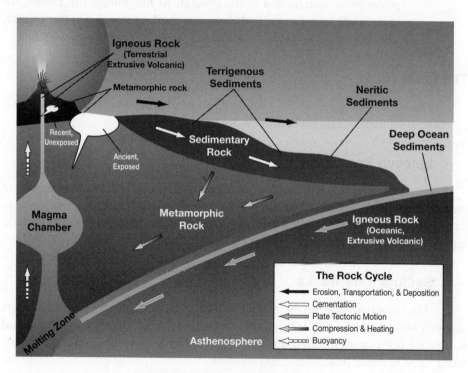

Figure 3.3 The Rock Cycle.

Soil

Soil Formation

Erosion creates small particles of rock and minerals. As those particles collide with one another in a stream bed or from the wind, they break into smaller and smaller pieces. When that inorganic material comes to rest, pieces of dead plant material and bugs contribute organic material to the inorganic material. Earthworms are important soil-builders as they take in the combination of organic and inorganic material, digest some of the organic material and bacteria, and then coat the inorganic material with the nutrients in their waste as they excrete the soil. The combination of all these processes takes time; good topsoil that is fertile and tillable takes about a year to build a 1.0 millimeter layer in favorable conditions.

Soil Composition

The following represent the major components found in most soils, the first four of which are listed in order of increasing particle size.

1. Clay: very fine particles with high adhesive ability due to mobile ionic attraction between the particles. Impermeable to water.

2. Silt: very fine particles (larger than clay and smaller than sand) that result from mechanical weathering. Silt particles will combine to form sedimentary rock when compressed.

3. Sand: coarse inorganic particles that result from mechanical weathering. The larger particles allow water to flow through easily.

4. Gravel: a mixture of macroscopic pebbles and fragments of rock.

5. Humus: a sticky composite of organic material made up of dead and partially decaying plants and animals. Humus provides the "glue" that holds together inorganic material and gives top soil a spongy texture.

6. Loam: a mixture of sand, silt, humus, and clay that is well-suited for growing crops. Varying the level of sand in the loam will vary its water-retaining capabilities.

7. Soil fauna: Algae, bacteria, and fungi flourish in the top layers of soil. Photosynthetic algae and cyanobacteria—collectively called **decomposers**—manufacture organic molecules from sunlight. Bacteria and fungi decompose organic material to process nutrients so that they are more usable by plants. Worms and microscopic insects form complete food webs that ultimately keep the soil vibrant for use by macroscopic plants.

Soil Profiles

As soil forms and a terrain evolves, it creates layers, or **horizons**, of different types of material. A cross-section of these horizons is called a **soil profile.** The following are the different soil horizons, starting with those closest to the surface.

1. O-horizon: the organically rich surface that contains leaves and partially decomposed organic material.

2. A-horizon: the top soil, which is composed of different combinations of weathered inorganic material and organic material.

3. E-horizon: the zone of leaching, where water percolates through and dissolves soluble ions and nutrients, moving them downward into the subsoil.

4. B-horizon: the subsoil, which usually has a lower organic content and higher levels of inorganic mineral deposits. Clay, silt, and soluble ions percolating down from upper layers tend to deposit in this horizon. This layer of soil often becomes compressed to the point of preventing roots from penetrating deeper.

5. C-horizon: weathered parent material with little or no organic material. Most of the parent material in the United States was transported to the site by glaciers and other geological events, rather than being composed of the underlying bedrock.

6. Bedrock: unweathered portions of the crust.

Soil Conservation

1. The Dust Bowl

    During the 1930s in Oklahoma, Kansas, and part of Texas, farmers had over-farmed the prairies and removed the natural organic material, creating a more sandy/silty soil. When drought and heavy winds came, the farmers paid the price for short-sighted practices and the top soil eroded away. The loss of natural wealth in the form of fertile farm land, combined with leveraged buying practices in the stock market, pushed America first into recession, and then into the Great Depression. As a result of the Dust Bowl, the Soil Conservation Act was passed in 1935 and established the Soil Erosion Service.

2. Erosion (See also Chapter Nine)

    Erosion occurs when soil is moved from its point of origin. There are three types of erosion: *rill erosion*, where water cuts small rivulets in the soil; *gully erosion*, where rill erosion escalates to form a large channel; and *sheet erosion*, where water removes a horizontal layer of soil. Erosion is caused by any of the following factors:

    a. Wind and water erosion can remove the top layer of nutrient-laden adhesive topsoil. Soil exposed to the open air, with few roots or little cover, is vulnerable to wind erosion. Water erosion can occur with flood or heavy rains, or if the land is simply tilled in such a manner as to create channels where the water runs downhill—taking the soil with it—as opposed to residing on the land and soaking into the groundwater. Planting crops in rows exposes the soil to the wind and also creates gullies for flowing water. Water erosion is also amplified by irrigation with too much water, so that it flows over and displaces the soil.

b. Chemical erosion is caused by nutrient depletion, acid deposition, or a process called salinization. **Salinization** is caused by irrigation with brackish (salty) water, which leaves mineral salts on the soil when the water is used or evaporated. The remaining salt makes it difficult to grow plants or becomes airborne and causes human respiratory problems.

c. Waterlogging results from excessive irrigation, poor drainage, or ocean encroachment.

3. Soil Conservation Practices (See also Chapter Nine)

a. Utilize no-till or low-till farming practices (leaving plant material lying on the ground after harvest)

b. Plant trees as wind breaks

c. Use contour farming or hill terracing to keep water from flowing downhill and washing away soil

d. Monitor soil nutrients so that the soil retains a high level of adhesive humus

Ocean

Ocean Currents

1. Currents Induced by Trade Winds

Trade winds created by the Coriolis Effect apply friction to the surface of the ocean and drag the surface water in the same direction as the wind. Consequently, the major ocean currents also move in circular paths that are clockwise in the Northern Hemisphere and counterclockwise in the Southern Hemisphere. When major currents hit continents, they tend to split and move in both directions along the coast of the continent. Trade winds and the placement of continents dictate the direction of the major surface currents in the oceans.

2. Thermohaline Currents

Both temperature and salinity can change the density of saltwater. When two masses of water of different densities collide, the water of lower density will move to the more shallow depth. Consequently, density differences created by temperature and salinity gradients can cause subsurface currents. Such currents are called **thermohaline** currents.

Tides

The gravitational pull of the moon on the water creates a bulge of water on the globe that faces the moon, and another bulge of water on the opposite side of the Earth. As the moon passes around the Earth roughly once per day, these two *tidal bulges* also cycle around the Earth—creating a high tide (and a low tide) twice per day. While tidal flow is not recognized in the middle of the ocean, it is much more visible along the coasts. The energy that is used to move water during tidal fluctuations ultimately comes from the gravitational potential energy of the moon acting on the Earth.

Atmosphere

Origin and Evolution

The original atmosphere was probably created in the first few hundred million years of the Earth's 4.6-billion-year history. It was composed of hydrogen and helium gases, but these gases could not be retained by the Earth's gravitational pull. Even now, these gases usually are not retained in the atmosphere and have enough kinetic energy to eventually diffuse into space. Volcanic activity produced water vapor, carbon dioxide and monoxide, sulfur dioxide, chlorine, nitrogen, ammonia, and methane.

Formation of Oxygen

1. Ultraviolet rays break up the water molecules to form ozone, which protects the Earth from UV light and facilitates the evolution of life.

2. Photosynthetic cyanobacteria produced the vast majority of current oxygen levels between 2.5 billion years and 400 million years ago. Oxygen production was later aided by plants.

Composition and Structure

Composition

- 78.08% Nitrogen (N_2)—Essential part of the nitrogen cycle and essential for life
- 20.95% Oxygen (O_2)
- 0-4% Water vapor (H_2O)
- 0.93% Argon (Ar)
- 0.036% Carbon dioxide (CO_2)
- 0.002% Neon (Ne)—Inert gas
- 0.0005% Helium (He)—Inert gas
- 0.0002% Methane (CH_4)—Produced by peat, anaerobic decomposition of organic material, and animal waste

- 0.00005% Hydrogen (H_2)—Remains from the early atmosphere; because of hydrogen's low molecular weight, it is held loosely by Earth's gravitational pull. The heavier molecules are held more closely to the Earth's surface, leaving only helium and hydrogen in the high-altitude exosphere.

- 0.00003% Nitrous oxide (N_2O)—Produced by burning fossil fuels, fertilizer use, and deforestation; a greenhouse gas.

- 0.000004% Ozone (O_3)—Absorbs UV radiation. Decreased by halogenated compounds (mostly fluorine and chlorine) in a catalytic reaction.

Layers of the Atmosphere

1. **Troposphere** (0–7 miles above surface)—Most of the Earth's weather takes place in the troposphere; 75% of the atmosphere exists here. Temperature decreases (to –60°C) with higher altitudes (and lower pressures; $PV = nRT$).

2. **Tropopause** (7–13 miles from surface)—The tropopause forms the boundary between the troposphere and the stratosphere. Jet streams occur here; temperature is constant.

3. **Stratosphere** (13–30 miles above surface)—Temperature increases with altitude because ozone (O) absorbs heat produced by solar UV rays. The ozone layer protects living organisms from damaging radiation. Most jet travel occurs here.

4. **Stratopause** (30–32 miles above surface)—Boundary between the stratosphere and the mesosphere; temperature is constant in this layer.

5. **Mesosphere** (31–50 miles above surface)—Temperature decreases with altitude; this is the coldest portion of the atmosphere and contains some ice-crystal clouds.

6. **Mesopause** (50–52 miles above surface)—Boundary between the mesosphere and the thermosphere; temperature is constant in this layer.

7. **Thermosphere**, which includes the **Ionosphere** (52–300 miles above surface). Temperature increases with altitude because of high energy solar radiation (gamma rays, UV radiation, X-rays).

8. **Exosphere** (300–6,000 miles above surface)—This region forms the transition to interstellar space. Most satellites orbit the Earth in this region. The atmosphere in this region is composed primarily of hydrogen and helium, but only at low partial pressures.

Weather

Physical Traits that Contribute to Weather

Three physical traits of the atmosphere contribute to weather: temperature, pressure, and moisture.

1. Temperature of the air is affected by convective cycles in the atmosphere, and air of different temperatures is moved by the trade winds, driven by the Coriolis effect. Because of the high specific heat of water, air temperature will vary more over land than over water. For example, when the sun goes down, the land will cool faster than the ocean, causing pressure to decrease faster there, and a wind will blow from land toward the ocean to equalize the pressure. In most cases, the daytime direction of the wind is reversed and blows from the ocean toward the land.

2. Pressure of the air is affected by altitude (as when air moves over a mountain) and by ascending air masses (lower pressure) and descending air masses (higher pressure) created by convection cells. Air pressure differences create wind on the surface of the Earth in addition to trade winds; air moves from a high pressure area to a low pressure area.

3. Moisture in the atmosphere is a form of stored heat. Water vapor is also a greenhouse gas; it holds heat in the atmosphere. Likewise, more moisture can be held in the atmosphere as vapor at higher temperatures and pressures. If either the temperature or pressure drop, the moisture will condense to form clouds first, and then precipitation. The **dew point** is the temperature at which condensation occurs at any given pressure.

Clouds

Clouds are composed of lightly condensed water vapor. Typically, clouds form by ascending air that carries moisture. As the pressure and temperature decreases, so does the atmosphere's ability to hold water in the vapor phase, and the condensation produces a cloud. The following are a number of different types of clouds.

1. Cloud Terms:
    a. Cumulus means "heap" and refers to a billowy cloud.
    b. Stratus means "layer."
    c. Cirrus means "curl of hair" and tends to be wispy.
    d. Nimbus means "rain."

2. Major Cloud Types

   a. High-level clouds: above 20,000 feet, tend to be mostly composed of ice crystals, are thin and wispy, and generate no precipitation (e.g., cirrus clouds).

   b. Middle-level clouds: between 6,500–20,000 feet, contain mostly water, but may contain ice in cold weather. These clouds may generate precipitation (e.g., altocumulus).

   c. Low-level clouds: under 6,500 feet, mostly contain moisture that is about to precipitate (e.g., stratocumulus).

   d. Vertical clouds: raised up through convection, particularly in a front, possible thunderstorm (e.g., cumulo-nimbus).

Weather Fronts

As mentioned earlier, weather fronts are created when an air mass of one temperature and pressure collides with an air mass of a different temperature and pressure. In North America, a common reason for such a collision is the movement of the cooler, arctic air in the circumpolar vortex to collide with the warmer, subtropical jet stream coming from the south.

A **cold front** exists when cool air moves into a zone previously occupied by warm air. The more dense, cooler air pushes itself under the warm air, forcing the warm air upward and causing it to cool. Cooling the warm air decreases the amount of water vapor the air can hold, increasing the chance of precipitation. Also, the difference between temperatures and pressures in the cold and warm air masses will cause air to move as wind from high pressure areas to low pressure areas. Cold fronts often bring severe weather, including thunderheads and tornados.

Hail is often associated with cold fronts. The water vapor in the rising warm air will precipitate and then freeze. As it falls to the Earth, the warm updraft catches it and moves it higher in the air, precipitating more frozen water around the hail stone. This cyclic movement of frozen water will continue until the weight of the hail stone overcomes the strength of the updraft, at which time the hail will fall to Earth. The size of the hail stone is an indication of the strength of the updraft and the overall energy of the storm. The movement of water polarizes electrical charges in the storm; lightning redistributes the electrical charge either within a cloud, or between the cloud and the surface of the Earth.

Warm fronts occur when warm air displaces cool air. Because of its lower density, the air in a warm front will slide above the cool air. This is a more gradual process than a cold front, and the resulting weather distributes the changes in energy over several layers in the atmosphere. Cirrus clouds will occupy the upper layers before precipitation reaches the ground, which comes as a steady drizzle when it does come.

Hurricanes

Low pressure regions that create rising air over warm tropical waters create strong, cyclonic storms. As the air rises, surrounding air on the surface rushes to replace it, turning as it approaches the center of the convection cell. If that air carries a lot of water vapor, the water will precipitate as the air cools, but the energy released increases the strength of the convection, which increases the amount of warm air and water vapor drawn into the rising air mass. This process—which is ultimately driven by the convection of air and the high heat of vaporization of water—is responsible for creating large cyclonic storms called hurricanes (in the Atlantic and eastern Pacific), cyclones (in the Indian Ocean), and typhoons (in the western Pacific).

Global Climate Changes

Milankovitch Cycles

As the Earth orbits around the sun, it both moves and spins. Like a top that wobbles with a characteristic period as it spins and travels along a tabletop, the Earth also contains a periodic wobble. The Earth's elliptical orbit lengthens and shortens every 100,000 years or so. Additionally, the angle of tilt fluctuates and causes a wobble in the Earth's axis over a 26,000-year period. Either shift would expose different latitudes to an amount of sunlight that is different from what now occurs, and may be responsible for shifts in climate over the centuries.

Volcanic Eruptions

A single volcanic eruption can put huge amounts of ash and crustal material into the atmosphere. The particulate matter blocks the sun's rays and can cause a drop in global temperatures.

El Niño/Southern Oscillation

Normally, the surface currents move equatorial waters westward from South America to Indonesia and Southeast Asia. As the water moves along the surface, it is heated, water evaporates, and the prevailing winds carry an increasing amount of water vapor as the current and air move west. By the time the air mass reaches Indonesia, it is laden with moisture and precipitates to quench the thirst of tropical rainforests in that region of the world. Likewise, as the water moves away from South America, nutrient-rich, more highly-oxygenated water wells up from the deep to replace the water that has left. This movement of nutrients has contributed to a strong fishing industry near the coast of Peru.

Peruvian fishermen were the first to notice warmer water and a decrease in their anchovy catches. They named the phenomenon El Niño ("boy child," referring to the infant Jesus) because these events would occur around Christmas every few years. During an El Niño event, the equatorial current begins to flow east instead of west. This brings the warm water and moisture to South America, rather than Asia. As a result, the cooler jet stream precipitates the moisture in the air after its Pacific trek and deposits the rain at various places in Central or North America. An El Niño will result in more precipitation in the Western United States or Gulf region, for example. Likewise, normally moist areas in Australia, Indonesia, and Southeast Asia experience droughts, crop failures, and fires.

Studies of tree rings, ice caps, and coral reefs suggest that El Niño events are becoming more frequent and stronger than they ever had before. The time when an El Niño event is not taking place is called La Niña, and the oscillation between the two conditions is called the El Niño Southern Oscillation (ENSO).

Greenhouse Effect

When solar radiation penetrates the Earth's atmosphere, it is either reflected or absorbed. The reflected radiation—about one-third of the incoming solar energy—returns to space; the absorbed radiation causes thermal motion in matter on Earth (conduction). When matter vibrates with greater motion, it emits low-energy, long-wavelength electromagnetic radiation. This long-wavelength radiation begins to be emitted back out into space under normal conditions. However, certain molecules in the air absorb that low-energy radiation and reconvert that energy to heat. The fact that this has happened has kept the temperature of the Earth warm and has actually facilitated the evolution of life. However, industrial processes have emitted more carbon dioxide and other gases that absorb low-energy electromagnetic radiation, which causes the Earth to warm up more than it has in previous eons.

Carbon dioxide production is accelerated by our use of energy—particularly when produced by burning fossil fuels. Using electricity that originates from a coal-fired power plant, driving to and from work in gas-thirsty automobiles, and using products and packaging that require energy to produce are practices that all increase the amount of carbon dioxide in the atmosphere.

Normally, carbon dioxide is absorbed by forests, which use the carbon dioxide to produce oxygen and carbohydrates through photosynthesis. However, global deforestation to provide fuel, shelter, or land space has reduced the acres of forests that would normally absorb the carbon dioxide. Therefore, not only is carbon dioxide production increasing, but the world's ability to absorb the carbon dioxide is rapidly decreasing—further amplifying the **greenhouse effect**.

The ability of carbon dioxide to hold in heat is similar to the workings of a greenhouse, which is warmer inside than outside because the glass holds in the low-energy radiation and warms the greenhouse. Consequently, the warming of the Earth by these gases is called the greenhouse effect.

Carbon dioxide is not the only pollutant that causes global warming. Methane, nitrous oxide, and chlorofluorocarbons (CFCs) also trap heat. While carbon dioxide is the highest volume greenhouse gas, methane is more effective at trapping heat. (CFCs are also involved in depleting ozone in the stratosphere.) CFCs are produced for foams, aerosol sprays, and refrigerants. Methane is produced from heated peat, melting polar ice, wetlands, sewage treatment, and livestock. Nitrous oxide is produced from fertilizer production and use and deforestation. All of these activities contribute to the greenhouse effect.

Effects of Climate Change

1. Climate change reduces Arctic and Antarctic sea ice. This increases the level of the sea, and also releases methane gas that is bound up in ancient ice, further increasing global heating.

2. Alpine glaciers are also melting, which results in a decrease in the freshwater runoff to local human populations. Already, measured decreases have been observed in the glaciers of the Cascades, Rockies, and the Alps. Decreased glacier runoff would result in less water available for drinking and power generation downstream.

3. Higher ocean levels will flood major population centers near the coast. A 45-cm rise in ocean level will flood much of Florida and coastal areas in Bangladesh and Pakistan. Major cities, such as New York, London, Calcutta, Jakarta, and Manila, would need to be abandoned or spend billions to protect buildings from the rising sea.

4. Global warming will spread the range of disease-carrying vectors that normally inhabit warm climates. Already, malaria and encephalitis vectors (i.e., mosquitoes) are emerging in parts of America where they have never before inhabited.

5. Each species of plant and animal evolves within a certain temperature range at a given location. Most species co-evolve in coordination with other species in subtle ecological relationships. To shift the temperature in a region alters a major biotic factor for all living things. They may not be able to shift their location together fast enough to survive.

6. Coral reefs undergo "bleaching" at higher water temperatures. Coral reefs are almost as biologically productive as tropical rainforests, and support a similarly high level of biodiversity.

7. Permafrost in tundra biomes have already begun to melt. Further melting will continue to damage roads and buildings and—like the melting ice caps—release methane hydrate that would further trap heat.

8. Peat releases carbon dioxide and methane when warmed, which further contributes to the greenhouse effect.

9. Warmer oceans would allow more water to evaporate and cause more energetic storms. For example, there are more severe hurricanes each year than previously measured.

Chapter Three: Earth Systems and Global Changes

Case Summaries

The Yellowstone Hotspot

Principles covered include:

- **Convection of molten material in the Earth's mantle**
- **Plate tectonics**
- **Groundwater heating and circulation**
- **The dissolving process**
- **Bacterial tolerance levels**
- **Global consequence (climate change) due to a cataclysmic event**

We might think that Mount St. Helens provided a large explosion and ash plume in 1980, but that event was minuscule compared to each of three major explosions that have occurred near what is currently Yellowstone National Park. The Mount St. Helens eruption produced about 0.24 cubic miles of ash that spread across North America. About 630,000 years ago, the Yellowstone caldera explosion emitted about 240 cubic miles of ash. About 1.3 million years ago, the same hotspot spewed forth about 67 cubic miles of ash. The largest of the three eruptions was about 2 million years ago, when the Yellowstone hotspot erupted with about 600 cubic miles of ash—about 2,500 times the volume of ash emitted by the Mount St. Helens eruption!

Most interesting is the fact that the three eruptions—2 million years ago, 1.3 million years ago, and 630,000 years ago—came out at different points on the Earth's surface, even though they came from the same lithospheric vent. The crust moves to the southwest over the hotspot at a rate of about one inch per year. Therefore, the previous eruptions are currently part of the Snake River Plain in Idaho.

At the time of eruption, molten basalt builds up under the crust and creates a bulge. Once enough pressure builds up, the molten rock pushes through and throws huge amounts of crustal material into the air. The remaining crust settles back into the remaining hole in the Earth and forms a caldera—or wide crater. The most recent Yellowstone caldera is about 45 miles long and 30 miles wide. The basaltic magma settled underneath the caldera and solidified into iron-rich igneous rock.

Rain runoff and groundwater seep into the fractured areas around the caldera, descending until it is heated by the heat rising from the basaltic magma that still exists about 25 miles underground. When the rising steam carries enough energy to overcome the water seeping downward, it erupts to the surface as a geyser. The hot water produced from this geothermal heating process easily dissolves marginally soluble sulfide compounds that later precipitate when the water cools on the surface of the Earth—giving some of the geysers a sulfurous, egg-like smell. Also, certain types of bacteria are the only organisms to live in water that is this hot, which gives many of the pools the beautiful colors. Around the edge of the pools, different colonies of bacteria—each with a different color—grow in different temperature "climates."

Massive Volcanism and Human History

Principles covered include:

- **Volcanism, subduction zones, tsunamis**
- **Air pollution as a result of volcanism**
- **Climate change as a result of a cataclysmic event**
- **The effect of climate change on human history**

The Pacific Rim is sometimes called the "Ring of Fire" because the Pacific plate descends into several other tectonic plates. When one plate descends underneath another, the friction created produces heat and spawns volcanic activity nearby. In 1815 on an obscure island in Indonesia, the volcano now called Tambora erupted. Approximately 100,000 people died as a result of the initial blast and as a result of the subsequent famine that resulted when nearby rice fields were destroyed from ash and acid rain. However, the long-term effects—like any major volcanic eruption—caused cooling on a global scale, decreasing average temperatures by about 3°C. The decrease in available sun shortened growing seasons and caused other famines in North America and Europe.

In 1883 in the same chain of islands, the more-famous Krakatau erupted for the same reasons. The blast blew about seven cubic kilometers into the air, formed a 5-mile-wide caldera, and created a 140-foot-high tsunami wave that killed approximately 34,000 people in low-lying islands throughout the South Pacific. Five hundred miles away, observers felt earthquakes that probably corresponded to the drop of the volcanic material back into the caldera.

Coastal Vulnerability to Rising Sea Levels

Principles mentioned in this case:

- **Relationship between sea level and global temperatures**
- **The economic and cultural consequences of rising sea levels**

One of the most likely outcomes from global warming is a rise in sea level. While small rises in water levels may not significantly decrease the overall land area on Earth, most of the world's population lives near a coast where a small increase in sea level will push the shoreline many miles inland. Therefore, local agencies need to take into account future possible shoreline changes to adequately plan for future needs of local communities. In addition, this type of planning will also make communities aware of future risks of erosion, storm damage, and flooding. Unfortunately, most coastal communities are being developed so quickly little plan is being made for these eventualities.

Recent models suggest that the sea level on the east coast of the United States will rise between 15 and 95 centimeters by 2100, which is more than double the rise in sea level in the previous century. Furthermore, the rise in sea level can be accelerated by catastrophic events—such as hurricanes—that will cause immediate, large-scale changes in shoreline due to beach displacement.

Coastal Vulnerability to Hurricanes

Principles mentioned in this case:

- **Ecological impact of hurricanes**
- **Role of barrier islands in protecting coastal wetlands**
- **Relationship between global warming and the frequency of severe hurricanes**

During two months in 1992, four category 4 hurricanes (winds between 131 and 155 mph) reached the coasts of Louisiana, Florida, Hawaii, and Guam. These natural events resulted in loss of life, severe coastal erosion, and many billions of dollars in property damage. Hurricane Andrew almost completely stripped vegetation from northern portions of the Florida Keys and removed over 50% of the sand from the barrier islands off the coast of Louisiana. Much of the sand transported from the barrier island was deposited on oyster beds between the islands and the coast, smothering and destroying the oysters. Many thousands of acres of coastal wetlands—normally protected from storms by the barrier islands—were also destroyed. In Hawaii, Hurricane Iniki caused massive beach erosion and a 25-foot storm surge that stripped the coastline of vegetation and homes.

It would be rare to have one hurricane of this magnitude hit American soil in a decade, but in 1992, there were four in a span of 60 days. Many scientists feel that global warming is responsible for more severe storms. Warmer equatorial oceans allow more water to evaporate. The evaporated water stores the latent heat energy and builds weather systems, such as hurricanes. One can consider a hurricane a heat engine that transforms solar heat into a machine that moves water and air with great force over many thousands of miles. Of the natural catastrophes that befall Americans, hurricanes cause most of the insured property loss. The question remains: Will global warming bring us more hurricanes that will deplete our finances and devastate our coastal ecosystems?

El Niño and Landslides

Principles mentioned in this case:

- **Relationship between climate change and land use practices**
- **Global consequences of El Niño**

El Niño brings a considerable amount of moisture eastward across the Pacific Ocean. When this moisture collides with cool, arctic air, higher levels of precipitation are experienced in North America. When that precipitation is absorbed by soil, the increased weight of the soil can often exceed the combined factors of foliage and angle of incline and the land will slide. For example, during the 1997–1998 El Niño, there were many more landslides in North America. The U.S. Geological Survey (USGS) produced maps that assessed the risk of landslides for the United States and established a Landslides Hazards Program to help local communities prepare for landslides. While El Niño is an event that begins in Asia, it may result in increased landslides in Idaho. With more frequent El Niño events, community planners and builders should carefully consider construction plans on or near a hillside.

The Biosphere

Chapter 4

The Biosphere

Productivity and Food Chains

Solar photons initiate the process of **photosynthesis**, during which **chloroplasts** in plant cells convert water and carbon dioxide to glucose and oxygen in the following reaction.

$$6\ CO_2 + 6\ H_2O \rightarrow C_6H_{12}O_6 + 6\ O_2$$

The amount of photosynthesis that takes place is one indicator of an ecosystem's **productivity**. But productivity of an ecosystem can be measured in other ways, too. For example, productivity can be measured at any level in a biological community where organic molecules are processed by cells to form living tissue or waste.

Primary productivity is the amount of biomass produced by photosynthetic organisms. It is a measure of how much solar energy is converted into chemical energy through photosynthesis. Different ecosystems have widely diverse primary productivities. For example, a tropical rainforest is able to convert massive amounts of carbon dioxide to plant carbohydrates because there is a significant amount of water available. Conversely, water is a limited resource in a desert ecosystem, so photosynthesis produces less plant carbohydrate and living tissue, and the ecosystem is less productive.

Secondary productivity is the amount of biomass produced by organisms that eat photosynthetic organisms. It can also be indirectly measured by the amount of waste that a consumer produces.

Food chains are sequences of organisms that begins with a primary producer and trace the movement of biomass through a series of predator/prey relationships. An example of a food chain is seen when grass is eaten by a rabbit, then a coyote eats the rabbit. There are three members of this food chain, or three **trophic levels**. The first step on any food chain is always a producer that converts solar energy, carbon dioxide, and water to organic compounds using photosynthesis. The second step on a food chain is a consumer that feeds on the producer—either an herbivore or an omnivore, or a decomposer. The third step on a food chain is a carnivore. There may be several steps on a food chain, but the last step is always a decomposer that recycles the nutrients stored in living tissue back into the inorganic nutrient cycles. Each step along a food chain is a trophic level. The amount of biomass at each trophic level in a food chain can be visualized with a biomass pyramid, where each trophic level contains about 10% of the biomass held in the previous trophic level.

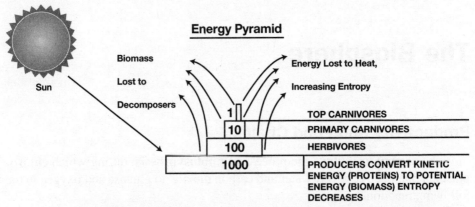

Figure 4.1 Biomass Pyramid — Three trophic levels, with energy that does not go into biomass in the next trophic level going into the environment as heat.

There is a different amount of volume associated with each tropic level because energy is lost to heat, metabolism, undigested food, or unconsumed food at each level of the pyramid. For example, a rabbit might not eat absolutely all of the lettuce in its surroundings; some might be consumed by decomposers or become part of geological sediments. Additionally, the lettuce the rabbit does eat is not all digested. Of the amount digested, much of those nutrients helps with the rabbit's metabolism and is expelled as unusable heat. Of the total biomass available to the rabbit, only a small portion actually becomes part of the rabbit's biomass. Likewise, only a small portion of the energy available from primary productivity becomes energy available from secondary productivity.

Food webs are interconnected food chains, where each organism is connected to one or more predator or prey relationships with other organisms. A food web that is highly interconnected and involves several species at each trophic level represents an ecosystem that is complex and can tolerate changes. The simplest food web—a linear food chain—represents a simple ecosystem where each species is highly dependent on the previous organisms in the food chain; such an ecosystem would not tolerate changes in the environment as well as a more complex ecosystem.

Population Interactions

Abundance and Diversity

Abundance is a measure of the total number of organisms of each species. For example, the mosquitoes are more abundant in a tundra ecosystem than are caribou.

Diversity is a measure of the number of different species, genetic variation, or habitats in the ecosystem. Biodiversity is more thoroughly discussed later in this chapter.

Niches and Competition

Habitat is the place in the environment where a particular organism, or community of organisms, lives. Habitat involves nonliving conditions, such as weather or terrain, as well as living conditions, such as plant cover or availability of food. For microorganisms, habitat may be described by a few parameters such as temperature and moisture; for more complex organisms, habitat involves many factors.

Niche is the combination of habitat and the ecological role that an organism plays in an ecosystem. Ecological role refers primarily to the method of obtaining food and the relationships an organism has with other organisms in the community. For example, where does an organism find its nutrients, what parasites does it host, and with what other organisms does it co-evolve?

Competition occurs when the niches of organisms sometimes overlap and two individuals need the same resources to survive. The more two niches overlap, the greater the level of competition; eventually, one or both species will engage in resource partitioning, or adapt a new niche in some way to reduce competition. There are two types of competition: intraspecific competition and interspecific competition. **Intraspecific competition** occurs when individuals of the same species compete for resources.

Population distribution suggests the level of intraspecific competition that takes place. Populations are typically a demonstration of one of three types of distribution: ordered, clustered, or random. An *ordered* population distribution suggests that members of the population are competing for resources. A *clustered* distribution suggests that members of a population clump together for protection, reproduction, or some other form of cooperation; intraspecific competition is low. A *random* distribution suggests that individuals are close to readily available resources.

Having different stages in the life cycle of an organism also reduces intraspecific competition between species. For example, many types of juvenile flies, or nymphs, grow and mature underwater in the beds of streams. Once they mature, they undergo metamorphosis, surface, and fly away. The adult never competes with the nymph for food, thus increasing the chance for both to survive.

Intraspecific competition tends to be more intense than interspecific competition because members of the same species compete for the same habitat and nutritional resources.

Interspecific competition occurs when individuals from different species compete for resources. Rainforest plants compete for available sunlight in a dense canopy by evolving large leaves, or conserve nutrients by growing up another plant's trunk rather than constructing their own. Interspecific competition pushes different species in a biological community to out-adapt other species through the process of natural selection.

Tolerance Limits

Tolerance limits are the minimum and maximum values for important environmental factors—such as temperature, space, and nutrients—that an organism needs in order to survive. Sometimes an organism's ability to survive is limited by a single environmental factor, which is called a critical factor.

Sometimes the existence of a species with specific tolerance limits and critical factors identify the health of a particular ecosystem; such a species is considered an indicator species. For example, mayfly nymphs require high levels of dissolved oxygen to survive in a stream ecosystem. Their presence is an indication that the level of oxygen dissolved in the water is high, so the mayfly is a good indicator species for high oxygen levels in a freshwater stream ecosystem.

Natural Selection

Members of a population with traits that survive in a particular environment will live long enough to produce offspring with those traits. Consequently, traits that allow individuals to survive better will tend to show up in succeeding generations. This process of selective survival is called **natural selection**.

There is a difference between adaptation of populations through natural selection and the adaptation of individuals through physiological adjustment. Individuals adapt to environmental changes because their physiology adjusts—like humans adapting our sleep schedule to a new time zone after traveling. Populations adapt over many generations through natural selection when a number of individuals with different traits experience varied reproductive success. New traits are introduced into populations through mutation, or a natural, random alteration of the gene that produces a particular protein.

Mutation introduces new traits in populations by slightly altering the DNA that carries the information needed to produce a particular protein. Most mutations are harmful, but if a mutation allows an individual to survive long enough for that individual to reproduce, then the trait will survive. Sometimes, random mutation brings an adjustment to the DNA that produces a protein that enables an organism to survive better.

Sometimes populations are divided by barriers and isolated from one another. **Speciation** occurs when each of the two subpopulations experience different combinations of mutation and selective pressures, and the traits that distinguish the two populations diverge enough so that they are not able to successfully interbreed and produce fertile offspring. Speciation is also an example of **divergent evolution**, where organisms with similar traits evolve so that they become less similar.

Convergent evolution occurs when different species evolve structures that are similar to one another, or occupy a similar niche in different ecosystems. For example, fish tend to have lighter underbellies that make them less visible from underneath, and coloration on top that blends with the substrate to make them less visible from above. Camouflaged coloration is a common theme of convergent evolution among both predators and prey; it helps predators more easily sneak up on prey and attack from above, and it allows prey to elude possible predators. The same trait is developed over many different species because all those species have the same need. Among the Galapagos finches studied by Charles Darwin, one species of finch has evolved a beak that looks more like a parrot's because that particular species of finch has assumed a niche which—like a parrot's—includes eating fruit. Two completely different species begin to look more similar—or converge—because they occupy a similar niche in different locations.

The concept of natural selection is important to this course because the world's current diversity of organisms evolved under specific selective pressures and critical factors. As human cultures alter those critical factors in the environment—temperature, pH, salinity, etc.—then many organisms will not survive, nor will the other species that are linked through symbiotic relationships.

Symbiosis

Symbiosis is a relationship between two organisms that co-evolve together. The three major types of symbiosis are parasitism, commensalism, and mutualism.

Parasitism is the type of symbiosis where one organism derives all of its nutrients from a host organism. A parasite is more successful if it does not kill its host. As its host evolves defenses against the parasite, the parasite evolves strategies to get around the defenses, but not so effectively as to threaten its own existence. Most macroscopic organisms are host to a number of parasites. **Ectoparasites** infect the outside of the host, such as fleas or ticks on a dog. **Endoparasites** live inside the host, such as tapeworms in people. The protist *plasmodium* is an example of an endoparasite that lives inside people and causes the disease malaria.

Commensalism is the type of symbiosis where one organism benefits and the other is neither benefited nor harmed. For example, a damselfish lives more safely among the protective stinging tentacles of a sea anemone, but the anemone gains no benefit from the presence of the damselfish. While *plasmodium* is a parasite for people, it has a commensalistic relationship with some mosquitoes, which act as vectors by transporting the *plasmodium* from host to host. The mosquito is not harmed by the *plasmodium*.

Mutualism is the type of symbiosis where both organisms in the relationship benefit. For example, bacteria find a home and receive necessary nutrients in human mouths. Such mutualistic bacteria compete with more harmful bacteria and help form a barrier of entry for pathogens. Nitrogen-fixing bacteria live in the nodules of plants called legumes; the plant obtains nitrogen, and the bacteria receive nutrients and a safe habitat. A tickbird grooms a rhinoceros and gains food and a place to live; in return, the rhinoceros is cleansed of parasites that might infect its rough skin. Coral reefs are composed of photosynthetic algae (a protist) and coral (an animal). The coral give algae a home, and the algae provide carbohydrates for the coral.

Traits of Biological Communities

Complexity

Complexity is the number of species at each trophic level and the number of trophic levels in the community. A biological community with a high level of complexity contains several trophic levels with several species at each trophic level. A complex community contains many different ecological connections between its members and an elaborate food web.

Stability

Biological communities remain stable because they have at least one of three traits: constancy, inertia, or renewal ability.

1. **Constancy** refers to a community's tendency to resist niche changes.

2. **Inertia** refers to a community's ability to resist changes that would damage or change the habitat.

3. **Renewal** refers to a community's ability to repair ecological damage. A highly complex community that has many species at each trophic level tends to be more stable. For example, if consumers have several food sources—as they would in a more complex ecosystem—they would still be able to survive if conditions shifted and one food source became extinct.

Succession

Succession is the process where niches and the composition of dominant species in a community change over time.

Primary succession occurs when species move into an unoccupied area. Organisms that are typically successful during primary succession reproduce quickly, survive in adverse conditions (have broad tolerance zones), and disperse easily (in order to get to the unoccupied area first). Species with these traits that colonize a new area during primary succession are called **pioneer species**.

In a rocky area, lichens may grow on the rocks and begin to build soil. Once enough soil has accumulated, grasses and herbs may be able to take root. Decomposers will break down dead, out-of-season grasses and further build the soil. Eventually, some trees may be able to grow. This entire process is considered primary succession, where an unoccupied area transforms into a biological community.

Secondary succession occurs when some change initiates a new biological community. This might occur after some disruption, such as a flood or fire. For example, after the pine forests were damaged from the great fires in the Yellowstone area in 1988, vast areas of ground were exposed to sunlight, which allowed for growth of plants that would normally only grow in meadows. Eventually, the pine forests will return, but only after a sequence of different plant populations rise, then give way to the next to eventually recreate the conditions needed for the forests to return.

Climax communities exist when succession has proceeded to the point where the biological community resists further change in composition. Different ecosystems are characterized by the climax communities they support. For example, the Yellowstone area succession seems to culminate with a pine forest until thunderstorm-ignited fires initiate secondary succession once again.

Biomes

Terrestrial Biomes

Tundra

Tundra is characterized by few if any trees, low moisture levels, extreme cold, and ground that is frozen (permafrost) during most of the year. There are two types of tundra. Arctic tundra is found at high latitudes in the polar region. Alpine tundra is found at high altitudes above the tree line. Plants are usually short because the growing season is very limited. Lichens are typical, particularly in alpine regions.

The cold temperatures freeze the little water that is available before it is absorbed into the ground, particularly in the arctic tundra. When this water thaws, it forms pools that allow short grasses to grow and insects to nest. These insects are food for migratory birds that arrive in the summer months to breed and lay eggs. Because of the very short growing season, the tundra is a fragile ecosystem with little top soil. Any environmental damage, such as an oil spill surrounding the Trans-Alaska pipeline or tracks from trucks, takes many years to recover from.

The Trans-Alaska pipeline was constructed in the 1970s to pipe oil from the oil-rich North Slope of the Alaskan coast down to warmer waters, where the oil can be piped onto a tanker and shipped to other places. Crude oil is very viscous and does not flow easily in the extreme cold, so it must be heated to high temperatures in order to flow through the pipeline. Periodic stations reheat the oil and pump it along. Occasionally, there is a breach or puncture in the pipeline. The extreme pressures cause the oil to shoot out and scar the land. The high heat of the oil melts the permafrost and creates a boggy, oily area that covers many acres and remains spoiled for many years.

Desert

Deserts are characterized by very low moisture—usually less than 25 cm per year. The low moisture level prevents top soil from accumulating, and the ground tends to contain less organic material and more rock and sand. The low specific heats of rock and sand (relative to water-filled leaves of other biomes) allow the desert to heat up faster during the day and also cool down faster at night. The rapid change in temperature creates more harsh air movement and higher winds during the transitional parts of the day. Although we think of deserts as always hot and sunny, some deserts—such as the Gobi Desert in Asia—exist at high latitudes and remain quite cool. In general, the hot deserts exist in areas on the earth where the convective cell tends to increase the temperature of the air with an earthward push of high pressure.

What little moisture there is available usually comes in the form of a quick thundershower and possibly a flash flood. Plants must adapt to quickly absorbing the water, and then retaining it for a long time. Many plants adapt by protecting their stored water with waxy leaves and a thorny barrier to animals that would take the plant's water for their own survival. Animals tend to be smaller, and often protect themselves from water loss by remaining underground during the day and coming out to feed and breed at night. Many animals obtain most or all of their water from their food, and have adapted some kind of structure in order to retain water.

Grassland

Grasslands—or prairies—receive about 25–75 cm of rain annually. The American Great Plains and broad regions in Africa and Russia are vast examples of this biome. Low rainfall, frequent fires, hot summers, and cold winters make it difficult for large trees to grow. Large herds of migratory grazing animals, such as horses, bison, or antelope, are typical. Most global grasslands have been converted to agricultural areas; the areas with the best soil grow crops, the areas with the weakest soil grow cattle and sheep. Desertification is a risk for grasslands that are either overgrazed or over harvested without concern for soil conservation techniques.

Savannah

Savannahs are similar to grasslands, but exist in more tropical latitudes and receive more rainfall (between 50–150 cm annually). The predominant mammals are grazers, but their high prevalence creates a niche for large carnivores.

Chaparral

The chaparral is a dry ecosystem characterized by wet winters and long, dry summers. Chaparral contains low, woody trees that are adapted to retaining water, such as mesquite, manzanita, oaks, and eucalyptus. These trees and other chaparral flora grow during the wet season, then shed their leaves to avoid transpiration during the dry season. These trees form a low canopy, below which is coarse, rocky ground and a light cover of decaying leaves from the previous season. Fire is an important aspect of the chaparral. The seeds of many plants open up only in the high temperatures of a fire, and secondary succession is rapid.

Temperate Deciduous Forests

Oak, maple, birch, beech, elm, ash, and other hardwood trees dominate temperate deciduous forests, where the leaves are shed in the fall. Underneath the tall trees, a wide diversity of shrubs and smaller trees usually also grow. Most of the original old-growth deciduous forests in the United States were harvested as the original European settlers moved west, and a second generation is now maturing. The largest temperate deciduous forests are in Siberia, but they are being logged at a high rate.

Taiga

Also called the **northern coniferous forests**, or **boreal forests**, the taiga is characterized by spruces and firs, with a short summer and long, snowy winter. There is slightly more moisture in a taiga than tundra, but when it falls as snow, it remains above the ground, and plants need to adapt to a cold, dry environment in order to survive. Coniferous trees are well suited to the taiga's climate. The waxy, long, thin needles reduce moisture loss. Flexible branches allow snow to accumulate and then fall off without hurting the tree.

Large mammals are able to act as grazers and top carnivores in taiga forests. Deer and elk graze in fields that are created from beaver activity or forest fires. Moose stand knee-deep in the many lakes and ponds to eat the tender shoots of aquatic grasses. Grizzly bears eat fish in the spring and berries in the fall, then hibernate through the winter months. Wolves hunt for rodents, marmots, and deer weakened by the cold.

Coniferous trees use needles, instead of leaves, because they are better adapted to cold, dry temperatures. The litter below the trees is composed primarily of dried needles, which create a highly acidic soil. Because of the cold temperatures and acidic soil, bacteria have a difficult time growing; various fungi tend to be the main decomposers. Some fungi live in a mutualist relationship with the tree roots. The fungi recapture nutrients from the needles, and the tree roots recapture the nutrients from the fungus.

Taiga forests have adapted to occasional lightning-ignited fire. The cones of lodgepole pines in the Greater Yellowstone ecosystem will only open and successfully germinate at the high temperatures caused by fire. After a fire, succession and the fertile top soil quickly restore a burned area. The long, straight trees that grow in taiga forests are highly prized for lumber, and deforestation is a major threat to this ecosystem.

Tropical and Temperate Rainforests

Low pressure zones at the equator allow for humid, tropical air to release water and form tropical rainforests. At high altitudes, these forests are cool and are called cloud forests. At lower altitudes, tropical rainforests receive about 200 cm of rainfall annually. Plants compete for sun in the lush growth by having broad leaves. The highly competitive biotic communities readily use decaying nutrients before they become a part of the soil; most of the nutrients are part of a living organism and the nutrients available for soil building are sparse. Plants adapt to the shallow top soil by having broad, shallow roots. The tallest trees in a tropical rainforest have buttressed roots that fan out at the base of the tree. The minimal top soil washes away easily when a portion of the forest is logged or is converted to agriculture, which makes it difficult to replant.

Temperate rainforests occur at higher latitudes where coastal mountains cause high precipitation—such as the Olympic Peninsula in Washington State. Mosses, ferns, and large conifer trees are dominant plants.

Marine Biomes

Marine biomes refer to the saltwater ecosystems in the ocean, which include the pelagic, benthic, and coastal ecosystems.

Pelagic Marine Ecosystems

The pelagic ecosystem is open ocean, where actively swimming creatures and suspended plankton exist and no land is available. It is composed of the **neritic** zone on the continental shelf, the **photic** zone in the open ocean regions that light can penetrate, and the **aphotic** zone in the open ocean where light does not penetrate.

The bottom of the pelagic food pyramid is characterized by plankton. **Phytoplankton** (which undergo photosynthesis) exist near the surface of the ocean. **Zooplankton** (microscopic larval forms of several invertebrates) usually live in deeper waters, then come to the surface at night to feed off the phytoplankton. Small fish eat the zooplankton, and large fish eat the small fish. Top carnivores include tuna and salmon, which are in turn eaten by the occasional shark or orca. The largest of the pelagic dwellers—the whales—eat from the lower levels of the food pyramid by taking in large volumes of plankton-rich water and then squeezing the water out through baleen filters around their enormous mouths.

While terrestrial ecosystems are often characterized by the amount of moisture and temperature, marine ecosystems are characterized by the amount of sunlight, nutrients, and oxygen dissolved in the water. Areas that exist in places where currents have churned up nutrients that have fallen to lower levels, or deposited by a river, tend to have greater productivity. Colder water can dissolve more oxygen, but warmer water allows chemical reactions to proceed and fosters richer plankton growth.

Benthic Marine Ecosystems

The benthos, or benthic, ecosystems exist near or at the bottom of the sea from the coastline to the deep abyss. The four most distinguishing characteristics that define the benthos are the constant rain of nutrients from above, the substrate of the bottom (sand, rocky, silty), and the depth (which determines both the amount of sun and pressure). There are different types of benthic ecosystems; the most common include the kelp forest, coral reef, mangrove swamp, and the abyss.

The depth of the water determines the amount of sun available for biological productivity, as well as exposure to physical challenges, such as wave action at the surface and water pressure at greater depths. For example, algae that grow on the rocks need to absorb sunlight in order to undergo photosynthesis. In shallow water, green algae are able to absorb a wide portion of the solar spectrum. In medium depths from 15–70 feet, brown algae are able to grow. The most abundant and ecologically significant of the brown alga is kelp, which attaches to the bottom and extends to the surface. Kelp forests provide homes for invertebrates in the convoluted holdfasts that attach the kelp to the bottom, and the towering leaves provide an important habitat for otters and seals. Deep water filters out the long wavelengths of light and the red algae are able to absorb the more energetic violet and ultraviolet light that penetrates the water. In warmer waters, coral reefs and mangrove swamps are two common benthic ecosystems.

1. Coral Reefs

    Coral reefs are built from a mutualistic relationship between photosynthetic algae and a marine invertebrate. The invertebrate provides a calcified home and feeds by extending small fan-like filters into the water, which catch food and capture sunlight. Other organisms find homes in and around the coral structure. Coral reefs are one of the most biologically productive ecosystems in the world, just behind tropical rainforests.

2. Mangrove Swamps

    Mangrove swamps occur near the shore where highly adapted, salt-secreting trees grow with their roots in the water. Like coral reefs and kelp holdfasts, the mangrove roots provide a home for marine invertebrates and small fish. Bacteria that live among the sediments trapped in the roots digest toxins that flow through, so mangrove swamps serve as a kind of filter in coastal areas. Eventually the trees grow close enough together where a terrestrial community is able to establish itself on top of the roots. Mangroves are found in the Caribbean, Southeast Asia, Africa, and tropical islands throughout the world.

3. The Abyss

    The **abyss** exists at great depths in the ocean. For biomass and energy, abyssal organisms either scavenge on the steady rain of decomposing biomass from above, or in some areas of the world, derive nutrients from the heat and sulfide gases of thermal vents.

4. Estuaries and Salt Marshes

Estuaries and **salt marshes** are unique because they represent the interface between a river ecosystem and a coastal marine ecosystem. Organisms must adapt to the salinity of the water that fluctuates with the tides. The river brings large amounts of nutrients into shallow water that allows photosynthetic algae to be highly productive. Young crustaceans and juvenile fish eat the algae and live among the grasses that grow along the shore. Like mangrove swamps, estuaries trap and filter sediment. As the sediment accumulates, the estuary becomes a salt marsh ecosystem, such as those found in the southeastern Atlantic seaboard. Both estuaries and salt marshes are important habitats for migratory birds.

In the salt marsh, a rich spongy soil builds up as the marsh snags detritus and decaying organisms. Near the water, mudflats are a microbial soup, with crabs and other invertebrates feeding from the organic material. Low wave action allows small particles to settle here, so little oxygen can penetrate into the mud. Bacteria undergo anaerobic digestion, which involves the reduction (adding hydrogen) of carbon or sulfur, rather than oxidation (adding oxygen). As a result, the marshes produce methane (natural gas) and hydrogen sulfide (which smells like rotten eggs).

5. Intertidal Ecosystem

The **intertidal** zone represents the region between low and high tides. Organisms must adapt to the forces of moving water and waves, and the periodic exposure to both open air and salt water immersion.

The food chain is based on organic nutrients that become washed ashore and are digested by decomposers, hearty algae that cling to rocky outcroppings, and invertebrates that are able to protect themselves from the changes in abiotic factors by either burrowing in sand or living behind a protective coating. Clams, for example, employ both strategies. Worms, protozoa, and bacteria live between the grains of sand on the intertidal beach, while filter-feeders, such as clams, mussels, and crabs, feed on them. Hardy species of starfish and sea anemones will inhabit tide pools and envelop a passing invertebrate.

Coastal Ecosystems

1. Supratidal Zone

Above the intertidal zone is the beach, which is either rocky or sandy. A sandy beach usually has a row of dunes above the high tide mark, where sand has blown and settled. Hardy dune grasses create a loose network of roots that help keep the dunes in place. Grasses and plants generally have small, sometimes waxy leaves to limit their water loss due to the exposure to salt spray, direct sun, and wind. Sand fleas feed on microbes, crabs feed on the fleas or small sand shrimp, and birds feed on the crabs.

2. Barrier Islands

Wave action over time has transported and deposited sand on the eastern continental shelf, creating an array of barrier islands, which in turn protect the mainland from storms. Seeds blow or drift to the island and pioneer new communities that mirror nearby coasts. The substrate is highly unstable; a strong hurricane can (and does, as in the case of Hurricane Andrew) annihilate whole islands.

Freshwater Ecosystems

Streams and Rivers

Because the flow of water in streams and rivers decreases the ability of photosynthetic plankton to remain, the major source of nutrients for this freshwater aquatic ecosystem is from terrestrial sources that fall into the water. Many insects develop during larval stages in the mud at the bottom of a stream (see Chapter Nine). Consequently, understanding which insects can only survive in oxygen-rich streams provides an indicator for the health of a stream. In general, fly nymphs are much more sensitive than mosquito or dragonfly nymphs.

Swamps, marshes, and other wetlands are a transitional ecosystem between freshwater streams or lakes and terrestrial ecosystems. Swamps contain trees in a permanently flooded flatland, while marshes are dominated by various grasses.

Lakes and Ponds

Lakes and ponds that are deep and tend to have little biological productivity are considered **oligotrophic**, while lakes and ponds that are shallow, warm, and nutrient rich are considered **eutrophic**. Specific portions of the lake are divided into zones. The portion of the lake near the surface, where photosynthetic plankton can live, is called the **euphtotic zone**. The **benthic zone** refers to the lake bottom. The shallow area of the lake that can sustain rooted vegetation is the **littoral zone**. The portion of the open water lake below the euphotic zone that cannot sustain rooted vegetation is called the **limnetic zone**.

Temperature zones in a lake shift with the seasons. In winter, the less dense ice rests on the more dense, warmer water. (Ice is most dense at 4°C.) In spring, the surface warms and, along with wind action, causes the surface water to sink and mix with deeper water in a process called **spring turnover**. In summer, the lake surface continues to warm and the difference between the surface and bottom temperatures increases. During this summer stratification, the upper layer of warm water is called the **epilimnion**. At the bottom of the lake is the colder, undisturbed **hypolimnion**. Between the warm epilimnion and the cold hypolimnion lies the **thermocline**—or temperature gradient—that lies in a zone called the **mesolimnion**.

Biodiversity

Genetic Biodiversity

Genetic diversity refers to the variety of different genes within the collective gene pool of a population. The importance of genetic diversity is based on an understanding of the role of genes and how they are expressed.

Genes are portions of DNA that contain the information needed to produce specific proteins. There is a separate gene for each protein produced by an organism. There may be many different versions of a gene within a population of a single species, and some genes will produce proteins that will be more useful to an individual than the proteins produced by genes in other individuals.

Gene expression is the cellular process of converting the information contained in the sequence of base pairs within a gene into a protein. Not all genes are expressed. For example, humans have two sets of each gene—one given by each parent, but often only one gene is expressed and is considered dominant. The nonexpressed gene is recessive.

Gene pool is the collective group of traits that exist in all the chromosomes of all the individuals in a population. It can also be considered the cumulative library of genes—expressed and nonexpressed—that occur in a population. While one set of genes may prove useful to individuals in a population for a time, a change in the environment might confer a survival advantage to another set of genes in the future. Therefore, the survival of a population in many different conditions depends on having a diverse pool of genes for each trait.

Mutations are random changes that change a gene that expresses one type of protein into a gene that expresses another type of protein. Mutation naturally introduces genetic diversity into a population's gene pool; although most mutations go unnoticed or are damaging to individuals, a single beneficial mutation may ultimately save an entire species.

Founder Effect

The **Founder Effect** is a limit to genetic diversity that is created when a small group of organisms begin a new population. For example, if a red-headed family began a new population on a desert island, all subsequent generations would have red hair. A corollary of the Founder Effect is the **Bottleneck Effect**, which is a limitation of gene pool diversity because some cataclysmic event has created only a few survivors, so that the gene pool of the subsequent population is limited by the available genes that the survivors transmit to the first generation of offspring.

Species Biodiversity

Species biodiversity refers to the number of available species in an ecosystem. If a community contains a broad number of species, it has a high level of species diversity. Broad species diversity depends on a population sustaining broad diversity at the genetic level. Without genetic biodiversity, a population would become extinct and thereby decrease species diversity.

Minimum Viable Populations

If the number of individuals in a breeding population drops below a critical value, then sexual reproduction of the adults will no longer provide the recombination of genes that is necessary to have the population survive over time. While the species may continue for several more generations, it is only a matter of time until the genetic diversity is not sufficient enough to provide the genes necessary to adjust to environmental changes in the surroundings.

Benefits of Species Diversity

An ecosystem benefits from having a broad species diversity. For example, if there are a number of different producers in a biological community, a change in climate and elimination of one producer may not pose as high a threat to all the species in the community.

Humans also benefit from the existence of broad species diversity. Much of what we do depends on adequately functioning ecological processes. Soil formation and nutrient cycling that we need to grow crops depend on microorganisms, for example. Species biodiversity means that humans have a broad number of species that are available for food and medicines. Our own survival might be compromised if we depended on a few species, and then a change in climate eliminated those essential species. Also, there is some aesthetic value for humans in appreciating broad biodiversity. This might take the form of enjoying a colorful garden, seeing wild elk in the wilderness, or enjoying the serenity of fishing in a remote stream. Some find value in simply knowing that a species exists somewhere in the world. For example, even though you might not have had—or want to have—an experience with a grizzly bear, many find comfort in knowing that such a magnificent creature exists.

Some would also say that there is value in having broad biodiversity beyond the functional value in holding an ecosystem together or making life better for humans. These people could be considered biocentric, and hold that living organisms have an intrinsic value that is quite apart from their functional value to humans or other species.

Threats to Species Abundance

Species can become extinct because of natural reasons, as well as human causes. Natural extinctions may occur from climate changes that result from cataclysmic events, such as volcanic activity or asteroid impact, or from long-term changes, such as continental drift.

Humans cause extinctions from over-hunting or a level of use of a species that outpaces its ability to reproduce itself. The most severe cause for human-imposed extinctions is from habitat destruction or fragmentation, where humans denude or develop an area and remove the habitat for species—sometimes even without knowing that a species has been affected. The spotted owl decimation as a result of logging in the Pacific Northwest is a prime example.

Sometimes human activity has caused extinction because a new species has been introduced that out-competes the species that originally evolved in that area. For example, kudzu vines were introduced in the 1930s by the U.S. Soil and Conservation Service to control erosion. By now, this pernicious Japanese vine has taken over and eliminated natural flora in large portions of the Southeastern United States.

While kudzu was purposefully introduced and carried an unintended outcome, many species are inadvertently spread long distances in this age of global business and jet travel. One example is the Asian long-horned beetle, which was probably brought to the United States on wood palettes or products. The beetles burrow into trees, inhibit the flow of sap, and eventually kill the trees.

Traits of Endangered Species (See also Chapter Nine)

Species that easily become endangered tend to have certain characteristics that either make it more vulnerable or decrease its resilience. The following are typical traits of species that are more easily endangered.

1. The species uses a logistic population growth strategy for one of the following reasons:

    a. Individuals of the species have a long life span and/or a long generation time; a lot of resources are invested in a single individual.

    b. The species has a large body; again, a lot of resources go into the success of a single individual.

    c. The species is a carnivore or top carnivore. This increases the species' dependence on the success of lower trophic levels, and increases the effect of bioaccumulation and biomagnification of toxic substances (see Chapter Six).

    d. The species has a low reproductive rate, which is consistent with the theme of investing large amounts of energy and materials into single individuals. Low reproductive rates allow for increased survival of individuals (e.g., bears), rather than producing many offspring over a short time in the hope that a few will survive (e.g., flies).

2. The species requires a large amount of land per animal for one of the following two reasons:

   a. The species is solitary and requires a large roaming area to find a mate or food (e.g., wolves).

   b. Seasonal migration requires multiple and intact biomes (e.g., migratory birds, such as geese).

3. It is a specialist species with a narrowly defined niche.

4. The species demonstrates one of the following reasons for having low genetic diversity:

   a. **Genetic bottlenecks** occur when a catastrophe has eliminated many individuals and the subsequent population has a gene pool that is limited by that of the original breeding pairs.

   b. **Genetic isolation** has occurred because a small number of individuals have been isolated; the gene pool of the subsequent population is limited by the genes available in the original breeding pair.

   c. **Genetic assimilation** threatens some endangered species when they crossbreed with closely related, more hardy species.

5. The endangered species competes for resources with a dominant, hardier species. Unfortunately, humans are one of those species that is out-competing many other species for global resources—such as competing with spotted owls for old growth forests.

6. The species has a low tolerance for pollution, and development or any association with human activity results in over-powering the organism's ability to survive.

Ecosystem Biodiversity

Ecosystems evolve and continue to shift over many millennia. They steadily provide essential habitats for a wide range of species upon which we depend. Diversity of ecosystems is important to support a variety of biome types for their intrinsic value, and also to provide the habitats necessary to support the species that depend on them.

Ecosystem biodiversity is connected to genetic and species diversity in that enough habitat is needed to sustain large enough populations so that there is significant genetic recombination with each generation. If the habitat is not large enough to sustain breeding requirements of a population, the number of genes being passed to the next generation decreases, and the population is less able to adapt to new changes. The Florida Panther is an example of a drop in species diversity that has come from a drop in genetic diversity, which in turn came from a drop in ecosystem diversity.

In addition, the habitat must be able to supply enough food to support a population. Decreased bamboo forests in China have endangered the survival of the panda, for example.

Fragmentation of a population occurs when human development has turned a large contiguous ecosystem into a patchwork of subpopulations that are unable to interbreed, or have a limited roaming region. When the breeding of populations is restricted, genetic isolation or bottlenecks can occur. Fragmentation also amplifies **edge effects**, or the effects experienced by an ecosystem when it borders another ecosystem. For example, the edge of a forest will experience greater light penetration than the core of a dense forest. As a result, different ground shrubbery and corresponding fauna will prevail at the edge.

Managing Biodiversity

Government Agencies and Public Lands

1. **The Endangered Species Act** (ESA) of 1973 has served to protect the habitats of endangered species. (*Endangered species* are those that are considered near extinction; *threatened species* are those that are nearly endangered.) The ESA limits harvesting, transporting, or selling endangered species, but it also limits the commercial use of habitats occupied by endangered species.

2. **National parks** are operated by the National Park Service, which is part of the Department of the Interior. The National Park Service was founded in 1916 to protect natural features and historic sites. Wildlife may be managed and studied in parks by rangers and ranger-sanctioned researchers, but wildlife may not be harvested.

3. **National forests** are operated by the U.S. Forest Service, which is part of the Department of the Interior. "Land of many uses" is the phrase that often accompanies the entry sign into national forests and encompasses the utilitarian mission of forest service land. National forests sometimes provide a buffer to national parks or wilderness areas, but the forests may be used for hunting, mining, logging, or other commercial enterprises at the discretion of the Forest Service.

4. **Wilderness areas**, also managed by the Forest Service, were established by the Wilderness Act of 1964. Wilderness areas are large, unspoiled areas where no mechanical or artificial devices are allowed to operate, no commercial ventures can be established, no hunting can take place, and no roads can be built. Most wilderness areas are in the west or Alaska.

5. **Wildlife refuges** are managed by the U.S. Fish and Wildlife Service, also in the Department of the Interior. Fifty-one refuges were established by Teddy Roosevelt in 1901; now there are about 511 refuges that represent every ecosystem in the United States. While every state has at least one refuge, most of the acreage involved with this program is in Alaska. Refuges were originally established to provide areas free from hunting, but some hunting, cattle grazing, oil drilling, and wilderness vehicle use is now allowed in refuges. The debate over allowing drilling for oil and gas in the Arctic National Wildlife Refuge on the north slope of Alaska characterizes the many pressures exerted on retaining true refuge for wildlife (see Chapter Eleven).

6. **Bureau of Land Management** (BLM) is also part of the Department of the Interior, and manages commercial use of the wilderness through logging and mining leases. Hunting and fishing, managed by the U.S. Fish and Wildlife Service, also takes place on BLM land.

7. **The Environmental Protection Agency** coordinates with the Department of Interior and the Department of Justice to enforce the Endangered Species Act (ESA) and related legislation that protects habitat and endangered species.

Management Techniques

1. **Gap analysis** by ecosystem policy makers helps to identify ways to cushion fragmented populations and provide public land to buffer core populations.

2. **Establishing corridors** between fragmented populations serve as genetic bridges between populations isolated by human development. Corridors also provide an escape route for endangered populations in the event of a local disaster, such as fire or flood.

3. **Wetland protection** through protecting flood plans, minimizing channelization (see Chapter Nine), minimizing lock and dam systems, decreasing beach erosion, preventing groundwater pollution, and refilling previously drained wetlands all protect this essential biome. Re-establishing wetlands provides a place for migratory birds to land en route to seasonal habitats. Wetlands also cleanse water as it passes through, keeping it cleaner for wildlife and humans.

4. **Range management** will help prevent overgrazing of public lands and desertification of grasslands. Range leases for public lands, such as national forests, can be issued at a rate that will allow the land to replenish itself between grazing seasons.

5. **Debt-for-nature swaps** allow financially destitute countries to trade development-related debt for an agreement to hold large tracts of land in reserve for wildlife and habitat preservation.

AP Environmental Science

6. **Fire management** has vastly improved—amidst sharp debate—in the last few years. There are drawbacks to both a "let burn" policy and an immediate extinguish policy in the public lands. Most policies are designed to protect public and private buildings and people. However, to some degree it is better for the environment to let fires in the wilderness burn.

7. **Land reclamation** is required after land has been used for mining since the enactment of the Surface Mining Control and Reclamation Act (SMCRA) of 1977. The SMCRA establishes "state primacy," which is state control over how land is reclaimed after it has been used for mining. Land reclamation returns commercially exploited land to an ecologically useful natural habitat.

8. **Land use mitigation** is a type of trade for developers that allows them to purchase and protect one tract of land in exchange for developing another tract of land. For example, coastal condominiums might go up near the beach, but developers can establish and protect wetlands nearby in an area that would be ecologically equivalent but less commercially desirable.

9. **Captive breeding programs and zoos** provide an essential rescue process for species on the brink of extinction. Although a species may never regain the genetic diversity once enjoyed, breeding in captivity may allow further generations of an endangered species to be reintroduced into the wild to establish new populations.

Case Summaries

Restoring Wolves to Yellowstone

Principles mentioned in this case:

- **Species biodiversity**
- **Biodiversity management**
- **Niche competition**

In the early portion of the twentieth century, ranchers and park officials were encouraged to shoot wolves so that they could not harm grazing animals in and around Yellowstone National Park. The steady removal of this single species eliminated the top carnivore of the largest contiguous wild ecosystem in the United States. Now biologists assert that wolves need to return to Yellowstone in order to have it be a complete and natural ecosystem.

Several misconceptions about wolves fueled the elimination of this majestic predator and made the re-introduction of the wolf controversial. Some were concerned that wolves compete with grizzlies; but they occupy completely different

Chapter Four: The Biosphere

niches because they seek different sources of food. Some were concerned that wolves would kill cattle and people, but studies show that very few cattle are lost to wolves, and there are virtually no reports of killings of humans by wolves. Some feared that wolves would deplete the celebrated herds of buffalo, elk, moose, deer, and antelope in Yellowstone; but wolf predation only removes the old and weak, allowing the gene pool to remain strong and conserving resources (especially during winter) for more highly adapted individuals.

Once re-introduced in Yellowstone Park, wolves tripled their population in just three years. Coyote populations seem to have suffered the greatest interspecific competiton, but the smaller mammals preyed upon by coyotes have flourished.

The re-introduction of wolves to Yellowstone has been a hotly contested debate and continues to receive court challenges. Ranchers, who now worry more about wolves attacking livestock, are among the strongest adversaries of the plan.

Seafood Markets Threaten Marine Biodiversity

Principle mentioned in this case:

- **Species biodiversity**

While seafood is a good source of protein that is low in certain fats, eating some seafood causes devastation to biodiversity that is unseen by a consumer. Sharks, swordfish, and marlins mature very slowly so that even low levels of commercial fishing have severely depleted these species. Harvesting shrimp and yellowfin ("chunk light") tuna also destroys endangered nontarget species. Albacore ("white") tuna, red snapper, and groupers are becoming very scarce. In addition, seafood is able to concentrate toxins from the environment in the meat. Seafood consumers should be aware of the food they eat and the effects that markets they encourage have on the environment.

Thermophilic Bacteria in Hotsprings: An Extreme Ecosystem in Miniature

Principles mentioned in this case:

- **Biotic and abiotic factors**
- **Reflection on the size of an ecosystem**

Each ecosystem is defined by its own specific set of biotic and abiotic factors. The abiotic factors are usually determined by temperature, moisture, and available nutrients; the extreme temperatures of Yellowstone hotsprings are no different. Within many hotsprings live colorful bacteria—a different color for each temperature range. Some of these bacteria have been found to contain chemicals that are used in research and medicine. Like any other ecosystems, the bacteria are limited by available nutrients and—at lower temperatures—can be prey for other organisms who take advantage of the bacteria's ability to convert dissolved nutrients into biological molecules.

Large Herbivores: The Ecological Impact of Eating from Low Trophic Levels

Principles mentioned in this case:

- **Trophic levels**
- **Ecological impact of food production**

Consider the largest organism on Earth, the mighty blue whale, which sustains itself with some of the smaller organisms of the sea, phyto- and zooplankton. By evolving this ability, the blue whale has taken advantage of an important ecological principle: Eating from trophic levels closer to producers minimizes the amount of energy that is lost to the ecosystem as waste heat going from one trophic level to another.

This principle is also seen with humans. For example, 16 pounds of grain are needed to make one pound of beef, and four pounds of grain are needed to make one pound of eggs. So 16 pounds of grain can feed one beef-eating human, four egg-eating humans, or 16 grain-eating humans. For an ecosystem that has a hard time supplying energy for large organisms—whether it be whales or people—it helps more organisms survive if they eat from a lower trophic level.

Silent Spring: Biomagnification of DDT

Principles mentioned in this case:

- **Biomagnification**
- **Bioaccumulation**
- **Vectors and pathogens**

In 1962, marine biologist Rachel Carson published the book *Silent Spring*. In this seminal work, Carson identified the environmental threat to our waterways, wildlife, and our own health from using fat-soluble pesticides—particularly DDT—to eliminate pests. For many years, Dutch Elm Disease plagued small towns and cities around America. Municipalities attempted to combat the disease by killing the insect vector with DDT, which did not apparently affect humans at first. However, DDT is fat-soluble and is easily retained in the bodies of organisms—particularly birds—that ingest the insects. As birds ate more and more insects, the DDT would continue to accumulate until it reached toxic levels for the birds. One sign of toxicity among birds was that the shells would be so soft that they could not sustain the life of young birds and would break under the weight of the incubating female. Within a single generation, entire bird species were at risk of being eliminated in our effort to eliminate Dutch Elm Disease with DDT. The principle of biomagnification demonstrated by DDT accumulation in predator species, and also in humans, contributed to legislation in the 1970s that regulated the use of fat-soluble pesticides like DDT.

Population Dynamics

Chapter 5

Population Dynamics

Fundamentals of Population Biology

Population Growth Dynamics

General Facts

1. In the year 2000, there was a net global gain of 3 humans per second, 226,000 births per day, or 78 million per year.

2. At the current 1.5% growth rate, the population will double in 48 years to over 12 billion people.

3. In the U.S., the population is expected to go from 293 million in 2004 to over 419 million in 2050.

4. About 20% of the Earth's population controls about 90% of the world's wealth.

Definitions

1. **Fertility** is a measure of the actual number of offspring produced. It is often expressed statistically as the **crude birth rate**, which is the number of individuals to be born per thousand people per year.

2. **Fecundity** is the physical ability of an organism to reproduce. Fecundity is what people actually refer to when saying "fertility" in casual conversation.

3. **Natality** is the production of new individuals.

4. **Morbidity** is the level of illness in a population.

5. **Mortality** is a measure of the actual number of individuals who die in a population. It is often expressed statistically as the **crude death rate**, which is the number of individuals who die per thousand per year.

6. **Survivorship** represents the number of people in a given age bracket who continue to remain alive each year.

7. **Life expectancy** is the most probable number of years an individual will survive.

8. **Life span** is the longest length of life reached by a given species.

9. **Total Growth Rate** is the sum of the increases to the population due to immigration and births, minus those individuals who have died or emigrated away.

10. **Natural Growth Rate** is the population growth due only to births and deaths, usually the crude birth rate minus the crude death rate.

Growth Rates

1. Arithmetic Growth

    Population growth that increases by a constant amount over time is considered **arithmetic**. An arithmetic series is one in which the same number is added to the previous number. For example, 0,4,8,12,16,20 is an arithmetic series. A town would experience arithmetic growth if it increased by 100 people every year—regardless of its size. On a population vs. time graph, arithmetic growth would be characterized by a line with a constant slope (see Figure 5.1). Arithmetic growth is not typical for populations unless other factors—such as birth and death rates—are in balance, or if there is some limitation of resources.

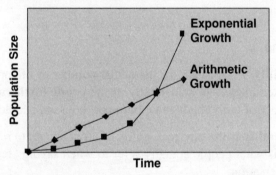

Figure 5.1 A comparison of arithmetic and exponential growth curves.

2. Exponential Growth

    More typically, populations experience **exponential** or logarithmic growth because each new generation of individuals produces more potential reproducers than the previous generation. In an exponential series, each value increases by a value that is itself increasing. For example, the series 2,4,8,16,32,62 demonstrates exponential growth. Exponential growth is particularly apparent in populations where many members of each generation survive long enough to have multiple progeny. The type of curve that demonstrates exponential growth is sometimes called a J-curve because of its shape.

3. Calculating Growth Rates

   a. Adding and Subtracting Rates

   To find the total population growth rate, find the difference of the birth rate and death rate, and add it to the difference between the immigration rate and the emigration rate; or

   $$N = B - D + I - E$$

   Where

   $N =$ Total population growth rate, per 1,000 people per year

   $B =$ Birth rate, number born per 1,000 people in one year

   $D =$ Death rate, number died per 1,000 people in one year

   $I =$ Rate of growth due to immigration, per 1,000 people per year

   $E =$ Rate of growth due to emigration, per 1,000 people per year

   Example problem 5.1: A town of 1,000 people experienced 16 births and 12 deaths. What is the natural annual rate of growth as a percentage? (Assume no immigration into or out of the town.)

   Solution: Use the formula $N = B - D$; where $B = 16$ per 1,000 and $D = 12$ per 1,000. Then multiply by 100 to get a percentage.

   $$\frac{16 - 12}{1,000} \times 100 = 0.40\%$$

   Example problem 5.2: A town of 20,000 people experienced a crude birth rate of 48 and a crude death rate of 18. Crude immigration and emigration rates were 12 and 3, respectively. What is the total annual rate of growth as a percentage?

   Solution: Use the formula $N = B - D + I - E$ where $B = 48$, $D = 18$, $I = 12$, and $E = 3$. Then multiply by 100 to get a percentage.

   $$\frac{48 - 18 + 12 - 3}{1,000} \times 100 = 3.9\%$$

b. Exponential Growth Simplified

While one can use the law of exponents to calculate growth rates and the number of individuals in a population at any particular time, questions in AP Environmental Science can usually be answered with a simplified approximation of the law of exponents, represented by the following equation:

Annual Percent Growth × Doubling Time = 70

Example problem 5.3: A country grows at an annual rate of 4%. How many years will it take to have the country reach a population that is 4 times as large?

Solution: For a country to reach a population that is four times as large means that the current population will have to double in size twice; it will undergo two doubling times. Use the equation,

Annual Percent Growth × Doubling Time = 70,

to find the doubling time. That amount of time is multiplied by two to arrive at the total number of years needed to quadruple the population size.

Doubling Time (in years) = $\frac{70}{4}$ = 17.5 years

17.5 years × 2 doubling times = 35 years needed to complete two doubling times, or quadruple.

Example 5.4: Calculate the annual percent growth of a population that doubles in size every 10 years.

Solution:

Annual Percent Growth = $\frac{70}{10}$ = 7 years

Biotic Potential

The **biotic potential** is the maximum rate of growth that a population can experience without any **environmental resistance**. Environmental resistance refers to forces that work to decrease population growth, such as limited food supply, disease, available habitat, or weather. The growth curve that defines the biotic potential is sometimes called a **J-curve**. Populations that have an **r-strategy** for reproductive growth tend to reproduce close to their biotic potential, then die back after using up too many resources. Populations that have a **k-strategy** for reproductive growth tend to respond more quickly to environmental resistance and experience a more sigmoidal, or **S-curve**, growth curve.

Effect of Environmental Resistance

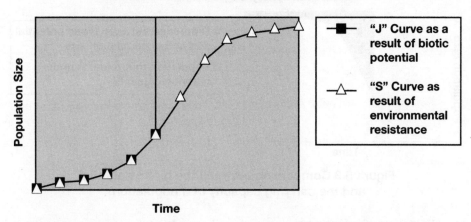

Figure 5.2 Comparison of an unrestrained biotic growth (J-curve) and a growth curve that experiences environmental resistance (S-curve).

Table 5.1 Comparison of the Traits of r-strategists and k-strategists

r-strategist species	k-strategist species
Tend to show irruptive growth	Tend to show logistic growth
Mature quickly, short generations	Mature slowly, longer generations
Tend to have short lives	Tend to have longer lives
Do not care for young	Young get more care, resources
High juvenile mortality	Low juvenile mortality
Tend to have many offspring	Tend to have very few offspring
Not sensitive to environmental resistance	More sensitive to environmental resistance

Carrying Capacity

Carrying capacity refers to the size of a population that can be supported with existing resources. The carrying capacity is the hypothetical limit to population size. The carrying capacity is determined by the amount of available resources, ability of the organisms to distribute the resources, available habitat, type of weather, the amount of disease, and the amount of competition with other organisms.

For k-strategists, population growth will slow as it experiences environmental resistance until it reaches zero population growth, just underneath its carrying capacity.

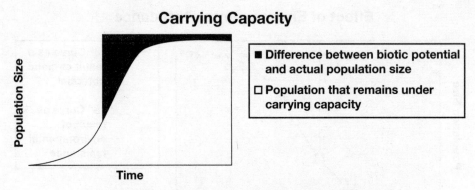

Figure 5.3 Comparison between the biotic potential and the carrying capacity of a population.

For r-strategists, population growth will **overshoot** the carrying capacity and then **dieback**. The population will tend to either be in a state of overshoot or dieback, and the growth curve will oscillate above and below the carrying capacity.

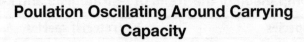
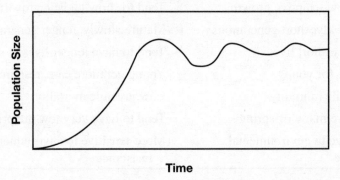

Figure 5.4 r-strategists will overshoot the carrying capacity and then dieback, oscillating above and below the carrying capacity of the environment.

Factors that Affect Population Size

Density-dependent Factors

In general, biotic factors that regulate population growth (those that have to do with other living things) tend to be density dependent.

1. Interspecific Factors

   When two species have a symbiotic relationship (parasitism, mutualism, or commensalism) then the population size—and therefore also the growth curves—will be related to one another. For example, an increase in the number of predators will exert greater environmental resistance on the prey, and the prey's population will decline.

The carrying capacity for the predator may be constantly shifting as the availability of its food source fluctuates. Which population, predator or prey, would tend to demonstrate r-strategies for population growth?

Wolf and Rabbit Population Oscillations

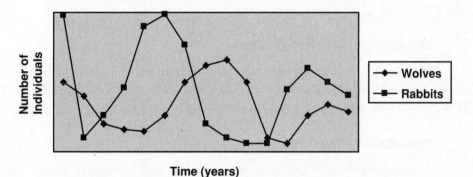

Figure 5.5 Relationship between the growth curve of a predator and that of its prey.

In mutualistic relationships, growth curves of the symbiotic species will tend to match one another. In commensalistic relationships, the growth curve of the dependent population will match the other species, but not vice versa. Also, growth curves will coincide in some manner if two species compete for resources (any portion of the niche overlaps).

2. Intraspecific Factors

When individuals in a population compete with one another, it suggests that there is a limiting resource, which is a form of environmental resistance. This may show up in the way individuals are distributed relative to one another.

A random distribution suggests that no competition takes place, but an evenly distributed set of individuals suggests that some competition is occurring for resources. For example, some desert grasses grow in clumps that are so predictable in their placement that they appear to be artificially planted. These grasses are competing for water, and the species as a whole has a better chance of surviving if the individuals are evenly distributed.

When organisms are overcrowded, they tend to experience stress. Stress is marked by a set of behavioral and physiological changes that take place when individuals are undergoing too much competition. Stress reactions in laboratory studies with mice include decreased fertility, decreased resistance to disease, hypo- and hyperactivity, aggression, lack of parental instincts, sexual deviance, and even cannibalism.

Density-independent Factors

In general, abiotic factors that regulate population growth tend to be density independent. Weather and climate tend to be the most prevalent density-independent factors. Volcanic eruptions, severe storms, and fire all occur without regard to the size of a population, and all have the capability of affecting the size of a population.

Immigration and Emigration

Immigration and emigration are more important in human populations because of our mobility. Therefore, human populations should be measured as a *total* growth rate, rather than a *natural* growth rate.

Pronatalist Pressures

Pronatalist pressures are those aspects in the environment or culture that increase the desire to have children. The following is a list of possible pronatalist pressures in human populations:

1. Social factors: companionship, pride, comfort, status
2. Financial factors: source of labor, security in old age
3. Fertility factors: high infant mortality
4. Cultural factors: producing an heir, etc.

Birth Reduction Pressures

The following are some of the factors that encourage people to have fewer children:

1. Personal freedom for women to find purpose in areas other than having children.
2. Materialism: the desire to spend resources on personal pleasure rather than sharing with a younger generation.
3. Socioeconomic status associated with wealth may encourage couples to have fewer children who would require resources.
4. Educational opportunities: as occupations require more and more time to prepare for the workforce, the childbearing years are postponed and the overall fertility of a population decreases.
5. Sense of financial security: if parents know that they will be well taken care of in old age (for example, do they have a nice retirement account?), they will be less likely to have more children in order to have a secure future.

Human Populations

Structure of Human Populations

The distribution of ages and gender in a population indicates some of the forces acting on the individuals and the population as a whole. There are two ways that the age structure of a population is summarized. The first is a survivorship curve, which summarizes the number of people at each age bracket who have survived. The second is an age-structure diagram, or histogram, which identifies the number of males and females in each age bracket that is alive in the current population.

Survivorship

Survivorship curves of different populations will reveal population growth trends and give hints about the social and environmental forces acting on a human population.

Four Survivorship Curves

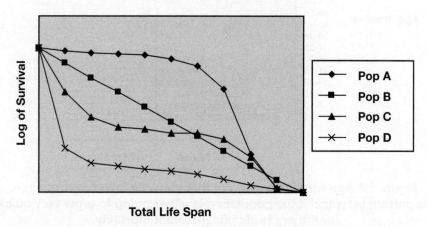

Figure 5.6 Comparison of the survivorship curves in four populations.

Population A: low infant mortality, high level of survivorship of most adults (indicative of a developed country).

Population B: steady mortality throughout life, independent of age.

Population C: medium level of infant mortality combined with high level of survivorship once people have survived childhood.

Population D: high level of infant mortality (indicative of r-strategist population or countries in the Third World).

AP Environmental Science

Age-Structure Diagrams

Age-structure diagrams are histograms that reveal the distribution of people at different ages within the population. The following represents three different types of age-structure diagrams.

The first age-structure diagram represents a population that would be similar to population D in Figure 5.6, with a high infant mortality. This type of age-structure diagram is typical of an r-strategist population, and suggests that either there is about to be very high population growth (when all those children have their own children) or high infant mortality.

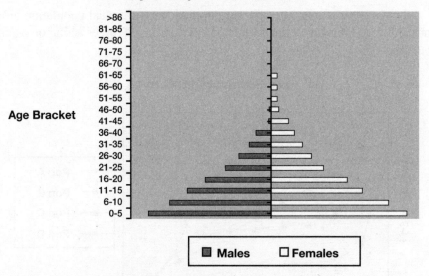

Figure 5.7 Age-structure diagram that shows a lower average age. This pattern is typical if the population is either going to grow very quickly, or if there is significant infant mortality.

The second diagram depicts a population that has a different structure. Children tend to live into adulthood, and adults tend to have long lives. The net result of this population structure is that it does not grow; it experiences **zero population growth** (zpg).

Chapter Five: Population Dynamics

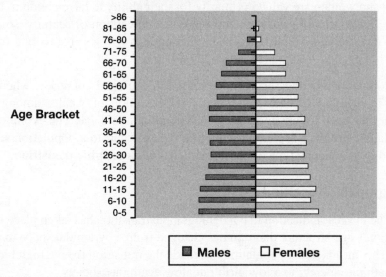

Figure 5.8 Age-structure diagram of a population that experiences zero population growth and low infant mortality.

The third age-structure diagram represents a population that is actually decreasing in size. Whether by infant mortality, decreased fertility, lack of interest in having children, starvation, or emigration, this population is experiencing fewer and fewer people who are about to enter the reproductive years.

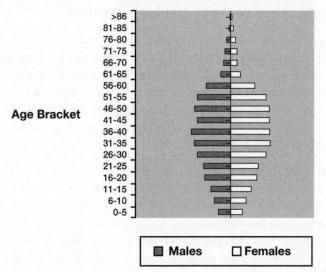

Figure 5.9 Age-structure diagram of a population that is declining. Fewer people are about to enter the reproductive years than did so in the previous generation.

Interestingly enough, populations of the first type tend to exist in Third World countries, where parents have children in order to provide for them in their old age, and women are valued primarily for their ability to have children. In these cultures, infant mortality is high and there is a perception of scarcity, so parents need to have several children to ensure that enough will survive to care for them in their old age.

Populations of the second type exist in developed countries, where more energy is invested in each child and the perception of wealth among adults precludes the need to rely upon the next generation in old age. The evolution from a population structure depicted by the first type of chart to a population structure depicted by the second type of chart is called a **demographic transition**.

Demographic Transitions

The clearest indication that a demographic transition has taken place is that a population's age-structure diagram has changed from a pyramidal shape to a block shape. This means that a population has made the transition from a high birth rate, high death rate society to a low birth rate, low death rate society.

Benefits of Demographic Transition

The largest benefit of a population undergoing a demographic transition is that it no longer grows. A government can better plan to meet the needs of its people if it has a stable population and sustainable methods of feeding them. Also, because the death rate is lower, there is less personal suffering associated with the death of children.

Characteristics of Demographic Transition

Countries that have not undergone a demographic transition tend to have less wealth, or at least the perception of less wealth. Families tend to rely on their children for social security in their older years. There tends to be a dominant culture that limits the value of women to that of childbearer. Population is still growing rapidly and there is a risk—similar to r-strategist populations—of overshooting the carrying capacity of the environment. Finally, there is a high infant mortality rate and the life expectancy is low.

Countries that have undergone a demographic transition tend to have more wealth, or at least the perception of enough wealth that families don't feel as though they have to depend on their children in their old age. Because women have fewer children, other norms for women exist in the culture—such as occupational and educational opportunities. The population takes on the traits of a K-strategist population, and more resources are invested in each child. There tends to be a higher quality of life because basic needs are met and there is less risk of the population overshooting its carrying capacity in the future.

Stages of Demographic Transition

The stages that a population undergoes in order to complete a demographic transition are summarized on this chart and below.

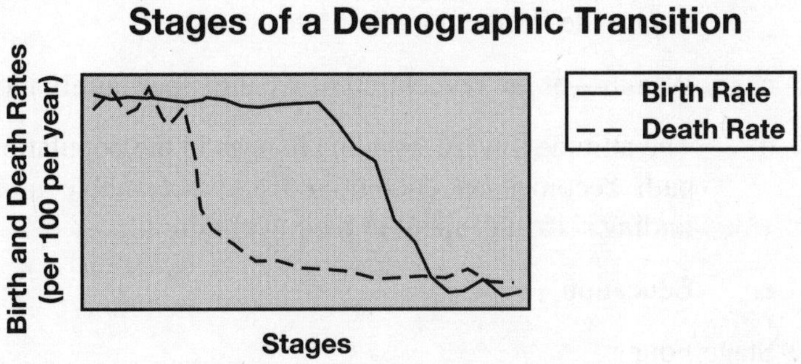

Stages of a Demographic Transition

[Graph showing Birth Rate (solid line) and Death Rate (dashed line) on y-axis labeled "Birth and Death Rates (per 100 per year)" versus "Stages" on x-axis]

Figure 5.10 The change in birth and death rates as a population undergoes a demographic transition.

1. Stage One

    Both birth and death rates are high and fluctuate with the following possible reasons:

    a. Little access to birth control

    b. High infant mortality

    c. Children are used as a "social security" to provide wealth for the parent generation.

    d. Cultural values encourage having children (pronatalist pressure) or discourage birth control.

    e. High death rates exist, especially among children. This may be due to poor medical support, infanticide, disease, famine, or poor hygiene.

2. Stage Two

    Birth rates remain high, but death rates begin to fall rapidly, which causes rapid population growth. This may be caused by any of the following:

    a. Improved medical care, sanitation, or water quality

    b. Food production and distribution improves or increases

    c. Some decrease in infant mortality

3. Stage Three

    Birth rates fall while death rates continue to fall. The total population growth rate becomes arithmetic (constant), rather than exponential. This may occur for any of the following reasons:

a. Contraception is used more often.

b. Parents sense less need to have large families and reset goals for family size. This probably is due to the parents' perception that fewer children need to be born in order to survive long enough to provide for them in their later years.

c. Wealth—or the perception of wealth—per family increases.

d. The attitude toward women changes in the population. A career path becomes an encouraged and acceptable alternative to finding self-fulfillment in having children.

e. Education

4. Stage Four

Both birth and death rates remain low, resulting in a steady population size. The age-structure diagram shifts to have a narrower base, perhaps even becomes rectangular. If the birth rate continues to drop below death rates, the age-structure diagram will take on an "inverted pyramid" shape and the population size would actually begin to decrease. This is already happening in some European countries.

History of Human Population Growth

Growth Strategies of Human Populations

Humans have undergone different growth strategies at different stages in our history. At the onset of human evolution, humans procured food by gathering and eventually hunting, which led to one type of growth strategy based on very limited resources. The advent of tools and fire helped to procure food—and especially protein—more easily, but we were still using a hunter-gatherer paradigm.

With the advent of agriculture during the Neolithic Revolution, increased productivity allowed human populations to grow faster and migrate throughout the world. The Industrial Revolution brought in a much higher level of human cooperation and specialization, which in turn has been able to accommodate unprecedented population growth. Each of these transitions significantly increased the carrying capacity, and each transition has brought a large jump in the global population of humans.

1. Hunters/Gatherers

In the course of human evolution, early hominids evolved a bipedal walking and running gait and a larger brain (see Table 5.2). These two characteristics made humans formidable consumers that could find plant foods and scavenge kills of other carnivores. With the development of tools and enough communication tools to build a social structure, humans became skillful hunters.

For millions of years humans roamed the terrestrial landscape in nomadic hunter/gathering tribes, steadily moving from resource-depleted areas to areas where more food resources were available. Even today, some groups of humans live a hunter-gatherer lifestyle. For example, bushmen of the Kalahari Desert depend on nomadic resource procurement.

The tribe's perception that working together would lead to a more constant food supply built social cooperation. The use of fire allowed tribes to cook meat and preserve unused portions by smoking.

Population was limited by a low population density, constantly confronting local carrying capacities, and low technological development. During this time up to the Neolithic Revolution about 10,000 years ago, the maximum population of humans was probably about 1–2 million people at its maximum.

Table 5.2 Hominid History

Approximate Onset	Genus, Species	Traits
4 million years ago	*Australopithecus*	Earliest hominid, originated in Africa
2 million years ago	*Homo erectus*	More upright, larger brain, developed tools, social structure allowed for cooperative hunting, migrated from Africa to Asia, Europe
250,000 years ago	*Homo sapiens*	Moved into cooler climates and developed new tools, moved to North America over land bridge about 25,000 years ago

2. The Neolithic Revolution

As populations began to grow and bands of people could not depend on a hunting/gathering lifestyle, some tribes settled in areas near food or rich soil. Over time, domestication of crops and animals made it easier for these tribes to obtain food. This shift to agriculture is called the **Neolithic Revolution**. As the amount of convenience and population density increased, so did a tribe's use of energy and need for social cooperation.

3. Industrial Revolution

The **Industrial Revolution** began in England in the eighteenth century, when agricultural technology, the availability of coal, and technology for making steel and using steam engines all coincided. The use of coal fueled the steam engines and provided carbon to make steel. The steam engines provided power to run more automated manufacturing—including pumping out mines to provide access to the coal. The steel provided a stronger, rust-free building material that could hold up trains and build larger ships. The better transportation also used coal to travel long distances. The increased commerce helped expand the British Empire and increased global awareness. All those activities were met with increased occupational specialization. People began moving to cities to earn higher wages in factories. And the improved agricultural methods (crop rotation, steel blades for plowing, for example) allowed the farmers to produce more crops outside the city and transport them to factory workers in the city.

However, there were environmental consequences for all this progress. The combustion of coal throughout the city created respiratory illnesses. The particulate matter blocked out the sun and prevented absorption of Vitamin D, which led to rickets. The high density of people in the city brought new diseases related to water quality and sanitation. Food was produced farther from the table, and the delay sometimes caused food poisoning. To a degree, we now deal with more complicated examples of the consequences brought on by the Industrial Revolution, although the advent of the automobile and the computer have each shifted the culture and distribution of people.

Overpopulation

Shifting Carrying Capacity

Carrying capacity is a shifting variable. Pollution works to decrease carrying capacity by reducing our ability to produce and transport food, survive disease, settle new places, and live close together. Technology works to increase carrying capacity by increasing crop yields, allowing the transport of food long distances, and improving our ability to fight disease. Other variables shift the carrying capacity. Food availability (remember the wolf and the rabbit earlier in this chapter) and weather disasters are just two additional ways in which carrying capacity can shift.

Case Summary

Easter Island

Principles mentioned in this case:

- **How will land appear many generations after it has been heavily degraded?**

Chapter Five: Population Dynamics

- **What factors lead to overpopulation?**
- **Shifting carrying capacity**

When Easter Island was found by European explorers in 1722, they discovered an island covered by sparse grasses, a very small poverty-stricken indigenous population, and a collection of ancient, monolithic stone busts dotting the landscape.

How did the ancient inhabitants erect, transport, and install the massive stone carvings? Anthropologists have reasoned that the only way the early inhabitants could have the time to create the statues is if they had considerable resource-based wealth and perhaps enough social stratification to provide a large workforce that could be devoted to a task that had little to do with food production. Archeologists have discovered the quarries that provided the huge stones for the statues, but they are many miles from the final resting place of the ominous faces. Technical experts agree that the only way the statues could have been transported and installed was to use many logs that would allow the statues to roll along the ground, then prop up the statue at its final resting place. The fascinating riddle is this: there are no trees on Easter Island.

Scientists now believe that Easter Island was once a lush island that was settled about 2,000 years ago. However, about 500 years ago, the advanced culture had grown unchecked to a population about ten times the size of the current population, but they eventually exhausted the island's resources. With the depletion of resources, the culture became war-like, hungry, and, in some cases, cannibalistic. Radioisotopic dating reveals that the statues depicting faces with the most sunken cheeks are the most recently erected, suggesting that toward the end of the climax, the population was starving. Over many generations without wood, the indigenous population ceased to build canoes and they lost the ability or knowledge to fish. There was no longer any wildlife, so the inhabitants engaged in meager subsistence farming on soil with few nutrients in it, just as they still do today.

The loss of natural resources severely decreased the carrying capacity of Easter Island and traumatic social upheaval resulted. Unfortunately, the ecological mechanisms that created the hearty soil and bountiful trees are no longer intact, and the island—along with its inhabitants—is locked in a state of minimal survival. In many respects, the Earth is an island. No new resources will flow to it from the outside. What will happen when the inhabitants have used up so many resources and misused the land that the regenerative ecological cycles shut down? How will we know when we are close to the point where we cannot steer ourselves away from such traumatic consequences?

Human Health

Chapter 6

Human Health

Morbidity and Mortality

Physical health is a state of **homeostasis**, or a balanced operation of all the systems of the body. Altering the body's homeostasis causes disease, or **morbidity**. If disease is severe enough, it causes death, or **mortality**. Disease can be caused by environmental health hazards, by lifestyle choices, for genetic reasons, or from some combination of these causes. This chapter covers different environmental hazards to human health, how these hazards are identified and measured, and the fate of toxins in the environment.

Environmental Health Hazards

Environmental sources of morbidity include pathogens, toxic and hazardous materials, diet, trauma and stress, although the first two causes will be outlined in greater detail.

Pathogens

Pathogens are disease-causing organisms or viruses. Every living kingdom contains some members that can potentially cause disease in humans. Also important for the life cycle and distribution of pathogens are those organisms that carry the pathogen from one host to another. Organisms that serve this function are called *vectors*. Quite often, efforts to eliminate pathogens have focused on controlling the vectors. The following is a list of the five living kingdoms and viruses, and examples of the types of diseases they cause.

Viruses

Viruses are so simple that most scientists consider them an organized assembly of organic molecules rather than a living organism. Instead of having a cell, they hold DNA or RNA inside a protein shell. Instead of metabolizing nutrients on their own, viruses take over the cellular machinery of their host in order to reproduce. Every living kingdom plays host to some virus or another. Typical viral human pathogens cause diseases such as flu, Four-Corners disease, HIV, herpes, and Ebola hemorrhagic fever.

Bacteria (Kingdom Monera)

Living organisms are divided into two different types of cell types, prokaryote cells and eukaryote cells. Prokaryote cells do not have an isolated nucleus, and are far more simple than eukaryote cells. Bacteria are single-celled organisms made up of prokaryote cells. Fungi, plants, and animals are composed of eukaryote cells.

Diseases that are caused by bacterial pathogens include tetanus, pneumonia, typhoid fever, Lyme disease, cholera, leprosy, trichinosis, botulism, gonorrhea, and syphilis. Not all bacteria are pathogens. Bacteria are useful in nutrient cycles, treating wastes, decomposing dead organisms, living inside our bodies to help us digest food, and making yogurt and cheeses.

Protists

Protists range in size and function more than members of the other kingdoms. Protists can be plant-like, fungus-like, or animal-like. Students might be most familiar with ciliated, animal-like paramecium found in pond water. SCUBA divers may be familiar with plant-like protists, kelp. Slime molds are sometimes mistaken for fungi. Diseases that are caused by protist pathogens include malaria, giardiasis, African amebic dysentery, and sleeping sickness. Many molds produce strong allergens.

Fungi

Fungi are typically decomposing organisms and include yeasts, many molds, and mushrooms. Diseases that are caused by fungi pathogens include athlete's foot, Valley fever, ring worm, and human yeast infections.

Animals and Plants

While most animals that are avoided because of their role in causing disease are done so because they are vectors, some families of Kingdom Animalia are pathogens. For example, worms are animals and cause diseases such as Guinea Worm disease.

Very few plants are considered pathogens. However, at this point, the definition of pathogen may become subtle. For example, food poisoning is a disease that is caused by the toxic metabolic wastes of bacteria. When a plant produces a toxin, we simply call it a poison, not a disease. Plant toxins are very prevalent. Sometimes plants evolve the production of toxins as part of their protection; sometimes the fact that a plant is poisonous to humans is coincidental.

Vaccines, Antibiotics, and Pesticides

Vaccines help prevent disease by introducing a foreign particle (**antigen**) into the system to stimulate the production of **antibodies**, which are then available should an actual infection take place. Some pathogens, such as flu virus, mutate frequently and a new vaccine must be established for each generation of flu.

Antibiotics and antifungals have been used for decades to eliminate internal bacterial and fungal pathogens. However, antibiotics that are prescribed to humans enter the environment through wastewater, and create a background level of antibiotics that leads to increased resistance to those antibiotics. Those organisms that survive the background level of antibiotics are then able to reproduce antibiotic-tolerant strains.

Pesticides are used to eliminate insect vectors, such as the mosquito vector for malaria. However, like the use of antibiotics, the use of pesticides to control vectors has also led to resistant strains of vectors. For example, malaria was thought to be almost entirely eradicated, but now malaria causes about 3 million deaths a year because enough mosquitoes are resistant to pesticides.

Resistance to antibiotics and pesticides is an example of natural selection, where only those individuals who are resistant are able to survive and reproduce. Since microorganisms and insects reproduce rapidly, a single resistant individual can quickly create an entire population of resistant individuals.

Emergent Diseases

Even though some diseases have posed threats to human populations for generations, other diseases are either new to a population or have been absent for some time. Such diseases are called **emergent diseases**. An emergent disease can be a potentially large threat to a human population because the population's gene pool may not be equipped to survive a new challenge.

Even the most virulent emergent diseases will decrease in severity over time. Lethal pathogens have a harder time continuing to survive after a host dies, unless the pathogen can move to another host before the host dies. As a result, lethal pathogens tend to either spread quickly, kill slowly, or decrease in virulence in subsequent generations.

Toxic and Hazardous Materials

Types of Toxicity

1. Neurotoxins attack nerve cells, which may be in the central nervous system (brain and spinal cord) or the peripheral nervous system, which affects every body system. Because the nervous system is in control of every aspect of human physiology, nerve toxins can be very damaging and fast acting. Nerve cells can be affected by toxins in many

ways. Heavy metals, such as mercury or lead, kill nerve cells. Some toxins block receptors so that neurotransmitters cannot bind to them. Some toxins, such as anticholinesterases in some pesticides, prevent neurotransmitters from being recycled.

2. Allergens are recognized by the human immune system as foreign objects; then, the immune system responds. One of the chemicals secreted by the immune system is **histamine**, which causes a set of symptoms similar to a cold to help remove the foreign particles. Sometimes the symptoms of the histamine are more bothersome than the foreign particles, and people take an **antihistamine** to minimize the discomfort of the symptoms.

The immune system can also be suppressed by toxins so that the body is less able to fight regular infections. There is growing evidence that this may occur as a result of long-term low-level exposure, but since the lag time for this effect is quite long, it is difficult to establish a clear cut cause/effect relationship.

3. Mutagens and Carcinogens

Mutagens alter the DNA in cells; **carcinogens** cause cancer. Since all cancer has a genetic component, most carcinogens are mutagens. A cancer is created when cells experience uncontrolled growth, or unregulated mitosis. Cancer can also start when cells do not undergo cell death at the end of their natural lifetime. Cancer-causing mutations might occur from any number of environmental reasons: toxins, hydrocarbons, nicotine, radiation, and ultraviolet light are a few examples. Because it only takes one out-of-control cell to begin a cancer, no level of exposure is considered entirely safe.

4. **Teratogens** cause birth defects. Perhaps the most infamous incident of widespread teratogen exposure occurred in the 1960s, when thalidomide was sold over-the-counter as a sleeping aid. Unfortunately, even a small amount of thalidomide in the early weeks of pregnancy prevented limb development, and children were born with a small hand or foot, but no arm or leg. Alcohol use during pregnancy can cause fetal alcohol syndrome, which is demonstrated by delayed development and mental defects.

Types of Hazardous Materials

Most nontoxic hazardous materials fall into one of the following categories, although many of these materials fall into more than one of these categories:

1. Ignitability

Materials such as gasoline, kerosene, or paint thinner present a hazard by being able to ignite and start fires. Sometimes only the fumes of these compounds are needed to start a fire. The hazard stems from the risk of burns from the fire and the risk of smoke inhalation.

2. Corrosiveness

Materials such as strong acids, bases, and oxidizing agents—such as chlorinated compounds—present a risk of corroding metals or oxidizing and burning skin. When disposing of these compounds, they must be isolated from other compounds and put into containers that will not themselves corrode and leak.

3. Reactivity

Materials such as gunpowder, carbides, sodium metal, nitroglycerine, and peroxides can react and explode. Different events will initiate an explosive reaction with different materials. Impact will ignite gunpowder, and water will react with sodium, for example.

Identifying Health Risks

Identifying Risks in a Population: Disease Clusters

It is sometimes difficult to make a connection between disease and an environmental pollutant, especially since it is unethical to test specific pollutants on a human population. One way to draw a correlation between diseases in a population and a pollutant is to look for disease clusters, or groups of people who have the same disease. A scatter plot may be used to show correlation between disease in the neighborhood surrounding a pollution source. Once a disease cluster is identified and some statistical correlation with environmental factors is established, further study may reveal an actual cause-effect relationship between the disease in a particular individual and specific pollutants.

Identifying Risks to Individuals: Dose-Response Curves

Identifying Risk with Bioassays

Over 100 years ago, coal miners would carry a canary with them into the mine. Canaries are very sensitive to carbon monoxide poisoning, which is a constant hazard in a coal mine. If a miner's canary passed out, he knew it was time for him to leave the mine. Even though coal miners no longer use canaries, the mindset of using animals to predict health hazards still continues. Each new medicine on the market must undergo animal testing before it is tested on people, and the toxicity of environmental pollutants can be assessed in the same way. Using the physiology of another species—whether it be animal, plant, fungus, or bacteria—to test a foreign substance is called **bioassay**.

Dose-Response Curves

With bioassay, the responses of a number of target organisms are measured at different concentrations of the medicine or toxin. Sometimes the response measured is muscle contraction or skin irritation; sometimes the response that is measured is the death of the test organism. The data accumulated in these tests is summarized on a dose-response curve, such as in Figure 6.1.

A dose-response curve places the independent variable (concentration of toxin) on the horizontal (x) axis and the dependent variable (response of target organisms) on the vertical (y) axis. When the response measured in the target organism is death, the concentration that causes death in 50% of the population is called the lethal dose for 50%, or **LD_{50}**. Each toxin and medicine has an LD_{50} with characteristic study organisms. In this way, scientists can compare the toxicity of different toxins by comparing the LD_{50} of the different toxins on the same organism.

The drawbacks of using LD_{50} to express toxicity include weighing the ethics of using large numbers of animals to calculate a reasonably accurate value; also, unrelated species can react quite differently to toxins, so the animal may not be an entirely appropriate model for all toxins. However, establishing an LD_{50} gives researchers some general idea of toxicity without using the unethical practice of deliberately exposing humans to toxins.

While the LD_{50} gives a crude indication of toxicity, observing other responses gives some idea of teratogenicity, carcinogenicity, or mutagenicity. For example, a group of human skin cells (fibroblasts) may be cultured independently in a medium and exposed to UV light to determine a dose that initiates abnormal cell growth. In the past two decades, cell culture technology has advanced significantly and has decreased the need to use as many animals for testing. However, toxins and medicines can still be more fully understood by eventually using animal models in testing.

Once a dose-response curve is established, regulatory agencies will determine the dose below which none of the test subjects was harmed; this level is called the threshold level, and becomes a guide for setting human tolerance levels.

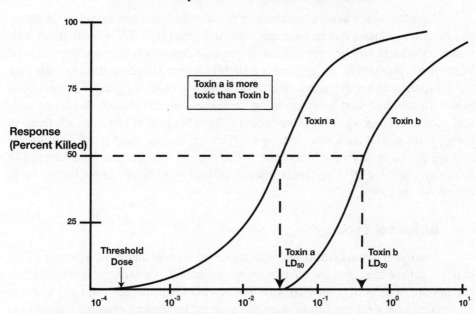

Figure 6.1 Typical sigmoidal dose-response curve. The horizontal axis is usually a logarithmic scale. The vertical axis represents the response—which could refer to anything from being healed from a disease to contracting cancer. When the response recorded on the vertical axis is death, the dose that kills 50% of the subjects is called the lethal dose 50%, or LD_{50}.

Factors That Affect Toxicity

Human Factors

Dose

Dose is determined by both amount and duration of exposure. Humans are exposed to a variety of toxins over a long enough period of time so that metabolism and excretion degrades or removes the toxin before it reaches a harmful dose. Humans have enzymes in the liver that can break down many toxins. Human kidneys eliminate soluble toxins up to a point, and the gastrointestinal system can eliminate solid toxins. However, if any of these systems are overwhelmed, the toxin may have a harmful effect.

A large dose that inflicts immediate harm on an organism demonstrates **acute toxicity**, whereas a smaller dose over a long period of time may cause **chronic toxicity**. It is easier to establish a study that outlines acute toxicity because the cause for poisoning can be easily ascribed to a particular effect. Chronic toxicity is harder to detect because the effects might not be seen for years, and it is difficult to ascribe toxicity to a specific cause.

Genetic Predisposition

An individual's genetic makeup, strength of immune and excretory systems, and overall hardiness due to nutrition, age, and general health will all affect how severely the toxin will be experienced. When cells reproduce, they undergo mitosis. With each reproductive cycle, there is a slightly higher chance that some cells may not reproduce correctly; abnormal, uncontrolled growth will cause a cancerous tumor. Irritation to frequently dividing cells increases the chance that those cells will initiate a tumor. Quickly reproducing cells in the human body include those in the gastrointestinal tract and the lungs, which are also exposed to foreign objects and are therefore susceptible to cancer. Fortunately, carcinogens that succeed in destroying cellular DNA are sometimes countered by enzymes in the human body that repair the DNA.

Chemical Synergy

Chemical synergism occurs when two toxins together have a greater effect than the sum of the effects of the two toxins separately. For example, even a moderate amount of alcohol combined with low levels of barbiturates can have a severe depressing effect on the central nervous system. In another example, while both smoking and asbestos can cause lung cancer, being exposed to both toxins causes cancer ten times more often than the sum of the two separate exposures.

Environmental Factors

Solubility and Persistence

Two physical properties of toxins greatly influence the effect the toxin has on the environment: solubility and persistence. Solubility is perhaps the most important physical property that determines how toxins can move through the environment. Water-soluble compounds tend to travel easily through human bodies and the environment. Fat-soluble compounds travel less easily through the environment, but they tend to be retained by cells (bioaccumulation) and magnify as biomass passes from one trophic level to another (biomagnification).

Some compounds retain toxicity long after they remain in the environment, while some toxins degrade quickly into harmless compounds. For example, DDT was a useful pesticide because it did not degrade quickly and did not need to be reapplied as frequently. However, its persistent toxicity meant that it would affect nontarget organisms in multiple trophic levels. If a compound is persistent and water soluble, it passes through the environment easily, spreading its toxicity throughout the environment.

Treatment of Toxins in the Atmosphere

Toxins can be broken down or removed from the atmosphere in one of three ways: photolysis, oxidation, or precipitation. Some toxins are broken down by high-energy ultraviolet rays from the sun, or photolysis. Solar rays may combine with the oxygen in the atmosphere to oxidize the toxin, reacting to form a less hazardous toxin. Sometimes toxins are simply washed out of the atmosphere by rain. For example, pollen is a potent allergen; a rain during the pollen season will bring brief relief to the allergic.

Treatment of Toxins in the Water

Toxins can be broken down or removed from aquatic and marine ecosystems by hydrolysis or microbial digestion, among other ways. Some toxins actually break down by reacting with water in a hydrolysis (or "water-breaking") reaction. Some toxins actually become food for bacteria that digest the toxin and secrete less hazardous by-products. Bacterial degradation is one of the most effective ways to remediate environmental catastrophes. Sometimes toxins in waterways simply sink to the bottom and become covered with sediment. For example, some waterways near industrial areas become more polluted after the area has been dredged, a process that removes the mud and kicks up any toxins that have been buried over time.

Treatment of Toxins in the Soil

Bacteria effectively digest many toxins that may percolate through the soil. For this reason, water in deep aquifers tends to be very clean; by the time water has percolated through hundreds of feet of soil, bacteria have been able to digest and metabolize the less persistent toxins.

Hazardous Waste Treatment and Disposal

The EPA established a hierarchy to reduce toxic waste that involves the following steps: pollution prevention, minimization of waste, recycling of waste, treatment of waste, disposal of waste.

Preventing Pollution

The best way to reduce toxic waste is to not produce it in the first place. With greater liability involved in taking care of toxic wastes, industries are looking into ways to not produce waste. Even just minimizing accidental leaks or spills can make a difference.

Waste Treatment

If toxic or hazardous waste cannot be prevented, minimized, or recycled, then the industry that produces the waste must treat the waste. Sometimes this can be done through neutralization or oxidation reactions. Sometimes the waste is treated by simply disposing of it in a safe manner.

1. Incineration

    Incineration can be used to reduce a variety of wastes. Only about 2% of the wastes in North America are reduced through incineration.

2. Air Stripping

A method called air stripping removes volatile toxins from aqueous solutions. Bubbling air or steam through the heated solution can force out the toxins, which can then be collected and treated.

3. Carbon Absorption

Toxic wastes in gaseous or liquid solutions can be passed through carbon filters, where the waste clings to the carbon and is removed from the solution. However, the carbon still contains the toxic materials and must then be disposed of.

4. Flocculation

Addition of aluminum sulfate, chlorine, or other compounds will cause some types of hazardous waste to form solid clumps, which can then be filtered out. However, like carbon absorption, the remaining solid waste must then be disposed of.

Waste Disposal

When wastes remain after the above steps have been exhausted, they are usually stored in a well or mine, put in a sludge pond, or taken to a land dump site. These methods of disposal all have a high probability of allowing toxic wastes to leach into the groundwater. All too often toxic wastes of this sort are surreptitiously dumped into rivers, or the ocean, or on vacant land.

Solid Waste Disposal

The Waste Stream

The waste stream is the steady flow of matter from raw materials, through manufacturing, product formation and marketing, to the consumer, and on to its final resting place—usually a solid waste dump. Sometimes the consumer experiences a very short time with the material between the time it is a pretty package and the time it becomes waste.

Some of our waste contains valuable resources that can be recycled, reprocessed, or at least combusted to produce energy. However, throwing all the waste together makes it very difficult to make use of these valuable resources. Below is a chart of the different categories of waste in the waste stream.

With each American producing about 4.5 pounds of solid waste per day, and 76% of that waste ending up in a landfill, managing our solid waste becomes a major land use issue.

Chapter Six: Human Health

Table 6.1 Components of Solid Waste

Type of Waste	Percent of U.S. Domestic Waste
Paper	38%
Yard clippings, trimmings	18%
Metal	8%
Plastic	8%
Glass	7%
Food	7%
Miscellaneous	14%

Methods of Waste Disposal

1. Dumps and Landfills

    While most developing countries use open, unlined dumps, which allow the water that seeps through the dump to contaminate the ground water, developed countries usually use lined, sanitary landfills. At a well-run landfill, metal objects (refrigerators, cars, and other large appliances), batteries, burnable trash, rubber tires, and various other types of recyclable or reusable wastes are sorted out. The leftover waste is then compacted and covered with dirt. The landfill is lined with clay or thick plastic. Wells drilled around the landfill allow monitoring of the groundwater nearby. Vents within the landfill allow methane gas—a natural by-product of bacterial anaerobic decomposition of carbon-containing waste—to escape. Some landfills even use the methane to generate energy.

2. Incineration

    Many municipalities generate electricity or steam by burning trash. This seems like an efficient use of material that would otherwise be useless, and it would certainly decrease the volume of trash that would need to be stored. However, burning trash in incinerators often produces a class of toxin known as dioxins. Lead, mercury, furans, and cadmium have also been found in the ash of incinerated trash. The EPA approximates that toxic emissions of dioxin from a typical incinerator causes less than one death per million over 70 years of operation—making it an "acceptable risk." Being sure to remove batteries will decrease the risk due to lead and other heavy metals, but it will be impossible to remove all the plastics from the combustion, which is what would be needed to eliminate dioxin production.

3. Selling Waste to Poor Countries

    As unethical as it may seem, several garbage producers have found that it is easiest to pay a poor country to accept its waste. For example, in 1986, the cargo ship *Khian Sea* left Philadelphia with a cargo of 28,000 pounds of incinerator ash. It was a last-ditch effort by the city because local states ceased to accept the waste. The *Khian Sea* dumped a portion of the cargo in Haiti, but the Haitian government ceased the offloading when Greenpeace alerted them of the toxic nature of the cargo. After leaving Haiti, the *Khian Sea* visited Senegal, Morocco, Yugoslavia, and other Third World ports, looking for a place poor enough to want cash more than it wanted to avoid the toxic liability of the waste. For two years, the *Khian Sea* roamed the seas until, finally, it mysteriously lost its cargo—presumably dumped in the sea.

In 1999, 600,000 pounds of incinerator waste from Taiwan was dumped in Cambodia after the owners paid a $3 million bribe to Cambodian officials. The real payment came when villagers near the dump site began dying of nerve and respiratory problems and eventually had to be evacuated from their homes. Unfortunately, there are enough unscrupulous officials to create a market for illegal waste—and present a significant hazard to people in Third World countries who suffer as a result.

Minimize the Waste Stream: Reduce, Reuse, Recycle

The real answer comes in reducing the total volume of waste in the waste stream. The total energy spent is minimized when consumers attempt—in order—to reduce consumption and generation of waste, reuse materials many times before they reach the waste stream, or recycle materials so they never enter into the waste stream.

1. **Conservation** and **Reduction**

    The best way to reduce the introduction of raw materials into the waste stream, and minimize pollution and energy use along the way, is to simply consume less. Here are some methods to reduce:

    a. Purchase foods with less packaging, or with none if purchased at a farmer's market.

    b. Don't buy a new car very often; make your cars last by maintaining them carefully.

    c. Ask cashiers to not give you a bag. Bring your own instead.

    d. Use books from a library rather than purchasing your own.

    e. Check the news on the Internet rather than purchasing a newspaper or news magazine.

    f. Don't make long trips for small reasons.

g. Instead of buying one appliance, then "upgrading" to be in style, accept the chic of being slightly out of style for the sake of not wasting materials and energy.

h. Consume less and save your money.

2. Types of Reuse

Reusing products that have already been manufactured requires more energy and raw materials than not purchasing the product in the first place, but far less energy than recycling. Reusing glass soft drink bottles is difficult to manage for large corporations, but is much more energy efficient than recycling aluminum containers. Here are a few items that a typical consumer can reuse and use less raw materials and energy.

a. Use your own container instead of bags at the store.

b. Use a refillable water bottle for personal refreshment.

c. Use old clothes for dust mops, then compost the material after years of use.

d. If some foods are purchased in durable containers, use them again and again.

e. Use aluminum foil repeatedly.

3. Recycling Materials

a. Recycling Solid Waste

Glass, metal, and plastic can be **recycled** in most metropolitan areas. Recycling material saves raw materials, to be sure, but mostly it saves the energy needed to extract, process, and transport the materials, and it saves the environmental degradation that is caused by searching for new materials. Below is a summary of a few of the benefits of recycling common materials.

Table 6.2 Savings by Recycling

Recycled Material	Raw Materials Saved
One *Sunday New York Times*	75,000 trees total, 280 gallons of water per newspaper
One aluminum can	95% of the energy and fresh water that would be needed to make a can from ore
Composting paper and yard trimmings	Eliminates over half of solid waste from a typical home

b. Recycling Nutrients: Composting

Composting uses microorganisms and decomposing invertebrates to accelerate the decay of organic material. Materials that can be composted include grass clippings, leaves, horse urine and manure, vegetable parts, wood, coffee, fireplace ash, and vacuum lint. Unacceptable materials include meats, oils, bones, plastic, glass, metal, pine needles, and pet feces. The decomposing organisms use carbon and nitrogen from the waste and oxygen from the air to reduce the waste to simple nutrients.

There are several variables to balance in managing a healthy compost pile. The ideal carbon:nitrogen ratio is 30:1. If the compost pile smells like ammonia, then there is too much nitrogen from sources like grass or food scraps. If the compost pile is too rich with carbon, it will decompose the waste very slowly. Oxygen is needed in order for the decomposers to undergo respiration and digest the waste quickly, and without odor. If the pile is too large or not turned over regularly, oxygen will not get into the center. If the pile is too small, the pile will not retain the optimal heat to allow the organisms to thrive. (A healthy compost pile becomes quite warm—often over 100°F.) A compost pile also needs water, but not so much as to reduce the exposure of the organisms to oxygen.

When a household composts yard trimmings and recycles glass, metal, paper, and plastic, very little waste would typically remain.

Case Summaries

From *Silent Spring* Flows Greater Understanding About DDT

Principles mentioned in this case:

- **Persistence of pesticides in the environment**
- **Bioaccumulation and biomagnification**
- **Effect on nontarget species**

DDT is a potent chlorinated hydrocarbon pesticide that began to be used soon after World War II. In the next ten years, the use of DDT expanded significantly because it did not show immediate toxicity to humans, it killed a broad spectrum of insects, and it was persistent enough to not need respraying as much as other pesticides.

After repeated efforts at writing magazine articles exposing the ecological effects of DDT, scientist Rachel Carson decided to write the book *Silent Spring*, which was first published as a serial in *The New Yorker* in 1962. Carson described how DDT not only killed a broad spectrum of insects, but the persistent chemical entered the food chain and—because it is fat-soluble—accumulated in the tissues of species who ate the insects, and magnified as it was passed to subsequent trophic levels.

Not only did *Silent Spring* identify the perils of using fat-soluble, persistent toxic compounds in the environment, some say that it escorted the environmental movement into public consciousness.

Love Canal

Principles mentioned in this case:

- **Social injustice**
- **Persistence of toxic wastes**
- **Superfund site**

In 1892, William T. Love began digging a canal between two sections of the Niagara River, but the project was abandoned soon after. In the 1930s, the partially dug canal was used as an industrial dump. In 1947, the Hooker Chemical Company purchased the site and used it to dump several thousand tons of toxic chemicals. When Hooker deeded the site over to the local school board, it was under the stipulation that it would be for surface uses only, such as a park, and no construction would pierce the clay lining of the dump. However, after the transaction, houses, roads, and an elementary school were built on the site. Soon after construction, people began to experience illnesses, and chemicals were seeping into the basements of homes. In 1978, the school closed down and over 200 families were relocated. Since this time, the Superfund has been used to clean the site, and in the 1990s, low-income families began to move back into the area, attracted by the low cost of homes.

Bhopal Crisis

Principles mentioned in this case:

- **Social justice in city planning and corporate responsibility**
- **Value of an effective crisis management plan**
- **Magnitude of a toxic spill in an urban area**

Union Carbide is a large chemical manufacturer that operates a plant in Bhopal, India, which produces pesticides to combat malaria in that country. On the evening of December 2, 1984, 45 tons of methyl isocyanate gas escaped from two underground storage tanks at this pesticide plant. At the time, the city was experiencing a temperature inversion that trapped the gas near the ground. By the next morning, over 2,000 people died; another 1,500 people would die in the following weeks from complications and illnesses. In the aftermath, thousands of people lost loved ones, could not keep jobs because of injuries, or experienced spontaneous miscarriages.

Methyl isocyanate undergoes an exothermic reaction when mixed with water. On this occasion, pipes were being cleaned with water, but the usual blocks that kept the water out of the methyl isocyanate tanks were not put in place. As a result, the water reacted with the methyl isocyanate and the tanks ruptured. Union Carbide was blamed for faulty facilities and procedures, but some also say the incident was made much worse by the government allowing dense slum growth so close to the plant, without adequate water, health care, electricity, or an evacuation plan.

Ebola and the Hot Zone

Principles mentioned in this case:

- **Emerging disease**
- **Role of CDC**

In 1976, a new disease emerged and infected over 200 people in Sudan and Zaire. The disease was named after the Ebola River, in Zaire, and fatally progressed in its victims in a gruesome manner. The Ebola outbreak among primates in Reston, Virginia, was the basis of the bestselling novel *The Hot Zone*. While there were no human deaths in the Reston outbreak, there have since been two other strains of Ebola that have surfaced—both in Africa, both claiming human lives.

Ebola is an example of an emergent viral disease, which is tracked by the Center for Disease Control (CDC) in Atlanta, Georgia. Epidemiologists from the CDC go quickly to clusters of new diseases to attempt to identify the reservoir (original vector), method of transmission, and—if possible—a potential cure. While there is no cure for Ebola, some comfort is taken by the fact that this disease usually kills its host so quickly that only less virulent forms of Ebola will probably persist into the future.

Agent Orange and the Vietnam Conflict

Principles mentioned in this case

- **Government use of toxic substances in war**
- **Toxicology of dioxins**

In the 1960s, the United States was involved in a prolonged military conflict in Vietnam. However, the jungles of Vietnam favored the guerrilla tactics of the Viet Cong rather than the type of open military conflict for which American troops were trained. So to destroy the foliage used for enemy cover, the military dropped about 19 million gallons of a herbicide called Agent Orange on the Vietnamese landscape between 1965 and 1970.

Agent Orange received its name because the barrels in which it was stored contained an orange stripe. It is composed of a 50:50 mixture of two fat-soluble pesticides, 2,4,D, and 2,4,5,T, although there was typically also a poisonous impurity in Agent Orange, commonly called TCDD, which is a type of dioxin. In addition to the deforestation of millions of acres, over 25 different cancers among army and marine servicemen have been connected to Agent Orange use in Vietnam. Today, veterans of Vietnam who were affected are cared for through the Veterans Administration.

The Flu Vaccine

Principles mentioned in this case:

- **Emergent diseases**
- **Importance of vaccines**
- **Bacterial resistance**

Every fall, families in the United States probably hear about a new flu vaccine. Flu is a general class of illness that is caused by a viral infection. The class of viruses that cause the flu mutate regularly, so it is difficult to determine which vaccine to produce and give. Usually the virus is generated among high-density farms in Southeast Asia where ducks, pigs, and people live in close quarters. Then the flu virus steadily spreads to the rest of the world. Scientists try to pinpoint the exact type of virus that will be most prevalent each year, but there is a lot of guesswork; many vaccines are actually combinations of different vaccines that attempt to build antibodies for the strains that will most likely spread.

A flu vaccine injects antigens into the body so that the body can produce antibodies for the flu before it arrives. It is estimated that flu vaccines decrease the chance of contracting the flu by 60–80%. If the flu is contracted, it can cause swelling and inflammation in the upper respiratory tract, which can in turn trap bacteria. A bacteria respiratory infection is a more serious infection, and can be treated with antibiotics. However, if every patient automatically takes antibiotics even when a bacterial infection has not set in, the heavy use of antibiotics in the population increases the chance that related bacteria become resistant to the antibiotics. For example, penicillin was an important antibiotic after World War II, but now several strains of staphylococcus and streptococcus bacteria are no longer affected by this powerful, broad-spectrum antibiotic. Resistance also spreads with new generations of bacteria. In 1987, only 0.02% of the pneumococcus strains surveyed by the Center for Disease Control were resistant to penicillin; today about 6.6% are resistant. So far, pharmaceuticals are generating new types of antibiotics, but it will continue to be a race between research teams and the evolutionary abilities of bacteria.

Atmosphere Resources

Chapter 7

Atmosphere Resources

Composition of the Atmosphere

- 78.08% Nitrogen (N_2)—Essential part of the nitrogen cycle and essential for life.
- 20.95% Oxygen (O_2)
- 0–4% Water vapor (H_2O)
- 0.93% Argon (Ar)
- 0.036% Carbon dioxide (CO_2)
- 0.002% Neon (Ne)—Inert gas.
- 0.0005% Helium (He)—Inert gas.
- 0.0002% Methane (CH_4)—Produced by peat, anaerobic decomposition of organic material, animal waste.
- 0.00005% Hydrogen (H_2)—Remains from the early atmosphere; because of hydrogen's low molecular weight, it is held loosely by Earth's gravitational pull. The heavier molecules are held more closely to the Earth's surface, leaving only helium and hydrogen in the high-altitude exosphere.
- 0.00003% Nitrous oxide (N_2O)—Produced by burning fossil fuels, fertilizer use, deforestation; a greenhouse gas.
- 0.000004% Ozone (O_3)—Absorbs UV radiation. Decreased by halogenated compounds (mostly fluorine and chlorine) in a catalytic reaction.

The above information shows that by far the greatest percentage of the atmosphere is composed of nitrogen gas (N_2), with oxygen gas (O_2) a distant second. Ozone gas (O_3) is essential to life on Earth because it blocks damaging UV rays from the sun, and yet it takes up a minuscule percentage of the atmosphere. Carbon dioxide (CO_2) also takes up a negligible percentage of the atmosphere, and yet increases in this gas threaten the health of the planet with global warming. The gas in greatest abundance has the potential of providing a nearly endless source of pollution, as high temperature combustion reactions that use air convert the nitrogen into nitrous oxide and nitrogen dioxide, both of which can react with water to form nitrous and nitric acids, and then precipitate as acid rain.

Air Pollution: Sources and Solutions

Natural Sources

Wind erosion, volcanoes, and waste from living organisms are common sources for the natural emission of sulfur compounds. Additionally, volcanoes give off an acidic plume mixed with caustic particulate matter. Trees and bushes can give off volatile organic compounds. Many people suffer from the natural emission of pollen. Forests fires give off extraordinary levels of carbon dioxide and particulate matter. These non-anthropogenic sources of air pollution provide the background of health hazards with which the anthropogenic—or human-caused—sources combine to have an adverse effect on the environment.

Anthropogenic Sources

Air Pollutant Categories

1. **Primary Pollutants**

    Primary pollutants are harmful to humans in the form in which they are initially released.

2. **Secondary Pollutants**

    Secondary pollutants are released in a form that is initially not harmful, but become toxic or hazardous after they are released. Photochemical smog is an example of a secondary pollutant that becomes damaging only after sunlight catalyzes the formation of oxidants and acids.

3. **Fugitive Emissions**

    Fugitive emissions do not come from a point-source, or smoke stack, but from a number of nonlocalized sources. Dust from soil erosion or mining operations is a good example.

Criteria Pollutants

The Clean Air Act of 1970 designated seven major pollutants as being the most damaging to human health, and established standards for those pollutants. These seven pollutants are called **criteria**, or **conventional pollutants**.

1. Oxides of Nitrogen

    a. Description

    Nitrogen oxides are formed when nitrogen from the atmosphere (N_2) is combusted at high temperatures to combine oxygen with the nitrogen and form the reddish-brown colored oxide, NO_2. This reaction occurs any time that air is used as a fuel source for combustion above 650°C. Nitrite (NO_2^{-1}) and nitrate (NO_3^{-1}) ions are formed when bacteria oxidize nitrogen from the atmosphere or ammonia as part of the nitrogen cycle. Nitrogen oxides and ions react with water to form nitrous acid (HNO_2) and nitric acid (HNO_3)—two major contributors to acid rain.

    b. Sources

    Over half of the nitrogen oxides are anthropogenic, mostly coming from power generation and internal combustion engines in cars, trucks, planes, and trains.

    c. Effects

    Acid rain from the hydrated oxides of nitrogen damage waterways and the wildlife they support, forests and plants, buildings, and people. For those who live in mountainous European towns, acid rain destruction of forests translates into soil erosion that threatens hillside villages. Oxides of nitrogen are responsible for about 30% of the acid deposition that takes place.

    The nitrate and nitrite ions, as well as acids of nitrogen, are essential nutrients for plants (see Chapter Eight). These nutrients promote photosynthesis in algae, which is then used as food by oxygen-demanding decomposing organisms. The overall effect of these nutrients entering a waterway is a reduction in dissolved oxygen in the water and an increase in eutrophication. For this reason, nutrients are considered *oxygen-demanding wastes*.

    d. Remediation

    Reduction is the best method for preventing pollution from oxides of nitrogen, but the next best method is to avoid getting combustion reactions hot enough to produce the oxides, or use pure oxygen rather than air to fuel combustion reactions.

    One way to run combustion reactions at lower temperatures is to use a staged burner. In the first stage, combustion takes place at high temperatures with little oxygen. In the second stage, combustion takes place in an oxygen-rich, low temperature environment. Both stages result in fewer oxides of nitrogen; combined, the two stages allow for complete combustion of the fuel (usually wood, coal, or liquid fuel). Catalytic converters also reduce nitrogen oxides, as well as carbon monoxide.

2. Oxides of Sulfur

   a. Description

   Sulfur dioxide is a corrosive gas that damages tissues of plants and animals. It can react with water to form sulfurous or sulfuric acid, which is a major component in acid rain.

   b. Sources

   Sulfur dioxide is produced primarily by electric utilities in the combustion of coal. The coal, which is primarily carbon, comes from the organic breakdown of plant material, which contains proteins. Proteins contain sulfur that form disulfide linkages to establish the shape of a protein molecule. The protein that is trapped with other organic materials decomposes, and leaves some type of sulfur compound when the coal is burned. Once burned, the sulfur combines with oxygen to form sulfur dioxide, which forms sulfurous or sulfuric acid when combined with water vapor in clouds. Non-anthropogenic sulfur dioxide is also emitted from volcanoes.

   c. Effects

   Sulfur dioxide causes breathing difficulties. Sulfates combine with other ions to form insoluble particulate matter that irritates the lungs. Sulfur dioxide combines with water vapor to produce sulfurous and sulfuric acids—the other major components along with the oxides of nitrogen that create acid rain. Oxides of sulfur are responsible for about 70% of acid deposition.

   d. Remediation

   Since the major anthropogenic source of sulfur oxides is the combustion of coal, the use of low-sulfur coal makes a large difference in reducing sulfur emissions. Using other fuel sources would eliminate sulfur emissions from power plants entirely.

   Fluidized bed combustion mixes crushed limestone with coal before combustion. This causes the sulfur dioxide to combine with the calcium to form a disposable solid, rather than go into the air. The products of this reaction are calcium sulfite ($CaSO_3$), calcium sulfate ($CaSO_4$), and gypsum ($CaSO_4 \cdot 2H_2O$). These same reactions are the principle behind *flue gas desulfurization*, or *wet scrubbing*. The post-combustion gas is sprayed with a liquid suspension of limestone. This method is effective, but very messy and difficult to maintain, and it results in exchanging an air pollution problem for a solid waste problem.

3. Oxides of Carbon

   a. Description

   Carbon dioxide and carbon monoxide are colorless, odorless gases.

   b. Sources

   Both carbon dioxide and carbon monoxide are produced from the combustion of carbon. Incomplete combustion of wood or hydrocarbons results in carbon monoxide. Respiration also produces carbon dioxide.

   c. Effects

   Carbon dioxide is not poisonous, but increased amounts in the atmosphere have most probably led to global warming which—if left unabated—could result in environmental calamity.

   Carbon monoxide is toxic to humans because it binds more readily to the oxygen-carrying hemoglobin molecule, thereby preventing the transport of oxygen to tissues.

4. **Volatile Organic Compounds (VOCs)**

   a. Description

   VOCs are organic compounds that are easily vaporized, or have a high vapor pressure (a high partial pressure of the compound exists at any given temperature above an open container of the liquid). Also, VOCs react to form ground-level ozone, which is a strong irritant.

   b. Sources

   Volatile hydrocarbons arise from evaporation of crude oil, air conditioning fluids, dry cleaning solvents, paints, adhesives, building materials, and other hydrocarbon sources.

   c. Effects

   Of the hundreds of VOCs that are used and found in the environment, different compounds have different effects. Different VOCs are carcinogens, irritants, neurotoxins, and liver and kidney toxins. VOCs form the largest category of the most highly toxic air pollutants.

   d. Remediation

   Evaporation of VOCs decreases when the source of the hydrocarbons are isolated so that gas may not escape. Automobile carburetors can be adjusted so that less gasoline evaporates.

Fewer VOCs will be emitted into the atmosphere if afterburners are installed so that more complete combustion completely oxidizes the hydrocarbon to form carbon dioxide and water. However, these afterburners require very high temperatures, and the trade-off with fewer VOCs is the production of more oxides of nitrogen.

5. Particulate Matter

   a. Description

   Another name for particulate matter is *aerosol*, which is a group of solid or liquid particles that are suspended in the atmosphere. *Suspended* refers to the fact that the particles are small enough to be held up by the kinetic energy of the surrounding gas molecules. Particulate matter includes ash (such as from a forest fire, or the fly ash from coal-fired power plants), dust, smoke, pollen, and mildew spores.

   b. Sources

   Non-anthropogenic sources include volcanoes, forest fires, leaf mildew, and pollen.

   c. Effects

   Of course, particulate matter reduces visibility, but it also irritates the lungs and carries the same toxic traits that the individual solid particles would have. However, the effects are magnified because the material is carried deep into the lungs, where it can diffuse into the blood or react with sensitive lung tissue.

   d. Remediation

   Particulate matter can be removed from air using bag filters or electrostatic precipitators. Bag filters are similar to a vacuum cleaner filter, where flue gas passes through a filter with holes smaller than the particles. The filter catches the particles, and they are later disposed of as solid waste.

   Electrostatic precipitators add a charge to particulate material in the flue or smoke stack. Then the gas is surrounded by a charged plate that attracts to it the charged particles and removes them from the air.

6. Metals and Halogens

   a. Description

   Metals enter the atmosphere as volatile gases (as with mercury), oxides, or particulate solids. Metals that pose the greatest concern are lead and mercury—both of which are neurotoxic.

Halogens are nonmetals that are very reactive, such as chlorine, fluorine, bromine, and iodine. Chlorinated and fluorinated hydrocarbons (CFCs) are particularly important because they migrate to the stratosphere and catalyze the conversion of atmospheric ozone into molecular oxygen, which is transparent to mutagenic ultraviolet rays. This process destroys the ozone layer that has been built up over many millennia and provides life on Earth with protection from damaging UV radiation.

b. Sources

Lead in the atmosphere typically comes from gasoline, to which it has been added to reduce the "knocking" and to catalyze more complete combustion. Mercury is emitted from coal-fired power plants, and to a smaller degree in waste incinerators. Mercury switches in home thermostats may emit low levels of mercury vapor.

Halogens are most often emitted as CFCs from propellants, coolants, foams, and dry-cleaning solvents.

c. Effects

Lead and mercury lead to decreased neurological functioning. Lead may decrease the body's ability to metabolize food and lead to nausea. In excess, both metals may lead to permanent debilitation and death.

The problem with any level of mercury in the environment is that it is metabolized by microorganisms to form highly toxic methyl mercury, which is a potent neurotoxin. Since methyl mercury is fat-soluble and persistent (does not break down further very quickly), it tends to undergo both bioaccumulation (residing in biological tissues) and biomagnification (concentrating in tissues as it passes from one trophic level to the next).

The decrease in ozone protection catalyzed by CFCs leads to skin cancer in the short term, but may have a drastic effect on global biology if allowed to continue.

d. Remediation

Lead levels have diminished with the use of unleaded gasoline; however, the use of leaded fuels continues at some level—and so does the problem. Stopping the use of leaded fuels entirely will all but eliminate atmospheric lead.

Some mercury is removed from fly ash by electrostatic precipitation, but then it becomes a solid waste problem. Since most of the anthropogenic mercury comes from coal-fired power plants, humans will continue to release mercury into the environment as long as we use coal as a source of fuel.

7. Photochemical Oxidants

   a. Description

   Photochemical oxidants are secondary pollutants that are synthesized with the aid of solar energy. The most common of the photochemical oxidants is ozone, which is produced when a single atom of oxygen is split off of the nitrogen dioxide molecule. The single atom of oxygen then combines with oxygen gas to form ozone.

   b. Sources

   Since the oxides of nitrogen are so critical to the creation of ozone, the sources and remediation of ozone and other photochemical oxidants are the same as those mentioned regarding the oxides of nitrogen.

   c. Effects

   While ozone has a beneficial effect when it exists in the stratosphere, it is highly reactive and caustic to painted surfaces and sensitive lung tissue when it exists around us in the troposphere.

Non-Criteria Pollutants

While the criteria pollutants contain strong mandates for monitoring and control from the federal government, the other air pollutants—or non-criteria pollutants—are not regulated. However, just because an airborne toxic substance is not regulated by the government does not mean that it is not harmful. Indoor air pollution contains several examples of non-criteria pollutants that can have a severe impact on human health.

Indoor Air Pollution

While the Clean Air Act establishes major criteria pollutants, and industries attempt to minimize large-scale anthropogenic sources of pollution, one of the major categories of air pollution occurs at home. Four general types of home pollutants that tend to cause the greatest number of health problems are *asthma triggers* (secondhand smoke, dust mites, pets, and mold), *toxic building materials* (VOCs, heavy metals, chromium, fiberglass, asbestos), *radon gas*, and *carbon monoxide*.

Asthma Triggers

1. Asthma afflicts millions of Americans, including five million children. It is the leading cause of childhood hospitalizations and school absenteeism. The EPA identifies secondhand smoke, dust mites, pets, molds, and cockroaches as the most prevalent household triggers of asthma.

2. Secondhand smoke contains over 40 substances that are linked to cancer, and it is thought to trigger asthma because it irritates bronchial passages. Children are particularly vulnerable to secondhand smoke because they are still developing and have higher respiratory rates than adults.

3. Dust mites, pets, molds, and cockroaches act as allergens to trigger asthma. Dust mites feed on skin flakes that remain on mattresses and fabric-covered items. The body parts and feces from both dust mites and cockroaches are strong allergens, as is the dander from cats and dogs.

4. Molds produce microscopic spores that circulate through homes. When the spores find a moist environment, they grow more mold. Controlling moisture is the best way to control the level of mold or mold spores in the home.

Building Materials

Many building materials that are commonly used in home construction may contain toxic compounds that adversely affect indoor air quality. A few of the common toxic compounds used in building construction include the following:

1. Volatile organic compounds (VOCs) are compounds that contain one or more carbon atoms. Formaldehyde is probably the most common VOC found in homes. Others include toluene, xylene, and turpentine. VOCs can be found in carpets, carpet padding, caulking, construction adhesives, particle board, plywood, and foam insulation. VOCs cause dizziness, nausea, fatigue, headaches, shortness of breath, dermatitis, and even cancer.

2. "Green-treated" waterproofed lumber contains toxic copper chromium arsenate. The chromium portion of this biocide cocktail may contain forms that are carcinogenic. The arsenate ion can dissolve in water and migrate through the environment, and it is carcinogenic to humans.

3. Fiberglass microfibers found in insulation are suspected of being a mutagen and causing abnormal cell growth in lung tissue.

4. Old ceiling tiles may contain asbestos.

5. Older plumbing systems may contain lead in the solder used to attach copper pipes.

6. Mercury switches may be used in thermostats.

Radon

1. Origin: Radon is a radioactive gas that is a by-product of the decay of uranium and other radioactive elements under the surface of the Earth. While the other radioactive elements remain in the Earth, the gaseous nature of radon allows it to rise to the surface and seep into the atmosphere. Normally this would not be a problem, but if a home is built over a spot where radon seeps out, the radon will accumulate in the home and—like any radioactive compound—cause cell damage and eventually cancer. In fact, radon is the second leading cause of lung cancer in the United States.

2. Entry into homes: Most homes contain a lower air pressure inside than outside. This reduced pressure draws subsoil radon gas into the home. Also, some building materials that come from the Earth, such as stone and concrete, may also contain some radioactive substances that decompose to create radon gas—although this is not the most prevalent source. It is important for homeowners to know the levels of radon in their home. The fact that a neighbor has not found radon is no guarantee that radon doesn't exist in a nearby home.

3. Reducing residential radon: Some methods of radon reduction involve preventing radon from entering the home, while other methods reduce radon after it has entered the home.

    a. Suction systems underneath or nearby the home will assure that the home is not the lowest-pressure area, and will draw radon into the system.

    b. Home pressurization will push radon out of the home and make sure that it is moving to another, lower-pressure area outside.

    c. Sealing cracks in foundation walls and floors may reduce the radon coming into the house, but the lack of cracks in a foundation is no guarantee that radon is not coming into the home.

    d. Increased ventilation, either through active or natural means, will remove radon so that it does not have a chance to build up.

Carbon Monoxide

1. Carbon monoxide is a colorless, odorless gas that competes with oxygen for the critical binding site on the hemoglobin molecule in human red blood cells. In other words, if carbon monoxide is present, our tissues cannot get oxygen.

2. Breathing low levels of carbon monoxide can cause fatigue; high levels cause headaches and dizziness, sleepiness, nausea, vomiting, and disorientation. Carbon monoxide is particularly dangerous because deadly levels can be reached without any warning to inhabitants.

3. Carbon monoxide is created from incomplete combustion in a fuel-burning appliance, such as a furnace, fireplace, grill, and all gas appliances.

4. Every home that uses some type of combustion appliance should have a carbon monoxide detector.

Other Indoor Air Pollutants

1. Para-dichlorobenzene is used in mothballs and air fresheners; carcinogenic.

2. Tetrachlorethylene is used as a dry-cleaning solvent for clothes. Some people are highly sensitive to this compound, others less so. Neurotoxic, liver toxic, carcinogen.

3. Methylene chloride is used as a paint stripper and thinner; it can cause nerve disorders and diabetes.

Effects of Air Pollution

Human Health

General Facts

1. Fifty thousand Americans a year die prematurely from air pollution-related illness.

2. Residents of polluted cities are 15 to 17 times more likely to die from air pollution-related illness.

3. In 2004, the EPA announced that 159 million Americans breathe air that is unhealthy.

4. Half of all autopsies demonstrate some degree of lung degeneration.

5. Other ailments, such as heart attacks and immunological disorders, are more likely in people who breathe polluted air.

6. Worldwide, at least 1.3 billion people live in dangerously polluted areas.

Lung Irritation

1. Irritation is caused by strong oxidizers in air pollution, such as sulfur dioxide, nitrogen dioxide, suspended particulate matter, and ozone.

2. Suspended particulates penetrate deep into the lungs and can trigger an inflammatory response and asthma. Asthma is usually caused by an allergic reaction—often to fine particulate matter. During an asthma attack, the bronchi fill with mucus and restrict breathing; thousands die each year as a result of severe asthma attacks.

3. Carbon monoxide blocks the ability of red blood cells to carry oxygen to tissues.

4. Bronchitis, or inflammation of the bronchial tubes, leads to mucus, coughing, and infection. Bronchitis can lead to emphysema, which breaks down the walls of alveolar sacs and decreases lung capacity.

5. Smoking is the largest cause of lung disease; about three million people die each year from tobacco-related illness.

Lung Cancer

Radon that has accumulated in homes is second only to smoking in the number of cases of lung cancer that it causes. Other mutagens and carcinogens include volatile organic compounds (VOCs) and asbestos.

Brain Function

Lead and mercury cause neural damage and decrease brain function, lead to learning difficulties, or cause paralysis or death. Pesticides can also decrease brain function.

Effects of Air Pollution on Ecosystems

Acid Rain

1. Effects on Aquatic Ecosystems

    a. Acid will cause leaching of metals from substrate, such as mercury and cadmium. These metals can exceed tolerance levels of aquatic species.

    b. Acid will change the pH of the aquatic environment, which may exceed the tolerance levels of aquatic species. A shift in the pH of a stream or lake will favor different types of organisms.

    [Reminder: pH is a measure of acidity. Since pH is a logarithmic scale, each pH level corresponds to a ten-fold increase in acid concentration. A pH of 3 corresponds to a hydrogen ion (acid) concentration of 10^{-3} moles acid per liter of solution. The pH is equivalent to the negative of the exponent of the concentration. pH 4 corresponds to 10^{-4}; pH 5 corresponds to 10^{-5}, and so forth.]

2. Effects on Plants and Soils

    a. As with aquatic organisms, a decrease in soil pH will increase the leaching of metals out of nearby substrate, which can have a deleterious effect on plant growth. Aluminum ions are particularly destructive.

    b. Nutrients essential for plant growth can leach out of the soil and enter the groundwater, thus depleting these essential nutrients for further plant growth.

    c. A change in soil pH changes a major abiotic factor (pH) that shifts a soil ecosystem to favor other plants, such as mosses, which remove air from soil and decrease soil health for other plants.

    d. Sensitive cells are damaged from the caustic action of acids. Leaves become discolored due to chlorosis (bleaching) and then die.

    e. Acid rain created from oxides of nitrogen will also bring nitrite and nitrate nutrients, which risk over-fertilizing plants.

    f. Plants damaged in any way from acid rain are more susceptible to pathogens and insects. This combination of pollution and pests can become synergistic; once pathogens and pests reach higher levels in a plant population, they spread more easily to neighboring plants.

The Greenhouse Effect

The greenhouse effect refers to the warming of the Earth that occurs when gas molecules absorb the low energy, long-wavelength infrared radiation that has been re-emitted by the Earth. The gases that cause such an effect are called greenhouse gases, and include carbon dioxide, methane, CFCs, nitrous oxide, and a few other trace gases. Of these, carbon dioxide is the largest culprit, and the largest anthropogenic source of carbon dioxide is the burning of fossil fuels.

Aerosol Effect

Particulate matter, or *aerosols*, from forest fires, urban pollution, and volcanoes has the opposite effect as greenhouse gases: they actually block the sun's rays and cool the Earth. For example, in 1815 the Indonesian volcano, Tambora, erupted. Approximately 100,000 people died as a result of the initial blast and as a result of the subsequent famine that resulted when nearby rice fields were destroyed from ash and acid rain. However, the long-term effects caused cooling on a global scale, decreasing average temperatures by about 3°C. The decrease in available sun shortened growing seasons and caused other famines in North America and Europe. Many scientists feel confident that sudden cooling was the reason for the disappearance of the dinosaurs. One theory suggests that a massive asteroid struck the Earth and kicked up enough particulate matter to have an effect similar to that which occurred from the Tambora eruption.

Ozone Depletion

First reported in 1985, the periodic creation of a hole in the ozone layer in the stratosphere was startling news. The ozone layer provides a shield for all living organisms from UV light that would otherwise cause burns, cancer, and genetic damage.

Ozone tends to be depleted in the cold Antarctic area because the cold temperatures create clouds of ice crystals in the stratosphere. Chlorine-containing compounds and ozone are brought together on the surface of these ice crystals and the third oxygen atom is removed to form oxygen gas. However, ozone depletion occurs to some degree worldwide, and preventing ozone depletion is a global effort.

The most likely culprits are hydrocarbon molecules that also contain chlorine or fluorine. They are particularly dangerous for two reasons: CFCs are very persistent and, since they have a catalytic role, are not used up in the reaction.

Oddly enough, ozone is also a greenhouse gas. Without it, the stratosphere gets even colder, which in turn leads to a higher rate of depletion. (Perhaps this is one greenhouse gas we can live with.)

Weather and Pollution

1. The Grasshopper Effect

    The grasshopper effect is driven by two physical principles: the convection cycles in the atmosphere and the variable solubility of toxins in water at different temperatures. The cumulative result of both of these principles is that toxins tend to be taken into the atmosphere in warmer climates, and released from the atmosphere—often with precipitation—in cooler climates. This results in a net transfer of pollutants from milder climates to cooler climates.

2. Inversions and Heat Domes

    a. Temperature inversions trap pollutants close to the ground when a blanket of warm air prevents mixing and dispersal of the cooler air underneath. There are two types of inversions: a subsidence inversion and a radiation inversion.

i. Subsidence inversions occur over a broad area when a less dense warm front moves over the top of cool air and traps it, like a bubble.

ii. Radiation inversions occur over a smaller area when, as the sun sets, the air near the ground cools faster than the air further up. The warmer air traps pollutants in the cooler air underneath.

b. Heat domes occur in urban areas where asphalt and concrete-covered land absorb large amounts of heat that is re-radiated at night. This heat retention creates an island of heat around a city that deflects weather that would otherwise disperse pollutants. As a result, pollutants are held close to the city.

Case Summaries

Volcanic Air Pollution in Paradise

Principles mentioned in this case:

- **Nonanthropogenic air pollution from a geological feature**
- **Composition of volcanic emissions**
- **Mid-plate location of a volcanic hot spot**
- **Physiological effects of air pollution**

When people think of Hawaii, they may think of beautiful beaches, surfing, a great restaurant in Waikiki, riding burros on Molakai, or the wonderful smell of a lei; but air pollution may not be on the top of the list. However, noxious gases are regularly emitted from the Kilauea volcano on the Island of Hawaii. Sulfur dioxide and acid aerosols produce volcanic smog (sometimes called vog) that damages local crops, initiates or aggravates respiratory ailments, and even leads to lead in local water supplies.

The Kilauea volcano is caused by a "hot spot" that steadily seeps asthenospheric magma through the oceanic plate. Over the millennia, the steady northwestern movement of the Pacific Plate over this hotspot has created the 1,500-mile-long Hawaiian chain, with the oldest island in the far northwest, and the youngest island—Hawaii—in the far southeastern tip of the chain.

Kilauea is best known for its spectacular lava fountains that tend to form two types of lava: slow-flowing, jagged, dense lava called *a'a*, and fast-flowing, smooth, *pahoihoe*. However, this seemingly tame, nonexplosive volcano also produces air pollution in one of two ways. The first occurs when hot lava reaches the sea; it vigorously reacts with the cold seawater to produce clouds of acidic steam. The second occurs at the site of the Kilauea crater, where about 2,000 tons of sulfur dioxide gas are emitted daily. The sulfur dioxide combines with suspended liquid and solid aerosol particles—which include mercury, arsenic, and iridium—to create the vog. Vog irritates the skin and the mucous membranes of the eyes, nose, and respiratory tract. This can induce asthma attacks and promote infection. The acidified steam in the atmosphere increases the level of acid rain on the island, which leaches out lead in roofing materials and is caught by rain-catchment systems that islanders use as a source of drinking water. As a result, some islanders are experiencing the symptoms of lead poisoning—nerve damage, learning issues, and chronic illness.

Water Resources

Chapter 8

Water Resources

Water Compartments and the Water Cycle

The Water Cycle

Water on earth circulates through different global water compartments, or places for storage, driven by solar heating and gravity. Energy from the sun evaporates water from oceans, inland bodies of water, leaves, and soil, then moves through the air as vapor. If the water vapor cools or decreases in pressure by moving to a higher altitude, the liquid water precipitates as rain or snow. If the rain lands on terrestrial ecosystems, it enters a river as runoff or infiltrates into the ground and enters the groundwater.

Water Compartments and the Availability of Fresh Water

The following chart identifies the percent of water on earth in each water compartment and the approximate length of time that water remains in that water compartment.

Table 8.1 Location of Water in Earth's Compartments

Water Compartment	% of Global Water Supply	Average Residence Time
Ocean	97.60	3,000–30,000 yrs.
Ice and snow	2.07	1–16,000 yrs.
Groundwater to 1 km	0.28	days – 1,000s of yrs.
Soil, animals, plant moisture	0.010	weeks
Lakes and reservoirs	0.009	1–100 yrs.
Saline lakes, inland seas	0.007	10–1,000 yrs.
Atmosphere	0.001	8–10 days
Swamps and marshes	0.003	months to years
Rivers and streams	0.0001	10–30 days

The fresh water that we use for all our needs comes predominantly from lakes, reservoirs, rivers, and groundwater. However, as the above chart shows, the combined percent of all freshwater that is available from these sources represents less than one-third of one percent of the global water supply.

Groundwater

As water is absorbed into the ground and moves downward, it eventually reaches an impervious layer of rock and forms an **aquifer**. The type of aquifer that is charged by percolation from above is called an *unconfined aquifer*. The upper boundary of an unconfined aquifer is called the *water table*. A *confined aquifer* is sandwiched between two layers of impermeable rock, and is also known as an *artesian well*. Confined aquifers are charged in *recharge zones* where the geologic layer of the sandwiched, porous rock absorbs water directly.

The water in aquifers is available for use by drilling a well. If too many wells are drilled, the aquifer will *subside*, and perhaps form a dip in water level around the well called a *cone of depression*. If this occurs near the coast, where an aquifer empties underwater, saltwater will diffuse up the aquifer. If a well near the coast uses water from the aquifer faster than it can be replaced from the freshwater uphill source, saltwater will be drawn toward the well and may make it unusable; this is called *saltwater intrusion*. As the water from an intruded well becomes more brackish, or saltier, a user may be tempted to use the water for irrigation. Doing so will leave a salt film, called *salinization*, on plants and soil that will eventually destroy the plants and degrade the land.

Geography and Rainfall

Rain Shadows

Trade winds tend to bring water vapor-laden air inland from the coasts. If there are mountains near the coast, the air must rise up the mountain in order to pass over it. As the air rises, both pressure and temperature drop and contribute to water vapor precipitating first into clouds, and then as rain or snow. By the time the air passes over the mountain, it is much drier. As the air descends on the other side, the air becomes more pressurized and warms ($PV = nRT$). The area on the downwind, or leeward, side of the mountain is then both drier and warmer than the air on the windward side of the mountain. The drier, warmer side is called a rain shadow.

For example, the Sierra Mountains in California capture considerable moisture in the form of snow as the air rises on the Pacific Ocean side of the mountains. On the downwind, or leeward, side of the range, the air descends into the much drier state of Nevada. The air is hottest in the zone where it has descended the most and reached the lowest altitude on the North American continent, Death Valley. Similarly, the rising Pacific air drops considerable moisture as it climbs the Olympic Mountains in Washington State. So much moisture falls that a temperate rain forest has formed between the mountains and the coast.

Desert Belts

Atmospheric convection cells create rising and falling air masses. The air pressure is lower where the air rises, and greater where the air descends. In the same way that descending high pressure air causes an arid rain shadow, it can also cause warm, dry zones at latitudes that experience descending air masses. Chapter Three described how air loses moisture and is heated and expands in the equatorial region, then moves north and south to descend at about 23 degrees from the equator. These zones are called the **Tropic of Cancer** in the north and the **Tropic of Capricorn** in the south. As the air descends, it comes under pressure and warms (PV = nRT), causing hot dry air in these regions. Because of this, these latitudes contain the largest deserts on each continent: Gobi in Asia, Kalahi in South Africa, Great Sandy Desert in Australia, and the Sahara in North Africa.

Trends in Water Use

Geography

In addition to latitude, varied rainfall and river locations lead some regions to have more water resources than others. For example, agriculture is limited in Kuwait, but this small energy-rich country obtains most of its water through *desalinization* of sea water through *reverse osmosis* plants, a very energy-intensive process.

Water use per person typically increases as countries develop. As each person uses more energy and manufactured goods in a culture of consumerism, both direct and indirect water use climbs. Currently, the United States uses about 1,300 gallons of water per day per person while Haiti uses about 30 gallons per day per person.

Water Use

Water is used by three general segments of society: municipal/residential, industrial, and agricultural. Although water use trends vary from country to country, the following chart gives approximate worldwide water use trends. Notice that the agricultural segment uses by far the most water.

Table 8.2 Water Use by Different Sectors in American Society

Water User	Percent of All Water Used
Municipal/residential	7%
Industrial	21%
Paper/pulp	10%
Wood/lumber	6%
Mining/oil/gas/chemical	3%
All other	2%
Agricultural	72%

Residential Water Use

As the following chart suggests, most residences use water primarily to wash away waste and water lawns; only a small portion (2%) is used for cooking and drinking.

Table 8.3 Residential Water Use Pattern in the United States

Water Use	Percent of All Water Used
Lawns and gardens	29%
Toilet	29%
Bathing	23%
Laundry	11%
Dishes	6%
Drinking, cooking	2%

Industrial Water Use

While it is clear when water is used directly by the average consumer, it is less clear when water is used indirectly. For example, we know we use a lot of water if we leave the sprinkler running, but we might be less sure about the amount of water that goes into the different foods we eat. A Sunday newspaper for a major metropolitan area requires nearly 300 gallons of water to be produced. Paper and pulp activities are the largest industrial water users. When we reduce paper use, or recycle paper, we are also conserving fresh water.

Agricultural Water Use

More water is used for agriculture than for any other sector. Many farmers are switching to low-use methods of irrigation. For example, instead of using large rotating sprinklers that lose most of the water to evaporation, some crops allow for the conversion to drip irrigation, which allows most of the water to go directly onto the plant and less is lost to evaporation. As world population increases, many regions may have to consider switching to less water-intensive crops. For example, wheat uses less water than potatoes. California, a state that struggles with having enough water resources, is the nation's largest producer of rice, the grain that demands the most water.

Managing Water Resources

Lock and Dam Projects

The Ohio River spans several states in the American Midwest. Along the river are 19 lock and dam combinations, 7 of which are also hydroelectric facilities, all built by the Army Corps of Engineers nearly 50 years ago. The combination of locks and dams allows large vessels to navigate up and down the river; in particular, it allows coal barges to cheaply transport coal from Pennsylvania and Kentucky downstream to the 47 coal-fired power plants along the river. In addition to navigability, the lock and dam system helps control flooding. However, some would argue that the increased channelization of the river has reduced wetlands along the bank, and thereby decreased the river's ability to flood without doing severe damage.

The lock and dam system has provided power and cheap transportation for industry and economic development. However, the power plants have caused thermal pollution in the water and the emissions are responsible for acid rain, mercury deposition, and high ozone levels in the area.

Hydroelectric Dam/Reservoir Projects

The Hoover Dam was built on the Colorado River in the 1930s. By allowing the river to back up behind it and form the immense Lake Mead, the Hoover Dam is able to control flooding downstream and produce a very large amount of power. However, the dam is over 700 feet high, and an adjacent lock to allow boats or fish to pass by is unreasonable. Therefore, one of the consequences is that fish are no longer able to pass through the dammed portion of the river.

In the Northwest along the Columbia and lower Snake rivers, many are advocating that the hydroelectric dams be breached—or broken open—to allow salmon to pass through and refurbish declining upstream populations. While these dams produce about 5% of the northwest's power needs, many feel that it would be better for business to pay more for power and allow the salmon to survive. As is the case with Lake Mead and the Hoover Dam, the Snake River dams create large reservoirs that are used for recreation and flood control.

In these type of projects, regional governments must weigh trade-offs between environmental and cultural costs (backed up waters often flood important cultural sites), and economic or recreational benefits.

Water Diversion Projects

In ancient Rome, the aqueducts that brought fresh water from the north into the city were a hallmark of Roman ingenuity and convenience. Moving water to areas of heavy use is still a big challenge and poses severe environmental consequences.

The Aral Sea in the former Soviet Union was used as a source of water and diverted to croplands to the south to grow cotton. However, so much water was diverted that the size of the inland sea diminished, causing local fishing villages to lose their income, and exposing vast, salty shores. Local winds blow the salt to nearby villages and increase respiratory illness. As a final insult, the briny water used for irrigation deposited salt in the soil (called salinization) surrounding the cotton and spoiling the land, which now requires even more water from an unidentified source to cleanse it. This water diversion plan turned into a cultural debacle, an air pollution problem that causes illness, and a soil conservation issue.

The California Water Project similarly used aqueducts to bring water from the Sacramento River near San Francisco Bay south 350 miles through hot, arid terrain and over a mountain range so that it could be used by communities in the Los Angeles area. A second portion of the project diverted runoff from the eastern slope of the Sierras with a similar aqueduct. While Los Angeles has been able to use more water, the mountain runoff was not able to enter its ancestral target, Mono Lake. As a result, Mono Lake has become an American Aral Sea: pillars of salt speckle the shore and a much smaller, brackish lake remains. Aside from the political conflicts this project has ignited between northern and southern California, there are severe health effects for the airborne salt from Mono Lake.

Steps Toward Water Conservation

Watershed Management

Clayton County, Georgia, developed a technique to supply more people with its limited water supply. The county implemented an innovative plan to use pine forests in the area to treat waste water, which would enter the groundwater and flow into the town's reservoir. From that point, the water would be treated, used again, and remain in the local hydrologic cycle. Such watershed management has helped communities use the local geography to clean and recycle water.

River Bank Management

After the Mississippi floods in the 1990s, many flooded areas remained open to allow the river to overflow, recharge groundwater, and supply local wetlands. A river with adjacent wetlands and less channelization tends to be cleaner and flow more steadily, with less flooding. See Case Studies in Chapter Nine for a story about planning for the levees of the Mississippi.

Reduce the Use of Water-intensive Products

If it really takes 280 gallons to produce a Sunday paper, perhaps reading the news from the Internet or listening to a radio would provide the same service. Water resources can be saved by purchasing fewer water intensive products. Paper recycling reduces the largest industrial user of fresh water.

Low Flow Utilities

Take shorter showers, or put a flow reducer on the shower head. Consider a low-flow toilet; some use less than half the water of traditional toilets. Some homes and public areas have installed composting toilets that use no water at all. As long as the waste is highly oxygenated, it does not smell. That would entirely eliminate our largest use of domestic water.

Decreased Spoilage

The most prevalent form of stream and river pollution is siltation. This can be reduced by limiting the runoff from construction sites, or other places where the land has been opened. Putting recharge zones in parking lots will reduce the speed and forcefulness of runoff after rains, and less silt will enter local creeks.

Water Pollution

Point Source vs. Nonpoint Source Pollutants

Point source pollutants discharge pollution from specific points, or locations. For example, drain pipes, sewer outfalls, and industrial drains are all considered point sources. Conversely, **nonpoint sources** of water pollution come from sources that are not limited to a discreet, identifiable location. For example, rain runoff in a city washes sediment, oils, and heavy metals into local streams and rivers, but it does not come from any one source. Pesticides entering the groundwater come from many fields; it is difficult to identify a single source. Acid rain is caused from reactions that occur in the atmosphere, then travel many miles before polluting a stream or river and is a major nonpoint source pollutant.

Types of Pollutants

Pathogens

Chapter Six identifies a sampling of pathogenic organisms, many of which are waterborne. Pathogenic pollution is the form of water pollution that poses the highest threat to human health. Most of these pathogens come either from a microbe contained in the life cycle of an insect or water organism, or from improperly treated human waste. Bacterial pathogens include those which cause cholera, dysentery, enteritis, and typhoid. Human coliform bacteria, such as *E. coli* and *giardia*—a protozoan—cause a severe intestinal reaction, but they do not cause death.

Viral pathogens cause Hepatitis A and polio. Schistosomiasis is one of the most prevalent aquatic pathogens and is caused by an animal (the blood fluke). Schistosomiasis has drastically increased in countries that have built very large water projects. The slower-moving impounded water allows the growth of small snails, which are the vectors for the fluke.

Oxygen-demanding Wastes

The addition of nutrients, such as nitrates and phosphates, or human waste will help certain types of algae to grow. Once the algae blooms, decomposing organisms break down the algal bodies and use up oxygen, which makes it difficult for organisms that undergo respiration to live. Therefore, anything that encourages the growth of algae robs a waterway of essential oxygen.

The **Biological Oxygen Demand** (BOD) is a good estimate of the load of oxygen-consuming organisms in a stream or river. The *Dissolved Oxygen* (DO) is a direct measure of the oxygen available for organisms in the water. When an oxygen-demanding waste is discarded into a river or stream, the levels of DO drop downstream and create *oxygen sag*. As the oxygen levels drop, so does the diversity and abundance of oxygen-demanding organisms. The *decomposition zone* is immediately downstream from the discharge, and will contain organisms like leeches and trash fish (such as catfish). The next zone is the *septic zone*, where the lowest DO exists (often as low as 2 ppm). Fish will not survive in the septic zone, and only highly tolerant organisms, such as worms and mosquito larvae, are hardy enough to exist.

As the water moves downstream, the natural turbulence in the water helps replenish dissolved oxygen in the water. The colder the water, the more easily oxygen will dissolve. After the septic zone, the *recovery zone* represents that area where the DO goes back to pre-discharge levels (usually about 8–10 ppm). When the water has fully recovered, the *clean zone* will once again be able to support sensitive, oxygen-demanding organisms, such as mayfly, stonefly, and cadisfly larvae. Like lakes, rivers that have clear water and low biological productivity are **oligotrophic**. **Eutrophic** rivers are nutrient-rich and carry a higher BOD.

Inorganic Wastes

There are two general types of inorganic wastes: suspended particles and dissolved ions—of which acid deposition is one type.

Suspended Particles

Suspended particles are undissolved solids, such as small objects and sediment, that spoil waterways. Sediment that has washed into a stream or river from runoff or erosion is responsible for spoiling more waterways than any other type of water pollution. Sediment spoils the water for drinking, and can cover or clog organisms that need a free flow of water or exposure to sunlight in order to survive. *Turbidity* is a measure of biotic and abiotic suspended solids, and is measured by determining the depth at which a black and white pattern on a secchi disk is still visible.

Dissolved Ions

Dissolve ions come from ionic compounds that dissolve in water (as described in Chapter Two). The most common problematic dissolved ions include heavy metal ions, calcium ions, iron ions, sulfide ions, chloride ions, and acidic hydrogen ions.

Heavy Metal Ions

Most heavy metal ions are toxic; some are extremely toxic. Mercury, lead, cadmium, and nickel can be fatal in the parts-per-million range. Metal ions already exist in their simplest form, so they do not break down further; therefore, they are extremely persistent. One of the most famous cases of metal poisoning occurred in Minamata, Japan, where residents ate fish that had ingested mercury and then metabolized it to form the more toxic methylmercury. Methylmercury poisoning caused birth defects and permanent brain damage. Methylmercury poisoning is now called **Minamata disease**.

The most prevalent form of metal ion water pollution exists as water passes through old plumbing and leaches out lead that was used for solder in incoming pipes and joints in waste pipes. Even if a particular household does not receive a dangerous dose, constant low levels of pollution can build up in the environment and cause illness somewhere else.

Calcium Ions

Calcium ions spoil the water by combining with soap to decrease its cleaning ability and create a bothersome film. Calcium ions can also precipitate with other ions inside appliances to form a calciferous crust that may eventually destroy the appliance. For these reasons, many families use a water softener to replace the calcium ions with sodium ions so that appliances and soap work better. The drawback to this practice is that the sodium in the water—like eating too much table salt—can increase the incidence of hypertension in some people.

Iron and Sulfide Ions

These ions are often found in areas where water is pumped from mineral-rich aquifers. Sulfide ions give water a rotten-egg smell, but can be removed with carbon filters. Iron ions affect the taste of water and can be harmful at high concentrations.

Chloride Ions

Chloride ions enter the water as runoff from salted roads and industrial processes. If combined with organic material, such as particulate material from sewage treatment facilities, the resulting compounds may be carcinogenic.

Acid Deposition

Acid runoff is perhaps the most damaging type of dissolved ion pollution. (Chapter Two outlined the definitions of acids and pH.) Acid deposition results from acid rain (Chapter Seven) or from runoff that has filtered through soil that contains acidic material, such as mine tailings. Acid deposition threatens the stable pH that aquatic organisms need in order to survive. The aquatic larval stages of insects are more sensitive to fluctuating pH; severe damage to these ecologically important organisms can result with a slight drop in pH.

Toxic Organic Wastes

Organic toxics can come from either natural or anthropogenic sources. One natural source of organic toxins is red tide, which is caused by the dinoflagellate *Pfiesteria piscicida*. Blooms of this organism are caused by nutrient wastes reaching marine pelagic environments. This microorganism can attack fish and be toxic to humans who eat the fish. The resulting disease is called paralytic shellfish poisoning.

Some artificially developed organic compounds represent the most toxic substances known; pollution of water by these compounds presents a serious threat to human and nonhuman life. Pesticides that wash off agricultural and residential land are one of the most prevalent culprits. Organic wastes that leach out of dumps or corroded fuel tanks into the groundwater often end up in streams and rivers. Dioxins, a type of toxin that is produced from burning trash, can rain down upon lakes and streams; exposure to dioxin at very low levels—a few parts per quadrillion—can lead to cancer and birth defects.

Thermal Pollution

Cool water is able to dissolve more oxygen, which can then be used by aquatic organisms. As water heats up, less oxygen can dissolve in the water. When power plants use river water to cool steam after it has passed through a turbine, then the cooling water is returned to a river a few degrees warmer and the river has a lowered DO. In metropolitan areas, thermal pollution is most pronounced when it occurs alongside nutrient pollution, both of which decrease the DO of a river or stream.

Wastewater Treatment

Pre-use Water Treatment

Before water is used by a municipality, it is generally filtered, flocculated, re-filtered, and disinfected.

Filtration and Re-filtration

Water is drawn through filters that contain gravel and sand of various diameters. At the end stage, carbon filters are used to remove the smallest particles and improve taste.

Flocculation

Aluminum sulfate may be added to water to bind small particles together so that they may be more easily filtered out.

Disinfection

Before entering the water supply system, water is frequently disinfected with chlorine, ozone, or ultraviolet light.

Wastewater Treatment

Primary Treatment

Primary treatment removes large particles through filtration, and then allows bacteria to partially digest carbon and nitrogen wastes in large settling tanks. However, this does not entirely degrade the carbon into methane, or the ammonia/nitrogen into atmospheric nitrogen, and pathogens or toxins may still exist in the water. Primary treatment is fundamentally a filtration and settling process.

Secondary Treatment

Secondary treatment holds the waste for a longer time in conditions that are favorable to bacterial digestion of the carbon and nitrogen wastes. The carbon-rich sludge tends to settle in the initial pond and can be removed to be digested by anaerobic methane-producing bacteria. The methane has a pungent odor, but it can be burned off or used as an energy source to operate the treatment facility.

The nitrogen-rich solution is skimmed off and sprayed over high surface area substrate to promote aerobic, or oxygen-using, bacteria growth. These bacteria will add oxygen to the ammonia form of nitrogen and convert it to nitrate ions. The carbon and nitrogen consuming bacteria can be easily filtered off before the waste is discharged. This solid waste is called sludge, and presents a major disposal problem for many metropolitan areas. Also, the nitrates remain in the water and, if left untreated, will lead to decreasing the available dissolved oxygen in the water after being discharged.

Pathogens can be killed by exposing the effluent to chlorine, UV light, or ozone before discharge. Secondary treatment removes the greatest percentage of waste and pathogens, but it does not remove nutrient or toxic wastes.

Tertiary Treatment

Tertiary treatment can involve many methods to remove nitrates, phosphates, and industrial pollutants. Nitrates and phosphates are fertilizers and can be removed by sprinkling the water on trees or fields, or running the water through marshes so the plants assimilate the nutrients. Exposing the water to denitrifying bacteria will convert nitrates into nitrogen gas. Passing the water between electrically charged plates will remove charged particles. Reverse osmosis will remove all dissolved and undissolved particles, but it is very energy intensive.

Tertiary treatment may also involve specific chemical treatment to disinfect the water or remove toxic wastes, depending on the needs of a community, but this is very expensive.

Removal of Toxic Wastes

Toxic wastes can be removed by physical, chemical, or biological methods, or some combination of those three.

Physical methods include vaporization (air stripping), filtration (especially using carbon, or reverse osmosis), UV oxidation/disinfection, or washing. For example, an underground oil spill can be remediated by using an air stripping method, which passes warm air through the ground to vaporize most of the oil fractions and allow them to rise to the surface of the ground.

Reverse osmosis is a common type of filtration that is able to remove suspended and dissolved particles and pathogens from water. This process uses energy to press water through layers of filters with microscopic pores. It is used on a small scale in homes and hospitals for water purification. On a large scale, it is a costly method of deriving fresh water from the ocean.

Chemical methods include acid neutralization, oxidation, reduction, precipitation, chelation, or hydrolysis. For example, chelation uses large molecules to combine with dissolved heavy metal ions so they may be more easily filtered from polluted water.

Biological methods often involved bioremediation. It may involve constructing wetlands, which have a natural filtering effect and use up nutrients that may otherwise cause a drop in dissolved oxygen. Duckweed can remove a large amount of organic nutrients from water. Duckweed lagoons are able to treat sewage at a fraction of the cost of traditional sewage treatment plants; then the duckweed can be harvested and fed to animals.

One noted example of bioremediation: After the *Exxon Valdez* oil spill in Alaskan waters, all the best artificial efforts were not as effective as allowing bacteria to digest the oil coating the coastal areas.

The Clean Water Act

The **Clean Water Act of 1972** was passed to "restore and maintain the chemical, physical, and biological integrity of the Nation's waters." This bill is one of the top five most important environmental legislative acts; it involves the following major objects:

1. Make all waters "fishable and swimmable."
2. Require discharge permits of major polluters.
3. Identify toxic pollutants and require the use of the best practical technology (BPT) to remove those pollutants.
4. Set goals for best available technology (BAT) to be developed in the future.

Water Pollution Control

Other than treatment, the best way to clean our waterways is to not pollute them in the first place. The following actions help avoid pollution altogether, but they involve expensive trade-offs that countries and local governments are rarely willing to make.

Reduce Emission of Sulfur and Nitrogen Oxides

These emissions come from point sources (power plants) and nonpoint sources (automobiles). This would reduce acid deposition of our waterways.

Modify Agricultural Practices

Reducing pesticide, nutrient, and sediment runoff will vastly improve the quality of our waterways. Using land practices that decrease runoff, or farming practices that reduce the need for water, pesticides, and fertilizers, would significantly increase available fresh water supplies. Golf courses use large amounts of fertilizer and weed control chemicals, and feedlots that produce nutrient, hormonal, and antibiotic waste also need to limit and/or treat runoff.

Treat Waste at Industrial Point Sources

If wastes are treated before it enters municipal sewage systems, those systems will not be overloaded. Municipal treatment facilities are not equipped to treat specialized industrial waste.

Separate Storm Runoff from Septic Treatment

Heavy rains can overwhelm wastewater treatment facilities and cause untreated waste to enter waterways.

Decrease Silt Runoff

Our largest water pollution problem could be solved if communities devote holding ponds that allow silt to settle before entering streams and rivers. However, numerous nonpoint sources would continue to exist from urban areas, parking lots, and construction sites. Using permeable pavers instead of asphalt or cement would allow water to soak into the groundwater and prevent silt from flowing into streams.

Seal landfills to prevent toxic substances from entering groundwater.

Stop ocean dumping of garbage or cleaning the bilges of large ships.

Reverse River Channelization

Wetlands adjacent to streams and rivers filter toxins and silt out of the water, decrease the flow of the river, and provide a buffer zone that decreases flooding.

When a river is engineered to be in a long, straight channel instead of winding through wetlands, all of these benefits are eliminated.

Case Summaries

Oxygen Demand on the Chattahoochee River

Principles mentioned in this case:

- **Combined effect of thermal and nutrient pollution**
- **Regional competition for water**

The Chattahoochee River flows through metropolitan Atlanta on its way to the Gulf of Mexico. It is the major source of water for domestic and industrial needs for this growing area. The water is in high demand as a repository for treated sewage, fresh water source, and coolant for power plants. However, local sewage treatment facilities must coordinate with regional power utilities in order to keep the dissolved oxygen in the river within acceptable levels.

As the population of Atlanta has increased, so has the burden of treated sewage. As a result, a major downstream coal-fire power plant must now reduce its use of river water as a coolant and must build large, unsightly cooling towers that will accomplish the same amount of cooling with less water. This is an example of how population growth puts increased stress on water resources.

The Diminishing Everglades

Principles mentioned in this case:

- **The effects of groundwater depletion on local ecology**
- **Toxic wastes and endangered species**
- **Channelization**
- **Restoring wetlands**

Steady use of Florida groundwater and ever-increasing diversion of water has reduced the high water table in and around the famed Everglades National Park. Steady development north of the park has converted many thousands of acres of wetland that previously drained into the Everglades into residential and farming uses. Not only does this decrease the flow of water into the Everglades, but runoff from these developed areas sends nutrient and toxic wastes into the Everglades, which also threatens the endangered species protected by the park.

One portion of the solution may lie with the "re-twisting" of the Kissimmee River, which was turned into a straight channel several years ago to promote navigation. Restoring the oxbow, circuitous nature of the river will revitalize former wetlands along its banks, which will in turn remove nutrients from the water before it reaches the Everglades.

The Ogallala Aquifer: A Hidden Treasure Dries Up

Principles mentioned in this case:

- **Water resource management**
- **Anatomy of an aquifer**
- **Subsidence**

The Ogallala aquifer is the largest aquifer in North America; it lies underneath the Central Plains states of Oklahoma, Kansas, Colorado, Nebraska, South Dakota, New Mexico, and parts of Texas. It ranges from dozens of feet to over 1,300-feet thick, and provides the water for America's "breadbasket."

Farmers and municipalities began to pump water from the Ogallala about 100 years ago, but use of the aquifer usually surpasses its ability to recharge itself. Some are concerned because it took millions of years to accumulate the water in Ogallala's porous rock, but all that water is being used within just a few generations. Careful attention is now being paid to use and recharge rates, water table levels, and groundwater pollution that spoils the aquifer.

The Gulf Hypoxic Zone: Symptom of a Larger Problem?

Principles mentioned in this case:

- **Oxygen-demanding wastes**
- **Relationship between global health and regional misuse**

The Mississippi River is the third-largest river in the world. It flows for approximately 2,300 miles, provides 23% of the nation's drinking water, and drains 40% of the United States. However, the mighty river also carries an enormous load of toxic waste—particularly agricultural waste. That waste continues to concentrate as the river approaches the Gulf of Mexico. Now there is a large zone in the Gulf—about 20,000 square kilometers, about the size of the state of Massachusetts—that is dead and cannot support life. The dead zone is hypoxic, which means that there is no oxygen in the water.

The Gulf hypoxic zone is a massive example of the consequence of oxygen-demanding waste mentioned in this chapter. Nutrient waste is emitted into a waterway. Algae populations bloom. Decomposer populations that feed off the algae then bloom—using up the oxygen in the water.

The Mississippi not only carries the bulk of the nation's water, but also a large proportion of the country's fertilizer runoff. Since the development of the United States, the land that drains into the Mississippi has steadily been converted from vast prairies to feedlots and croplands. The nitrates and phosphates that are produced on these farms run off into the Mississippi and ultimately create the hypoxic zone in the Gulf of Mexico. An estimated 56% of the nutrient pollution enters the river above the Ohio River confluence. Iowa alone contributes an estimated 51 million tons of animal waste per year to the river.

Compounding the effect is the increased channelization of the great river. (See Case Studies, Chapter Nine.) The vast wetlands that used to assimilate the pollutants of the river have been turned into agricultural lands—which instead of assimilating runoff fertilizer, only contribute more.

When this happens in a stream, the natural mixing and turbulence of the stream helps to dissolve more oxygen into the water; the stream has a chance to recover. However, the Gulf water does not experience enough mixing, and the hypoxic zone only increases. The shrimp fisheries in the Gulf must entirely avoid the hypoxic zone, which has pushed shrimp populations further from shore—requiring more energy to procure them.

As the river moves along, the water becomes increasingly polluted. Near the mouth of the river, there is considerable poverty. To some people the mismanagement of the Mississippi becomes an issue of social justice.

As a family in Baltimore or Newark or whatever city sits at the table to have dinner, they probably do not think about the consequences to people and the environment for growing that food in America's Breadbasket.

The Three Gorges Project: The Largest Dam Project in the World

Principles mentioned in this case:

- **Effects of a large water project**
- **Social justice of community displacement**
- **External costs**

The Three Gorges area along the Yangtze River is the construction site for what will be the largest hydroelectric dam in the world. It will span a mile-long canyon and rise 575 feet above the Yangtze. Its reservoir will extend 350 miles upstream, and it will produce a behemoth 18,000 megawatts of electricity with 26 turbines. The project is scheduled to be completed in 2009.

However, the environmental and social cost is commensurate with the size of the project. It will displace nearly 2 million people, many of whom have no place to go. The Chinese government's web site reports that about 850,000 people will need to be resettled, and they are doing so happily because of a generous re-settlement stipend. However, interviews of indigenous villagers by human rights organizations suggest that relocation stipends are being embezzled, and villagers are being sent to jail if they ask too many questions.

The Chinese government feels that electrical power is critical to continued development in the country that contains over one-sixth of the world's population. The dam will also help prevent flooding along a river famous for catastrophic floods.

Land Resources

Chapter 9

Land Resources

Principles of Land Use

Abuse of the Land

Tragedy of the Commons

In 1968, Garrett Hardin wrote an essay titled *Tragedy of the Commons*. In this seminal essay, Hardin cites those aspects of human nature that caused the overgrazing of "the commons" in early New England towns when local herdsmen would add just one more head of cattle to feed on the communal grazing area. With little effort on the herder's part, he could gain the benefit of having an additional animal. However, the cumulative effect of all the herders "just adding one more" to graze on the commons is that the land becomes overgrazed, denuded, and unable to support any animals. Tragedy results by spreading out among many the responsibility for the common area, or, as Hardin writes, "Therein is the tragedy. Each man is locked into a system that compels him to increase his herd without limits—in a world that is limited." Exercise of the freedom of choice by many individuals leads to the lack of choice for all.

The same nature in humans is also at play today as we dip from "the commons" to use not only grazing land, but forests, cropland, and waterways, and develop real estate to build cities. While the economics of the system favor the person who grabs the resources first, disaster results when everyone attempts to build wealth. As a result, the worldwide community is faced with the following land-use crises:

1. **Deforestation** of old-growth and rainforests in order to provide fuel and building material, and create space to grow food and cash crops or raise cattle for profit.

2. **Desertification** of formerly arable farm or grazing land because nutrients and moisture have been depleted.

3. **Waterways and wetlands** become altered so that they no longer provide ecological cleansing or biodiversity. Instead, the over-engineering of the waterways creates frequent floods, which themselves create environmental havoc.

4. **Urbanization** stems from people moving from self-sufficient lifestyles in the country to high-density populations in the city, further removed from sources for food and energy, and residing in pools of pollution. Huge amounts of energy are needed to pave the landscape, heat and cool artificial environments, transport needed raw materials used for economic growth, and—to a much smaller degree—meet basic human needs.

5. **Solid waste** that results from disconnected lifestyles and exponential population growth must also find a place. The consumer who purchases an item from the store and throws away the packaging material does not ultimately pay the price to store that packaging in future years. Like the herders in the commons, each individual seeks gain, but the consequences are paid later by everyone.

This chapter summarizes the basic principles of land restoration, and then summarizes current land use practices in residential and commercial building, urban growth and city planning, agriculture and forestry, flood control, mineral extraction, and solid waste dumping.

Land Restoration and Reuse

In looking at specific land use management techniques, it is important to keep in mind that land restoration and reuse can be maximized through conservation of land and nutrient resources, preservation of land for use of wildlife, restoration of denuded land, remediation of contaminated land, reclaimed formerly degraded land, and mitigation to minimize the overall damage to the land.

Conservation

Conservation refers to simply not using and protecting resources that might otherwise be expended with a less responsible pattern of use. For example, utility companies are finding that it is less expensive to educate consumers about using less power than it is to find new sources of power for energy-ignorant consumers. In this case, the policy that makes the most economic and environmental sense is conservation, which can be achieved by educating the consumers and using incentives to decrease consumption.

Preservation

Preservation refers to providing an ample reserve of resources so that they might be enjoyed by others in the future. It is different from conservation in that preservation protects actual resources, whereas conservation simply decreases use of resources—whether or not they have been identified, isolated, or protected for others. For example, the Park Service Act of 1916 sought to preserve natural features, unique populations, and historical objects "by such means as will leave them unimpaired for the enjoyment of future generations." Similar acts to preserve nature are the 1964 Wilderness Act, which established wilderness areas, and the establishment of wildlife refuges, which began under Theodore Roosevelt in 1901. Currently, about 1% of the United States is preserved as a wildlife refuge. The most recent and massive addition to this block of land is the 1980 Alaska National Interest Land Act, which nearly tripled the total acreage preserved as a wildlife refuge.

Restoration

Restoration refers to bringing a damaged ecosystem back to its unspoiled, natural condition. For example, the Nature Conservancy is restoring a 40,000-acre prairie in Kansas by using periodic fires and introducing bison. The fires tend to kill exotic plants that have not adapted to fire and help natural prairie grasses reseed themselves. The bison serve as seed dispersal agents as they eat the grass in one area and eliminate the nutrient-rich, seed-laden waste as they roam. The gentle disturbances from bison hooves also provide opportunities for other organisms to build the soil and help establish a more complete, complex prairie biome. The hope is that the area will be completely restored to its prior prairie grandeur.

Remediation

Remediation refers to using chemical, biological, or physical methods to remove chemically active (either hazardous or toxic) pollutants. Chemical treatment may include neutralization of acids or oxidants. Examples of biological treatment include bacterial digestion of oil or nutrients and using water plants to remove nutrients from wastewater. An example of a physical method includes the vaporization of hydrocarbons in an underground plume from a broken oil transport pipeline. Another example, CERCLA (1980; modified in 1984) was passed to establish a "superfund" that would begin cleaning up toxic waste dump sites across the country. Taxation of a large number of people provided the funds for these massive cleanups. In this case, the government took an active role in remediation.

Reclamation

Reclamation used to refer to using large water projects to bring water to otherwise un-arable land. However, now this term usually refers to earth movement that returns massively scarred, denuded, or devastated land to a condition that is environmentally useful and socially or politically acceptable. For example, when a mining operation wishes to create an open-pit mine, they are required by the Surface Mining Control and Reclamation Act (enforced by the Office of Surface Mining Reclamation, passed in 1977) to put into escrow enough money to pay for total reclamation of the land after mining operations are complete. Reclamation might include reburying mine tailings so that surface runoff does not carry soluble toxic ions into waterways, refilling open pits to allow plants to grow and wildlife to thrive, and digging out "high walls"—or the cliff-like edge of an open-pit mine—so that it does not present a hazard to hikers and the terrain may take on a more natural topography.

Mitigation

Mitigation is a general term that refers to finding a solution to a problem. In the context of environmental degradation, mitigation usually refers to establishing another ecosystem elsewhere of comparable health and magnitude in exchange for damage done as a result of developing a nearby area. For example, the Fish and Wildlife Conservation Act of 1980 prompted states to develop plans to conserve wildlife resources. One such state plan occurred in California where a developer had to establish a remote preserve to protect the habitat of a rare frog before the state and county officials would grant a permit to undergo construction that would decrease the habitat of the frog in another area. Such mitigation hopefully accomplishes a greater benefit for both the frog and humans by providing housing that is necessary for a growing population.

Management of Specific Land Uses

Residential and Commercial

Building Construction

1. Current Building Practices

    a. Heavy Dependence on Toxic Compounds

    Many current building materials contain toxic resins or organic compounds to make them easy to install or durable. However, over time, some portion of those toxic compounds moves into the air and is taken in by the residents. The following is a chart of some common building materials that contain toxic compounds.

Table 9.1 A sampling of indoor pollutants, where they are found, and some of their physiological effects.

Toxic Compound	Found in	Physiological Effects
Fiberglass	Wall, ceiling, and floor insulation	Irritant, suspected carcinogen
Lead	Plumbing solder	Neurotoxin, leads to learning disorders
Volatile organic compounds	Carpet fibers, wood stain, adhesives, insulation foam, caulking, plywood	Dizziness, nausea, headaches, fatigue, carcinogens
Mercury	Thermostat switch	Neurotoxin
Radon	Cement, wallboard, soil underneath the home	Carcinogen
Chromium	Green fungicide used to pressure-treat lumber	Carcinogen, liver and kidney toxic

b. Heavy Dependence on Energy

In addition to heavy dependence on toxic compounds, current building practices are very energy dependent. Consider the wood for the framing alone. The trees collected solar energy for years to produce the wood (a typical, three-bedroom home requires about one acre of forest to provide the necessary lumber). Then it had to be harvested and transported many miles—using trucks that burn gasoline—to a mill, which used energy to turn a large tree into a few boards and a lot of saw dust. Then those boards were wrapped and shipped to your local lumber yard, where they were stored. Then they came to your site, and contractors used electric or pneumatic guns to push fasteners and nails into the boards to hold them together. The contractor got sore and tired and needed to eat a larger dinner, then relax on the weekend by water skiing at the local lake, which was dammed to create the power to run the nearby sawmill. While this description is a little "tongue-in-cheek," the fact remains that every step of building a home or commercial building requires energy, and lots of it.

Once the house is built, it will need energy to operate. The design of the house will determine its energy consumption level over time. Does the house use ambient sun well? Is the house well insulated? Does air flow carry heat effectively from one portion of the house to another? These are a few of the questions a prospective homeowner should ask.

c. Heavy Input into Waste Stream

The typical three-bedroom home produces about four to seven tons of solid waste, which must then be disposed of at a local dump. Some of the waste is much more toxic to the environment than the material it protected. For example, a toilet fixture might be made of ceramic, but it may have been packed in Styrofoam, which will continue to pollute the environment from the landfill for years to come.

It seems improbable that our society will be able to sustain current construction practices, so what type of practices are more sustainable?

2. Sustainable Practices

a. Use of Nontoxic and Recycled Materials

To a large degree, American home buyers may be addicted to the "new house smell," so that most families expect to have the background smell of VOCs as a consequence of purchasing all new materials and traditional materials. To use less toxic materials and to reuse materials will eliminate that smell, and some people will feel as though they haven't purchased something "new." However, for those who want to decrease the levels of toxic substances in the home, have home construction use less energy, or produce less solid waste, there are some options.

i. Use nontoxic finishes on wood surfaces, such as molding and siding.

ii. Site the location near places of business and shopping so that less energy is expended in daily activities.

iii. Recycle materials from older buildings. Some companies have made a business out of manufacturing doors, molding, and timber supports out of old structures, such as barns, railroad trestles, and mills. Use shingles made from recycled paper. Use "glu-laminated" beams that are constructed from scrap lumber, rather than large solid beams that are whittled from a whole tree.

iv. Use cellulose insulation rather than fiberglass.

v. Use ecologically efficient carpeting. Carpeting is one of the longest lasting, most prevalent artifacts in landfills. Some companies (such as Interface, Inc.) recycle carpet materials, or manufacture carpet in smaller squares so that the whole carpet does not need to be replaced when a small portion wears out. Recycled carpet not only requires less energy and takes up far less space in landfills, but it does not contain toxic VOCs.

vi. After construction, use nontoxic cleaning fluids: club soda for upholstery, kill bugs with soap water or brewer's yeast, vinegar for windows and chrome, baking soda solution for tile, etc.

vii. Purchase materials from local sources to decrease the energy needed to get the materials to the construction site.

viii. After construction, landscape the area with drought-resistant plants that need little water. Be sure to plant deciduous trees on the sunny side of the house to block the sun in the summer and allow the sun to hit the house in the winter.

ix. Consider composting waste.

x. Consider a "grey water system" that recycles water from dishes and showers to gardens (although this is legal in only some states).

b. Design for Decreased Energy Use

The total energy consumption of a residential or commercial building may be decreased by using the alternative energy sources described in Chapter Ten.

Urbanization

1. Trends Toward Urbanization

In 1800, 6% of the United States population lived in an urban environment and 94% lived in a rural environment. In 1990, 75% lived in a city and 25% lived in the country. Not only is our population growing, but an ever-increasing percentage of it is living in a city.

There are some environmental advantages to living in a city. Less land is used for each person. Mass transit means that less energy is expended by each person. Recycling is more efficient. Sanitary systems are more centralized and efficient. More educational and cultural opportunities better meet the needs of city dwellers. However, there are some drawbacks to urban growth. People are disconnected with the real cost of the energy they demand and the food they eat. Consumerism reaches far beyond subsidence. The magnitude of solid waste and wastewater is such that natural processes are not capable of remediating the pollution; it must then be artificially treated, stored, or released into the environment. The artificial environments inside malls and buildings require massive amounts of energy for cooling and heating in areas where people would normally be comfortable by simply opening a window.

2. Causes for Urban Growth

Urban areas grow in size and population because more people are being born there, and more people are immigrating to the city. Reasons for immigration can be divided into "push" factors that encourage people to leave rural areas, and "pull" factors that draw people toward cities.

    a. Immigration "Pull" Factors

        i. Improved sanitation

        ii. Improved access and choice of food and other consumer products

        iii. Improved medical care

        iv. Greater opportunities for employment and upward mobility

        v. Higher salaries

    vi. Opportunities for recreation and cultural events

    vii. Diversity of lifestyles, more luxurious accommodations

  b. Immigration "Push" Factors

    i. Decreased employment, or economic downturn in specialized rural economy

    ii. Specialized industrial monoculture makes it more difficult for subsistence farmers to compete

    iii. Cultural conflict; homogenous village or rural local culture may create discomfort

    iv. Increased need for medical care

    v. Land division has reached a point where the plots available to an individual may not be enough to support a family

3. Current Urban Problems

  a. Traffic and Transportation

  Cities struggle with transportation issues. Some American cities, such as New York, Chicago, San Francisco, and Seattle, have developed public transportation to such a level that traffic is reasonable. Some cities have experienced such rapid growth that they have not had a chance to develop a suitable transportation infrastructure to accommodate the growth, and chaotic traffic problems result. Along with traffic comes additional air pollution and personal conflict and stress—both of which are already higher in cities.

  b. Air Pollution

  The industrial practices around cities, traffic, energy production (from coal plants, for example), and the lack of trees all contribute to additional air pollution around cities. Air pollution is made worse when local climate conditions create either an air inversion or heat dome effect (see Chapter Seven: Atmosphere Resources), which will hold the pollution toward the ground and keep it from blowing away. In Third World cities, lenient air quality standards, corrupt government, and lack of funds make urban air quality worse.

  c. Water Pollution

  Even in developed countries, it is a challenge to treat all the human waste that is produced in a large metropolitan area. Local rivers carry a burden of nutrient pollution from treatment facilities that are not perfectly effective all the time. Flow of waste is difficult to estimate in growing metropolitan areas. In Third World countries, the problem is far worse. Approximately two percent of the urban septic waste in Latin American cities receives any treatment at all. Poorly treated water makes it more difficult for authorities to provide fresh drinking water, so personal health in cities is further marginalized.

    d.   Housing

Construction cannot keep pace with the population influxes of rapidly growing Third World cities. Even cities in developed countries provide cramped quarters—or no housing at all—for the poor and destitute.

    e.   Crime

With a continual heavy flow of people into cities, people become increasingly desperate for some level of comfort, or just survival. Robberies and prostitution to gain money for food or drugs, and violent actions to vent frustrations or injustices—made easier because of the anonymity of the city—both contribute to higher crime levels in urban areas.

    f.   Mental Illness and Substance Abuse

When people have supportive families with money who can afford to treat relatives with personality disorders, neurosis, psychoses, alcoholism, or drug addictions, or when they are elderly, then society does not see them so easily as when these ailments occur to the poor and destitute. With a drop in national trends toward state-supported and mandated institutionalization, these people tend to migrate toward the streets to survive in homeless shelters and by begging, whereas they would not survive at all in rural areas.

4.   Urban Renewal

In the 1950s and 1960s, government-subsidized housing was funded through taxes and built in many American cities. However, the cement-block, industrialized architecture depersonalized people in these communities—which were already struggling for identity. Many of these government "projects" have since been demolished. When residents are involved in the planning of the projects and take responsibility for their upkeep and safety, then a sense of community grows and urban areas become vibrant cultural centers.

In the past, to save real estate costs, some of these government subsidized projects were constructed near industrial sites that contained toxic dump sites and air polluting industries.

AP Environmental Science

5. City Planning

    a. Typical Evolution of an Urban Area

        i. Early pre-industrial cities were centered around religious centers, palaces, or universities.

        ii. Early-industrialized cities tend to be centered around railroad junctions, as rail lines are needed to bring in raw materials for manufacturing.

        iii. With automobiles, cities tend to be at the junction of major roads; wealthier individuals live in the suburbs, further from the noisier, more polluted city center. In-city real estate prices drop and only the poor remain.

        iv. Well-established roads and freeways eventually divert shoppers to malls; industries find cheaper land out of the city core; core becomes financial center and housing is revitalized by wealthy urban-dwelling professionals.

        v. Eventually, perimeter villages themselves become urban centers, each with a full range of industry, residential opportunities, shopping, and services.

    b. Transportation

    Mass transportation—busses, light rail, bikes—is vital to keep down traffic and accommodate the movements of an ever-increasing population. Mobility of people also allows for social mobility and greater opportunity to gain employment. Also, by using public transportation and planning for "foot-traffic only" areas, fewer parking places will be needed and people will interact with each other and with nearby shopping and cultural opportunities.

    c. Multiuse Design and Clustering

    Creating multiuse space—with retail, office, and residential uses in the same building—decreases transportation costs and fosters a sense of community in an urban area. Also, more open space may be obtained by clustering buildings together as villages, with common green space between.

    d. Social Equity in Placement of Subsidized Housing

    When government-subsidized urban housing is not sited in an unhealthy environment, then the least affluent are not asked to pay the health cost for industries—whose increased profit margin results from environmental irresponsibility. Also, building users should be involved with design, maintenance, and supervision so that there is a stronger sense of community.

e. Open Space

Parks and greenbelts reduce air and noise pollution, establish a sense of serenity and pride in a community, and provide habitat for wildlife and migratory birds. Building tops may also be used for green spaces, which would allow food to be grown more locally and decrease the reflection of heat from the building, thus decreasing the "heat dome" effect of many cities.

Agriculture and Forestry

Human Nutritional Needs

If the land must contain the nutrients that are needed to grow the food, which in turn provides nutrition for humans, it is important to be aware of those human nutritional needs.

1. Undernourishment

   a. Humans need 2,000–2,500 calories each day; to get less is considered **undernourishment**.

   b. Humans need 22 different amino acids in order to build all the proteins that are needed by the body. The body uses proteins for critical functions in every cell. Of these 22 amino acids, the body can synthesize all but eight. These eight amino acids are the **essential amino acids** that must be in the diet if the person is to survive. From these eight, the other 14 critical amino acids can be synthesized. If a diet is missing any one or more of the essential amino acids, then the building blocks necessary for healthy living are not present. Meats and cheeses contain all eight essential amino acids. They can also be obtained from combinations of plants, such as a legume (beans, for example) and a grain (such as wheat). The essential amino acids are tryptophane, lysine, methionine, phenylalanine, threonine, valine, leucine, and isoleucine. Agriculture in different cultures has focused primarily on supplying these amino acids, and secondarily on providing other essential vitamins and minerals.

   c. **Famines** are massive, acute incidences of undernourishment that are usually catalyzed by political or economic upheaval (as in war), or environmental devastation (often due to a major storm). In either case, while the acute nature of a famine may pass quickly, the ability of the group affected to return to full productivity is severely weakened. Usually, large groups of people are displaced as they seek refuge in aid camps.

The Bangladesh famine of 1974 is an example of the combined effects of environmental and economic disaster. Severe flooding after the spring monsoons destroyed rice fields and the ability of families to support themselves—either nutritionally or financially. The perception of scarcity caused panic buying by those hoarding the remaining rice, which caused prices to escalate. It turned out that the harvest was not highly affected, but the combination of people out of work and high rice prices meant that people did not eat. Also at that time (during the Cold War), the United States cut off food aid because Bangladesh sold commodities to Cuba. The reduced aid fueled further panic buying and amplified the rising cost of rice—and the inability of families to feed themselves.

2. Overnutrition

In the United States and many other developed countries, many people eat too much food. The average per capita intake of food in the United States is 3,500 calories per day—1,000–1,500 calories more than necessary. About 1.1 billion people in the world are overweight, a condition that results in higher risk of heart disease, high blood pressure, diabetes, and stroke.

3. Malnourishment

Malnourishment is the inability to acquire adequate vitamins and nutrients. A shortage of any of the essential vitamins and minerals leads to an imbalance, which eventually leads to a disease.

   a. **Kwashiorkor** is easily recognized in children with swollen abdomens and reddish orange hair; it results from a lack of protein.

   b. **Marasmus** results from a lack of both protein and total caloric intake and causes skeletal thinness and wrinkled skin. Marasmus will weaken bones and increase a victim's susceptibility to disease.

   c. **Anemia** results from a lack of iron, which often comes as a secondary result of a lack of protein. Anemia prohibits oxygen from traveling to tissues and results in low energy and fatigue.

   d. **Ariboflavinosis** results in a deficiency of Vitamin B2, and is one of the most common nutritional disorders in the United States. Symptoms include skin problems, sore lips and mouth, and anemia.

   e. **Goiter** and hyperthyroidism result from an iodine deficiency.

   f. **Rickets** results from the body not having enough calcium in the bones, which is brought on by a deficiency of Vitamin D.

g. **Pellagra** results in a deficiency of niacin, which can produce skin problems, learning difficulties, or even death.

h. **Vitamin A deficiency** results in poor vision.

i. **Scurvy** results from Vitamin C deficiency. Symptoms include loose teeth, black-and-blue spots on the skin, and swollen gums.

j. **Beriberi** results from thiamine (Vitamin B1) deficiency. Symptoms include a loss of appetite and cramps.

Types of Land Degradation Due to Agriculture and Forestry

1. Desertification

    When land has been over-farmed, the nutrients and organic material in the land become depleted. Sometimes the only difference between good soil and poor, sandy soil is the organic material that holds soil together. When the soil becomes like sand, it is more susceptible to erosion by wind and water. The process of converting farmable grassland into nonarable desert is called **desertification**. For example, overgrazing by animals in the Sahel savannah in sub-Sahara Africa is quickly expanding the boundaries of the Sahara Desert. Ironically, chopping down rain forests in South America removes most of the nutrients from the ecosystem (which is mostly bound up in the plants, not in the soil); the resulting soil is low in nutrients and erodes away easily.

2. Erosion

    Erosion occurs when soil is moved from its point of origin. There are three types of erosion: *rill erosion*, where water cuts small rivulets in the soil; *gully erosion*, where rill erosion escalates to form a large channel; and *sheet erosion*, where water removes a horizontal layer of soil. Erosion is caused by any of the following factors:

    a. Wind and water, which can both remove the top layer of organic and nutrient-laden, adhesive topsoil. Soil exposed to the open air, with few roots or little cover, is vulnerable to wind erosion. Water erosion can occur with flood or heavy rains, or if the land is simply tilled in such a manner as to create channels where the water runs downhill—taking the soil with it—as opposed to residing on the land and soaking into the groundwater. Planting crops in rows exposes the soil to the wind and also creates gullies for flowing water. Water erosion is also amplified by irrigation with too much water, so that it flows over and displaces the soil.

    b. Chemical, which is caused by nutrient depletion, acid deposition, or a process called salinization. Salinization is caused by irrigation with brackish (salty) water, which leaves mineral salts on the soil when the water is used or evaporated. The remaining salt makes it difficult to grow plants or becomes airborne and causes human respiratory problems.

c. Physical, which includes compaction by cattle or machines. This occurs in industrial monoculture when the fields are trampled by excessive tractor use, or when grazed land contains too many animals.

d. Waterlogging, which results from excessive irrigation, poor drainage, or ocean encroachment.

3. Pesticide Use

Pesticide use becomes a land management issue because it greatly influences the type of agriculture that is practiced on the land, particularly with respect to the toxicity to non-target species and to pest resistance and resurgence. The toxicity of pesticide use is discussed further in Chapter Six: Human Health.

a. Toxicity to Nontarget Species

The largest drawback to pesticide use is the effect that pesticides have on nontarget species, whether directly or indirectly. Some nontarget species are destroyed by direct application of the pesticides (honeybees, for example). Some nontarget species are affected by eating other poisoned nontarget species. The pesticide then accumulates and eventually reaches toxic levels in the carnivore.

Humans can experience toxic effects from either acute (short-term) or chronic (long-term) exposure. Acute exposure at high levels may cause nerve problems, vomiting, bleeding, or death at high doses. Chronic exposure is suspected to cause a drop in immunity, cancer, birth defects, and other diseases.

b. Pest Resistance and Resurgence

Even with heavy use of pesticides, the differences in tolerance by individual insects result in at least a few hardy insects who survive. When these insects survive, they reproduce to give rise to a generation that is more resistant to the pesticide. Given the rapid reproduction levels of insects, this new population of pesticide-resistant individuals can quickly bloom to create a new menace that is not solved by using pesticides. This blooming of a resistance population is called pest resurgence. To manage resurgence by using more and stronger pesticides increases the collateral health consequences of pesticide use and decreases the effectiveness of the pesticides on the intended target organisms.

c. Types of Pesticides

    i. **Inorganic pesticides**, such as arsenic, copper, mercury, or lead. Highly toxic and persistent.

    ii. **Chlorinated hydrocarbons**, such as DDT, aldrin, lindane, and toxaphene. Block nerve membrane ion transport and the nerve signal; tend to bioaccumulate and biomagnify.

    iii. **Organophosphates**, such as parathion and malathion. Effective neurotoxin, but not persistent.

    iv. **Carbamates**, such as carbofuran and aldicarb; behave much like organophosphates.

    v. **Botanical pesticides**, such as pyrethrum, which is extracted from chrysanthemums.

d. Integrated Pest Management (IPM)

In the wake of the Green Revolution, growers reconsidered the benefits of nonchemical solutions to pests and sustaining high yields, under the catchall phrase **Integrated Pest Management (IPM)**. IPM involves a combination of pest control strategies—chemical and nonchemical—that are unique to the crop and location. Nonchemical strategies include the use of natural predators, using sex pheromones to attract bugs to a place away from the crops, introducing sterile breeding partners, and crop rotation to decrease pest momentum. Chemical pesticides can still be part of an IPM program, but with greater deliberation and specific targeting so that a minimum is used. For example, planting a trap crop so that it matures slightly before the rest of the field will attract pests to it. Spraying and destroying the trap crop then eliminates the bulk of pests— and the health risk experienced by farm workers due to pesticide exposure when they work in the other fields.

4. Fertilizer Use

Plants need inorganic minerals to grow, in addition to water, carbon dioxide, and sunlight. The most important nutrients are nitrogen, phosphorus, and potassium, although magnesium, calcium, and sulfur may also be needed in some situations. Farmers can ensure that plants reach their full growth potential by adding these nutrients in fertilizer. While much of the improved crop yields around the world are due to fertilizer use, it is very possible to overuse fertilizer; what is not used up by plants sometimes causes nutrient pollution in the environment. (See Chapter Eight: Water Resources.)

Crop rotation is one alternative to the application of chemical fertilizers. For example, farmers could rotate planting nitrogen-demanding crops with nitrogen-producing crops. Peas, alfalfa, and clover are examples of crops that fix nitrogen into the soil. Such plants have nodules on their roots that contain symbiotic bacteria, which convert atmospheric nitrogen into ammonia—the form of nitrogen that remains in the soil and can be used by other plants.

5. Energy Use

    When farmers used horses, the horses were powered by hay—a renewable resource. Current farming practices depend on a much greater amount of fossil fuels—a nonrenewable resource—for tractors, fertilizer and pesticide production, transport, storage, manufacture of final product, and disposal of waste. Consequently, meeting the expanding nutritional needs of the world's population is not just a land use issue, it is also an energy issue.

Sustainable Agriculture vs. Industrial Monoculture

1. Subsistence Farming

    Subsistence farming produces only what is needed to support the nutritional needs of the grower and is not designed to produce food for market. More than 65% of the global population is involved in subsistence farming—nomadic herding of livestock, migratory farming (such as swidden agriculture—where a farmer burns an area, then lives off of plants that are a part of the natural succession of the area), and stationary crop cultivation.

2. Industrial Monoculture

    In an effort to increase both profit and yield, and to cover the large investment in energy, chemicals, machinery, land, and labor, the economics of agriculture has favored **industrialized monoculture**. Large tracts of land tend to be planted with a similar crop, and the same maintenance techniques are implemented over a broad region. While this trend has resulted in both profits and higher yields, the same crop over a large geographical area allows for large accumulations of one type of pest that would devastate the whole crop. Pest resistance and resurgence is more likely. Also, small family operations that bring diversity of crops and management styles cannot compete economically.

3. Sustainable Farming

    a. Uses low- or no-till farming and contour farming to decrease wind and water erosion.

    b. Uses crop rotation and simultaneous polyculture (for example, planting a pest deterrent species or nitrogen-fixing species between food production species).

c. Uses natural fertilizers as much as possible, rather than chemical fertilizers.

d. Minimizes pesticide use by using Integrated Pest Management strategies.

e. Minimizes the use of fossil fuels by selling to local markets, using local materials, limiting the number of passes over the field, and decreasing the use of chemicals.

f. Reduces dependence upon irrigated water.

4. Farming vs. Ranching

a. Ecological Aspects of Meat Production

i. Total Energy Input

It takes approximately 16 pounds of grain to produce one pound of beef, and three pounds of grain to produce one pound of chicken. From an ecological standpoint, eating beef requires far more energy and agricultural land. In fact, there are more livestock than people in the United States, and those animals consume about 70% of the grain produced in the country.

ii. Feedlot Pollution

Most of the cattle raised in the United States spend at least some of their lives on a high-density feedlot. Here they are fed with grain—usually corn—that is grown on a farm. Feedlots make economic sense in a cultural climate of escalating real estate costs and range fees.

However, since the cattle are in such tight quarters and their hooves are usually in manure-strewn mud, they are given antibiotics to decrease disease. They are also given growth-hormones to maximize meat production. Consequently, the natural runoff from feedlots contains high levels of antibiotics and hormones (that have been excreted) and nutrients (from the manure). The nutrients cause eutrophication of streams, rivers, and lakes (see Chapter Eight: Water Resources). The hormones and antibiotics create background levels of those substances in the environment which has other consequences. The antibiotics, for example, contribute to bacterial resistance to antibiotics in our society.

b. Overgrazing Public Lands

Using public lands (National Forest and BLM land) for cattle grazing is a major component of agricultural economics. In the 11 western states, 75% of the public land is available for grazing. The grazing permits cost about 3–5% of the true cost of grazing land, so leasing public land helps ranchers minimize the cost of beef production. Consequently, the metaphor of the "Tragedy of the Commons" becomes reality; the tendency to add "just one more" animal unit (a steer or cow/calf pair) to the land is a strong temptation that leads to land degradation. Indeed, an estimated 85% of the government-owned rangeland is considered poor in quality.

Agricultural "Revolutions"

1. **Green Revolution**

   In the 1950s, scientists began developing strains of crops that would provide higher yields in hopes of solving world hunger problems. One of the most significant new strains was a dwarf rice strain developed by the International Rice Institute. This rice was planted throughout the Philippines and, for a time, yields rose sharply. This increase in crop yields as a result of producing and mass planting high-yielding strains was known as the Green Revolution.

   However, the Green Revolution in the Philippines used only a few genetic strains of rice where once there had been several—all with different levels of resistance to different pests and diseases. The single strains became vulnerable to diseases and insects and increasing amounts of pesticides were needed to sustain the high yields. In the Philippines and in other places, the Green Revolution brought greater dependence on expensive seeds and higher dependence on chemical fertilizers and pesticides. In some years, lower yields result when high usage levels of pesticides backfires in the form of environmental contamination or pest resistance and resurgence. Overall, however, many would credit the technology coming from the Green Revolution as responsible for allowing worldwide crop yields to keep pace with world population growth.

2. **Genetically Modified Organisms (GMOs)**

   In the 1970s, scientists began to understand more about moving genes from one species to another. Species that carry such a hybrid combination of genes from other species are called transgenic species or strains, or genetically modified organisms. Agricultural scientists began to experiment with this technique in crops in hopes of producing strains that are more resistant to pests and adverse environments, remain more durable during shipping, yield better nutrients or more protein, or grow at different times of the year. Some of these benefits have been realized, but commercial scientists have also used genetic engineering techniques to create strains that simply work better with their pesticides. Today, transgenic crops represent about 70% of the food grown and sold in the United States and are planted on about 75% of the global cropland.

Scientists and consumer advocates are concerned because it is difficult to predict the health and environmental drawbacks of GMOs. Some feel that we have reached a crisis in labeling. After all, how can a consumer know that they are not eating a plant that contains a gene that produces a protein to which they are allergic?

Multiple Use of Forests

Forests occupy about 30% of the entire land area of the world, and are important in absorbing precipitation, controlling climate, providing oxygen and purifying the air, producing usable resources (fuel, lumber, and paper), and creating habitat for wildlife. About 40% of the forests are old growth forests—those where disturbance is low enough that the forest is able to undergo natural ecological cycles.

Current Forest Harvesting Practices

1. Selective Cutting vs. Clear-Cutting

    Clear-cutting a stand of trees involves cutting down every tree, regardless of species or size. Large machines can be used to cut and then drag (or "skid") the logs to an access road, but the smaller trees are wasted, the soil is exposed to erosion by wind and water, and the habitat is severely disrupted. Large "slag piles" of the waste wood and stumps are pushed up, although now some logging operations are attempting to protect from soil erosion and encourage nutrient return to the soil by leaving waste distributed over wide patches of ground. Some species do not reseed easily unless by natural conditions.

    Selective cutting involves a less disruptive harvesting of a portion of the mature trees, which results in better growth, a more stable habitat for wildlife, and protection from erosion. However, selective cutting is not usually cost-effective.

2. Swidden Agriculture

    Swidden, or "milpa," agriculture refers to a practice used by indigenous people in tropical rainforest areas. In swidden agriculture, the farmer clears a small plot of land by cutting and, in some cases, burning. The ashes from the fire provide essential nutrients, which—because the plot is small and surrounded by rooted trees—do not erode away. Some crops, such as bananas, are planted immediately for an instant food crop to prevent erosion and to provide shade for useful root crops, such as sweet potatoes. During the natural succession of the forest, a number of other crops are planted or naturally harvested as that portion of the forest steadily returns to its climax condition. Swidden agriculture is sustainable as long as the density of farmers does not exceed the forest's ability to regenerate. However, if the density of farmers is too high, or if outside companies open up large tracts of rainforest, then deforestation occurs.

3. Rainforest Deforestation

Tropical rainforests are a rich resource that contains about two-thirds of the global biomass and more than half of the flora and fauna biodiversity worldwide. However, this rich resource is threatened as world population increases. Initially, deforestation occurred to grow hardwood for furniture and commercial food crops, such as coffee and sugar. Currently, about one percent of Brazil's rainforest is destroyed annually to make room for settlers, charcoal producers, developers, and farming or ranching by international corporations. This continued reduction of tropical rainforest threatens the rich biodiversity, climate stabilization, flood control, and oxygen production available from rainforests.

4. Forestry as Agriculture

In an effort to provide lumber products long into the future, some forest supply firms have replanted many millions of acres with dense single species stands. Such monoculture forestry increases yield and ease of harvesting, but it also encourages disease and pest infestations. Monoculture forests cannot support the same level of biological diversity necessary to build healthy soil and establish a hearty level of ecological complexity in the forest biome.

Fire Management in Forests

For many years, the policy of the Forest Service was to put out all fires. However, this policy allowed the undergrowth and brush under the tree canopy to grow unabated, which resulted in hotter and more damaging fires when they finally did come. In addition, eliminating all fire from some forests, particularly the taiga forests in the Northern Rockies, also eliminated the benefits that fire provided to many species. For example, a fire opens the cones of the Lodgepole Pine and facilitates re-seeding. Fires open up areas and help with meadow formation, which allows grazing wildlife to move to new areas. The combination of wildlife and ash from the fire returns necessary nutrients to the soil. Now the Park and Forest Service have a "let burn" policy that uses fire-fighting resources only when lives or property are threatened. This new policy attempts to allow smaller fires to remove the undergrowth and revitalize the forest community. Part of the pressure on the Forest Service to fight fires prematurely comes from a public that sees a burned forest as "ruined," rather than being an essential part of a natural and healthy cycle.

Wildlife

Traits of Endangered Species

Species that easily become endangered tend to have certain characteristics that either make them more vulnerable or decrease their resilience. The following are typical traits of species that are more easily endangered.

1. The species uses a logistic population growth strategy for one of the following reasons:

    a. Individuals of the species have a long life span and/or a long generation time; a lot of resources are invested in a single individual.

    b. The species has a large body; again, a lot of resources go into the success of a single individual.

    c. The species is a carnivore or top carnivore. This increases the species' dependence on the success of lower trophic levels, and increases the effect of bioaccumulation and biomagnification of toxic substances.

    d. The species has a low reproductive rate, which is consistent with the theme of investing large amounts of energy and materials into single individuals. Low reproductive rates allow for increased survival of individuals (e.g., bears), rather than producing many offspring over a short time in the hope that a few will survive (e.g., flies).

2. The species requires a large amount of land per animal for one of the following two reasons:

    a. The species is solitary and requires a large roaming area to find a mate or food (e.g., grizzlies).

    b. Seasonal migration requires multiple and intact biomes (e.g., migratory birds, such as geese).

3. It is a specialist species with a narrowly defined niche.

4. The species demonstrates one of the following reasons for having low genetic diversity (or diversity of gene pool). Low genetic diversity limits the ability of the population to respond through natural selection to a wide range of challenges in future years. For example, because of the low genetic diversity of the Florida panther, a single disease for which the panther is genetically unequipped could wipe out the few remaining individuals.

    a. **Genetic bottlenecks** occur when a catastrophe has eliminated many individuals and the subsequent population has a gene pool that is limited by that of the original breeding pairs.

b. **Genetic isolation** has occurred because a small number of individuals have been isolated; the gene pool of the subsequent population is limited by the genes available in the original breeding pair.

   c. **Genetic assimilation** threatens some endangered species when they crossbreed with closely related, hardier species.

5. The endangered species competes for resources with a dominant, hardier species. Unfortunately, humans are one of those species that is out-competing many other species for global resources.

6. The species has a low tolerance for pollution, and development or any association with human activity results in over-powering the organism's ability to survive.

Causes of Extinction

1. Loss of habitat is the largest and most common reason for species extinction. Critical habitat may be diminished for any of the following reasons:

   a. Human development and pollution will eliminate species. For example, road-building limits the range of grizzly bears; DDT accumulated in the food chain and began to eliminate predatory birds; clearing land for a housing development will eliminate indigenous plants.

   b. Competition with dominant or exotic species for the same habitat will eliminate species. For example, the introduction of kudzu vines in the American southeast choked out local plant species; humans use intelligence to out compete just about any species for any resource we choose.

   c. Change in climate or other abiotic factor will shift the environment and make it uninhabitable. For example, a stream is warmed by the effluent of a power plant and can no longer support a population of mayfly nymphs. Some scientists believe that a climate change is responsible for the elimination of dinosaurs.

2. Hunting has eliminated or endangered many dozens of species. If there is a market for some portion of an animal (whether for food, profit, or sport), or there is a perception that it presents a nuisance to human society, the species will be hunted—whether legally or not.

3. Loss of genetic diversity, such as can happen as a result of genetic drift, genetic isolation, a genetic bottleneck, or genetic assimilation (described above).

4. Normal population fluctuations experience a downward swing (from predation or climate changes, for example) that results in total elimination of the population in question. If the population is the only existing population for that species, it would result in the demise of the species.

Mitigation

1. Monitoring Markets for Endangered Species

    As long as markets exist for an animal, someone will figure out a way to remove it. If cultural practices (folk medicines, exotic pets, gourmet dishes, etc.) are adjusted to not include endangered species, then market forces will not pressure people whose desire for short-term profit overides their concern for biodiversity.

2. Hunting and Fishing

    Ironically, hunting and fishing have created a market for habitat protection of waterways and woodlands, which has in turn allowed for populations of game species to flourish. Seasonal harvesting of these animals tends to remove those most easily obtained, or the weakest, and a hardier population results. For example, there were approximately 500,000 white-tailed deer in the United States in the early 1900s; presently there are approximately 14 million.

3. Legislation

    The Endangered Species Act (ESA, 1973) represents a significant change in endangered species protection. ESA designates categories of concern for species: *vulnerable* (early identification of species at risk), *threatened* (most likely to become endangered), and *endangered* (in imminent danger of extinction). Perhaps the most powerful tool of ESA is the protection that it provides to critical habitats from economic development. Habitat Conservation Plans are one attempt at a solution.

4. Habitat Conservation Plans

    The Fish and Wildlife Service now negotiates Habitat Conservation Plans with private landowners so that they are allowed to use natural resources as long as the species experiences some overall benefit. For example, in some areas developers arrange trades with permitting agencies to create protected areas in exchange for the opportunity to develop wildlife habitat. This provides a new protected habitat and creates a secondary real estate market for critical habitat to be used as such.

5. *In Situ* Management

    In situ management protects species and habitats by protecting wildlife in existing parks, wilderness areas, and preserves. This method is most successful when the necessary habitat is available to be protected by private donation or government action.

6. *Ex Situ* Management

Ex situ management protects species by removing individuals from a natural habitat and preserving them in zoos or captive breeding programs with the hope of returning offspring to a natural setting. This method is most successful when the natural habitat has become hostile in some way (hunting, fire, etc.) or if the habitat has been destroyed and a new one needs to be developed in another place.

Flood Control

While flood control has much to do with managing water resources (see also Chapter Eight: Water Resources), it also affects the management of surrounding land resources. In particular, changes in flood control and waterways policies affect the availability of wetlands, which are ecologically important.

Importance of Wetlands

1. Wetlands provide essential habitat for endangered species, breeding freshwater organisms, and migratory birds.

2. Wetlands provide a reservoir for flood waters and prevent downstream flooding and bank erosion during periods of high river flow.

3. Wetlands provide a natural water treatment process by removing nutrients from the water and decreasing eutrophication.

4. Coastal wetlands provide a buffer to saltwater intrusion that would otherwise contaminate inland wells.

5. Wetlands provide recreational opportunities and retreat space for outdoor-lovers.

Channelization

During the last century, the Army Corps of Engineers created straight, deep channels in many rivers and coastal wetlands in order to make the waterway more navigable. However, creating a straight channel with firm boundaries increases the flow of water moving down the channel, which increases the force exerted on the shore and increases siltation. Additionally, channelization decreases the flow of water to surrounding wetlands, thereby drying them out and diminishing their benefit; there is less of a nursery available for important species, water-dependent plants die, flood waters rise more quickly, and the water is not cleansed by the plants and slow-moving current.

Mineral Resources and Mining

Common Mineral Resources

1. Aluminum: usually mined as bauxite ore, requires large amounts of energy to process into metal. Multiple industrial uses include construction, food packaging, transportation, and electronics.

2. Chromium: used as additive to create alloys

3. Copper: building construction and electronics

4. Gold: hard currency, jewelry, medical and electronics industries

5. Iron: steel production

6. Lead: gasoline, batteries, paints

7. Manganese: alloys

8. Nickel: alloys, batteries

9. Platinum: catalyst for catalytic converters in automobiles, electronics

10. Silver: photography, jewelry, electronics

11. Uranium: radioactive metal used to create fuel pellets for nuclear reactors

12. Nonmetal minerals

    a. Silicates: used for glass production

    b. Sand: road construction

    c. Gravel: construction; most prevalent mineral used

    d. Limestone: used in cement, paint, glass

    e. Evaporites (formerly dissolved minerals left behind after the water solvent has evaporated): include halite, which is rock salt that is used for making ice cream and refined to make table salt, soften water, and melt ice on winter roads; gypsum, which is used for making wallboard in building construction; potash, which is predominantly potassium chloride and sometimes used as a fertilizer

    f. Sulfur: used for sulfuric acid production

    g. Coal: used as an energy source

    h. Oil: used to produce carbon-based industrial molecules, fuels (natural gas, gasoline, kerosene, heating oil), lubricants, road construction (tar), waxes, and other hydrocarbon products

The Mining Operation

1. Due Diligence

   a. Site Analysis

   A mining company will use chemical analysis, tests of well drilling or core samples, geological surveys from independent companies, and seismic surveys in order to assess whether or not a site will yield a cost-effective amount of mineral resources.

   b. Leases, Licensing, and Permits

   Mining operators must register with the state agency that implements the SMCRA in that state. This process includes setting up an escrow account with mining company funds to pay for land reclamation after mining has ceased, and also ratify the site cleanup plan. For public land, leases with the federal government must be negotiated.

2. Extraction

   a. Surface Mining

   To dig an open mine, the overburden is first removed and stored. Then the mineral seam is mined. In some cases, whole mountains are displaced, or open pits several kilometers across are created.

   b. Subsurface Mining

   Underground mines are used to extract deeper mineral deposits, although this method contains many more risks to miners—gases, lack of oxygen, fires, explosions, tunnel collapse. Also, mine tailings tend to be removed and discarded on the surface. Water percolating through the tailings creates a toxic, acidic runoff that pollutes streams and waterways.

   c. Wells

   For deposits in the gaseous or liquid phases (water, oil, natural gas), a well may be the best way to extract the resource.

3. Processing

   The United States already imports about 50% of the mineral resources it needs. As mineral resources become depleted, lower quality ores are mined, which require more elaborate and pollution-producing processing. Below are three different—but very common—types of mineral processing.

a. Heap-Leach Extraction for Gold

In this process, large piles of gold ore are placed over a basin that catches the runoff created by spraying the piles with a solvent. Gold is not easily dissolved (which is why it is a precious metal), so the ore is sprayed by an acidified cyanide solution to coax the gold to dissolve. Electrolytic reduction (adding electrons to the gold) is then used to extract the metal from the gold-laden solution. The remaining cyanide solution can then be re-used.

One difficulty with heap-leach extraction is that when mining operations are completed, the operators often leave the highly toxic cyanic solution on site. In one well-publicized example, a Colorado mining company extracted $98 million in gold, then went bankrupt and left the EPA with a $100 million cleanup bill.

b. Uranium Processing

Mining and processing uranium is one of the most environmentally devastating types of resource acquisition. For example, for every 100 tons of ore mined, about 0.8 tons of milled ore move to the next level of purification. The remaining 99.2 tons of mill tailings represent high-level nuclear waste that must be handled and stored. The 0.8 tons is further processed through an enrichment process, and finally the enriched material is fabricated into pellets that can be used in a nuclear reactor. At each step of the process, nuclear waste is produced.

c. Aluminum Electrolytic Extraction

Processing aluminum from bauxite ore requires a very large amount of energy. Because of this, the benefits of recycling aluminum is most apparent in the energy savings of processing ore to get "virgin" aluminum. In order to retrieve aluminum metal from bauxite ore, the ore must be crushed, melted, and then put in large vats with two electrodes (anode and cathode) in it. The aluminum ions in the ore are reduced (electrons added to the positive ion) at the cathode.

4. Reclamation

The Surface Mining Control and Reclamation Act (SMCRA) requires that coal mining operations put into an escrow account sufficient funds to reclaim mining sites after mining operations have ceased. The law also empowers states to oversee this process ("State Primacy") and provides funds to clean up sites in existence before the law was passed.

Reclamation involves the return of the overburden, or the land that was taken off to access the mineral deposits. Then topsoil must be put down and plants grown. "Highwalls," or the upper edge of the mine site, must be sculpted to appear natural. Tailings must be removed and placed in such a manner as to prohibit the leaching of toxic and acidic runoff.

5. Environmental Consequences of Mining

There are many environmental consequences for recovering natural resources from the land. The land can be deformed by the digging, or eroded from exposure to wind and water. Mine tailings exposed to rain create a toxic runoff that spoils local soil, destroys streams and rivers, or seeps into the groundwater and shows up in wells many miles away. Particulate matter from mining operations (particularly in open mines) stirs into the air and creates an air pollution problem—sometimes causing respiratory problems in populations many miles away. There needs to be enough top soil to reclaim the land and allow plants to grow on the previously mined site.

Solid Waste Disposal

The Waste Stream

The **waste stream** is the steady flow of matter from raw materials, through manufacturing, product formation and marketing, to the consumer, and on to its final resting place—usually a solid waste dump; this is sometimes referred to as a "cradle-to-grave" concern for waste management. Sometimes the consumer experiences a very short time with the material between the time it is a pretty package and the time it becomes waste.

Some of our waste contains valuable resources that can be recycled, reprocessed, or at least combusted to produce energy. However, throwing all the waste together makes it very difficult to make use of these valuable resources.

With each American producing about 4.5 pounds of solid waste per day, and 76% of that waste ending up in a landfill, managing our solid waste becomes a major land use issue. See Chapter Six: Human Health for a more complete discussion about waste and waste treatment.

Case Summaries

The Mississippi Flood of 1927

Principles mentioned in this case:

- **The historical background of channelization**
- **The flooding consequences of channelization**
- **Relationship between land use planning, personal pride, and socioeconomic politics**

In his seminal work, *Rising Tide*, historian John Barry documents the power struggles associated with managing the banks of the Mississippi River. As early as 1816, a raging debate began to grow about how to protect towns when the river flooded. Two sides emerged: one advocated higher levees; the other, spillways and reservoirs. By 1850, the debate about how to handle the river was personified in the pride and egos of three engineers, Andrew Atkinson Humphreys, James Buchanan Eads, and Charles Ellet.

At the start, both Ellet and Humphreys were conscripted to study the river and publish their findings about whether levees or spillways provided the best solution. Ellet was brilliant and intuitive, but his recommendation to use spillways was not well researched. Humphreys was an intensely driven man who had a breakdown just as the due date for the reports approached, so Ellet's ideas were the only ones standing. Humphreys recovered and published his work by the time the Civil War broke out; in it, his research supported Ellet's conclusion, but Humphreys could not allow himself to agree with Ellet and pointed his conclusions toward building levees. Ellet was killed in the war, which left Humphreys as the one who was most knowledgeable about the river. Humphreys became the head of the Army Corps of Engineers, and—saving face by not changing his conclusions—went ahead with the building of the levees and increased channelization of the Mississippi.

In 1927, a massive series of storms dumped water over states that drained into the Mississippi. As flood waters rose, so did the tempers of regional political adversaries. When New Orleans was ultimately threatened, a portion of a levee in a predominantly African-American area upstream was dynamited—flooding that area and destroying the land, but saving New Orleans. In the aftermath, the displaced African-Americans were severely mistreated, which exposed the political stripes of regional and national leaders. It greatly accelerated the Great Migration of blacks out of the South and into the northern cities, sending nearly a million to Chicago. The flood was one of the natural disasters that helped usher in the Great Depression, and it defined the terms of two presidents. Some feel that it is the single most influential event in American history. Some wonder if it could have been prevented if a single engineer had been willing to agree with his adversary.

Today there are over 1,600 miles of levees on the Mississippi. The river defies control. Some towns have changed states as the water boundary shifted over the years. The mouth of the river has moved many times. The current plan to control flooding involves not only the levees, but the shortcuts advocated by Eads and spillways promoted by Ellet.

Real Estate Development in Hurricane-Prone Coastal Areas

Principles mentioned in this case:

- **Land erosion due to natural disaster**
- **Land erosion due to community mismanagement**

- **Conflict between rights of landowners and ecological impact**
- **Limits to federal disaster aid**
- **Trends in coastal land management**

Folly Island is an aptly-named barrier island in South Carolina that is experiencing a real estate boom because of a new trend in American home-ownership. A *USA TODAY* survey estimates that one in seven Americans now lives in a county that borders the Atlantic Ocean or the Gulf of Mexico. In many of these communities, natural features and the threat of hurricanes makes many of these counties poor choices for building, but a frenzy for high-end ocean-view homes overshadows these concerns and the real estate market is booming.

On Folly Island, as with many other coastal islands and beaches, erosion from natural sources and poor ecosystem management is a major problem. A lighthouse that was built on one tip of the island in the 1870s is now 2,000 feet from shore. The Army Corps of Engineers routinely restores sand to local beaches, but they are regularly washed away. In 1989, Hurricane Hugo flattened the island community. However, the real estate boom was only encouraged when the federal disaster assistance and flood insurance responded with such generosity that they made it profitable for homeowners to replace damaged small cottages with large, high-end homes. Attempts to artificially fortify the beach only led to further erosion. In one case, sand was brought in by the Army Corps with the intention of lasting 10 years, but 80% of it was washed away by a storm a few months later. If one resident builds a sea wall—which may or may not withstand shifting sands or the impact of a hurricane—the presence of the wall only accelerates beach erosion on nearby lots.

Anti-growth groups attempt to halt development on unstable Folly Island beaches, but the strong property-rights state and local judges tend to rule in favor of the property owners' desire to develop. The assurance of federal aid and the expected guarantee of flood insurance gives landowners a sense of invulnerability regarding assets they invest in beach homes, so building continues. However, the expectation of flood insurance may be about as sturdy as a foundation built on sand. In 1999, a House subcommittee met to consider the problems associated with insuring areas that are heavily hit by natural disasters. Many insurers have stopped underwriting insurance in high-risk areas, such as earthquake or slide-prone counties in California, Hawaiian neighborhoods in the path of flowing lava, and coastal Floridian counties. In particular, insurance claims after Hurricane Andrew and the Northridge earthquake in California resulted in huge losses for insurance companies. However, government subsidies and insurance companies continue to underwrite sensitive coastal areas such as Folly Island.

Even though there doesn't appear to be any solution on the horizon that favors careful environmental stewardship and storm safety over unabated construction, perhaps local residents will eventually re-learn the meaning behind the name of the island on which they live.

Energy Resources

Chapter 10

Energy Resources

Understanding Energy

Energy Sources

Fusion: Our Original Source of Energy

Much of the energy that reaches Earth originates from **fusion** reactions in the sun. Fusion reactions combine lighter elements to form heavier elements. In the combination of these elements, some mass is converted into light energy that is carried through space by photons.

Energy that we use from coal and oil also comes from solar energy. **Photosynthesis** converts atmospheric carbon dioxide into an organic molecule, glucose. Dead organic material is compressed over time and either produces carbon alone (coal) or a mixture of hydrocarbon compounds (oil). When we burn fossil fuels, such as coal and oil, we are using solar energy that was captured by photosynthesis millions of years ago.

Solar energy can also be captured by water during the **hydrologic cycle**. When water evaporates upon being heated, then precipitates and falls into a reservoir, it can turn a turbine in a hydroelectric power station.

The energy we derive from nuclear sources came from a star, but not our own sun. During a previous generation of planets, a star exploded and created large, unstable elements, such as uranium and plutonium. These elements give back the energy of the star's explosion as they undergo radioactive decay.

Some energy is made available to us as tidal energy, which originates by the gravitational potential energy of the moon as it pulls against the seas.

Photons and Light Absorption

When photons reach the Earth, they are either reflected back into space or they are absorbed. Photons are absorbed because they are intercepted by a molecule that has a bond in it that vibrates when struck by the photon. In this way, light energy is converted into **kinetic energy** in the molecule. The increased motion within the molecule gives it heat.

For example, sunlight carries energy through the atmosphere and strikes the ocean. The bonds in ocean water molecules absorb the light energy and convert it into kinetic energy. If the water molecule has sufficient kinetic energy, liquid water will evaporate. The movement of the atmosphere carries the water vapor up a mountain. As the altitude increases, the pressure drops and the air releases the water as rain. As the water moves downhill in a river and through a power plant, some of its potential energy converts to kinetic energy and pushes the turbine to make electricity. This electrical energy is available because the water molecule originally absorbed light energy from the sun.

In the formation of coal, light was originally absorbed by the **chlorophyll** molecules in a plant. The chlorophyll initiates photosynthesis in the plant's cells, which convert carbon dioxide and water into glucose and oxygen gas. The glucose and other biological molecules in the plant fall off, decay, and become compressed under great heat and pressure over many years to become coal. The coal can then be burned to warm a building or create steam that turns a turbine that generates electricity. The coal provides energy because the chlorophyll molecules originally absorbed light energy from the sun. In both these examples, different mechanisms stored solar energy, which was used in the form of electricity some time later.

How Electric Power Is Made

Much of the energy humans currently use in daily life is in the form of electricity, which ultimately comes from the above-mentioned sources. Electricity is made by using some mechanical force to turn a turbine, which turns a generator to make electricity. The needed force can come from flowing water (hydroelectric) or high pressure steam that comes from water that has been boiled by the heat produced from either a chemical or nuclear reaction. The turning generator moves a set of magnets relative to a conducting wire. Through electromagnetic induction, the moving magnetic field exerts a force on the mobile electrons in the wire and causes the electrons to flow. Flowing electrons represent electrical current, which contains voltage.

The Second Law of Thermodynamics states that some energy will be lost as heat when energy is transformed from one type of energy into another, such as with the generation of electrical energy. Figure 10.1 shows a sequence of the forms energy takes as it is transformed from sunlight to electrical energy via hydroelectric power generation. In each transition, some energy will be lost as heat. The percent of energy available to do useful work after each step is called the *efficiency* of that energy transfer. The total energy efficiency for a number of steps equals the product of the efficiencies of each step.

For example, the series of energy transformations in Figure 10.1 represents six steps. If each step allows 80% of the energy to move to the next step (20% would then be lost as heat in each step), then the total efficiency for the process would be

$0.8 \times 0.8 \times 0.8 \times 0.8 \times 0.8 \times 0.8 = 0.8^6 = 0.26$, or 26% efficient.

Figure 10.1 Energy Transformations Needed to Make Electricity

Sun
⇓
Evaporated water
⇓
Precipitated water
⇓
Falling water
⇓
Water pushing turbine
⇓
Turning generator magnets relative to conductor
⇓
Electrical energy

Energy Units and Calculations

Units of Energy

The fundamental metric units of energy are: kg • meter^2/seconds^2. The derived metric unit for all of these forms of energy is the **joule**. Therefore,

$$1 \text{ joule} = 1 \text{ kg} \cdot \text{m}^2/\text{sec}^2$$

Power is the amount of energy exerted in a given time and is expressed in units called **watts**; therefore,

$$1 \text{ watt} = 1 \text{ joule/sec}$$

Electrical energy is often expressed in terms of watts instead of joules by multiplying both sides by seconds; so the following relationship becomes important:

$$1 \text{ joule} = 1 \text{ watt} \cdot \text{sec}$$

Example 10.1

Question: How many joules are in a kW-hr?

Answer:

$$3.6 \times 10^6 \text{ J} = 1 \text{ kW-hr} \times \frac{1 \text{ J}}{\text{W-sec}} \times \frac{1,000 \text{ W}}{1 \text{ kW}} \times \frac{3,600 \text{ sec}}{1 \text{ hour}}$$

Sometimes energy is expressed in terms of nonmetric units, such as **British Thermal Units** (BTU) where 1.0 BTU = the amount of heat needed to raise 1.0 pound of water by 1°F; or calories, where 1.0 cal = the amount of heat needed to raise 1.0 grams of water by 1°C. Therefore:

$$1 \text{ kW-hr} = 3,400 \text{ BTUs} = 3.6 \times 10^6 \text{ J} = 1.5 \times 10^7 \text{ cal}$$

Unit Conversions

Many problems simply involve converting one set of units to another. The following steps help solve unit conversion problems:

1. Write down the units of the answer.
2. On the other side of the equals sign, write the given information.
3. Use conversion factors to cancel the units of the given information so that only the units of the answer remain.

Example 10.2

Question: How many BTUs are in 14.0 kW-hr?

Answer:

$$4.76 \times 10^4 \text{ BTU} = 14.0 \text{ kW-hr} \times \frac{3,400 \text{ BTUs}}{1 \text{ kW-hr}}$$

Example 10.3

Question: Assuming one pound of coal used by a power plant yields 5,000 BTUs of heat energy, how many BTUs are produced by 2,500 pounds of coal?

Answer:

$$1.25 \times 10^7 \text{ BTU} = 2,500 \text{ lbs. coal} \times \frac{3,400 \text{ BTUs}}{1.0 \text{ lb. coal}}$$

Example 10.4

Question: How many joules of heat are produced from 2,000 lbs. coal? (There are 1,059 J per BTU; 1.0 lb. coal produces 5,000 BTUs.)

Answer:

$$1.1 \times 10^{10} \text{ J} = 2,000 \text{ lbs. coal} \times \frac{5,000 \text{ BTUs}}{1.0 \text{ lb. coal}} \times \frac{1,059 \text{ J}}{1 \text{ BTU}}$$

Specific Heat Calculations

Specific heat capacity refers to a material's ability to absorb heat. For example, it takes more heat to raise the temperature of water by 1.0°C than it does to increase the temperature of sand by the same amount. Each unit of heat energy is connected to a different definition of specific heat, each of which carries meaning about units that can be used to solve problems.

Chapter Ten: Energy Resources

Table 10.1 Units of Energy and Specific Heat Definitions

Unit of Energy	Amount of Unit Needed to Heat Water
Joule	Heat needed to raise temperature of 1.0 gram water by 4.2°C
Calorie	Heat needed to raise temperature of 1.0 gram water by 1.0°C
BTU	Heat needed to raise temperature of 1.0 pound water by 1.0°F

Each different material has a different specific heat capacity, or the amount of heat contained in the material per degree. In general,

Heat needed = mass of material × specific heat × ΔTemp

Or $q = mC\Delta T$

where q = amount of heat absorbed by the material

m = mass of the material

C = specific heat capacity of the material

ΔT = change in temperature of the material
(final temperature – initial temperature)

Example 10.5

Question: How many joules of heat are needed to heat 5.0 g water by 10.0°C? (C_{water} = 4.2 J/g • °C)

Answer: Using $q = mC\Delta T$,

$$210 \text{ J} = 5.0 \text{ g water} \times \frac{4.2 \text{ J}}{\text{g water} \times °C} \times 10.0°C$$

Example 10.6

Question: How many BTUs of heat are needed to heat 10.0 lbs. water by 10.0°F? (C_{water} = 1.0 BTU/lb. • °F)

Answer: Using $q = mC\Delta T$,

$$100 \text{ BTUs} = 10.0 \text{ lbs. water} \times \frac{1.0 \text{ BTU}}{\text{lb. water} \times °F} \times 10.0°F$$

Example 10.7

Question: How many calories of heat are needed to heat 25.0 g of water by 10.0°C? (C_{water} = 1.0 J/g • °C)

Answer: Using $q = mC\Delta T$,

$$250 \text{ cal} = 25.0 \text{ g water} \times \frac{1.0 \text{ cal}}{\text{g water} \times °C} \times 10.0°C$$

Electrical Energy Calculations

Some calculations take advantage of the time component in the common electrical units of energy, kW-hr. Don't forget that when you divide by units, you turn them upside-down and multiply. For example, to divide by miles-per-hour, you simply multiply by hours-per-mile. Also, even though you might cancel the units of kW in kW-hr, the time unit (hr) remains.

Example 10.8

Question: How long will a 100 W bulb shine with an energy input of 1.0 kW-hr?

Answer:

$$10 \text{ hr} = 1.0 \text{ kW-hr} \times \frac{1,000 \text{ W}}{1.0 \text{ kW}} \times \frac{1}{100 \text{ W}}$$

Example 10.9

Question: How many kW-hrs of energy are needed to light a 75 W bulb for 100 hours?

Answer:

$$7.5 \text{ kW-hr} = 100 \text{ hr} \times 75 \text{ W} \times \frac{1.0 \text{ kW}}{1,000 \text{ W}}$$

Non-Renewable Energy Sources

Formation and Use of Fossil Fuels

Coal

During the carboniferous period about 300 million years ago, conditions on earth favored freshwater swamp ecosystems, which in turn produced a significant amount of plant material. Upon dying, the plants decayed underwater, became compacted, and formed peat—which is about 5% carbon. As peat was covered by sediments, it became further compressed and formed lignite coal—which is about 60% carbon. Further compression yielded bituminous coal (about 75% carbon), which is the type of coal most often used in electric power generation. The most compressed form of coal is called anthracite—which is over 90% carbon. As a general rule, older coal has a higher carbon content, which provides more heat when burned.

Oil and Gas

In the same manner that coal was formed from plants, crude oil and natural gas were formed by the decomposition of microorganisms. Oil typically exists within the pores of sandstone. The sediment that compresses the decaying organisms is itself compressed and forms shale, which absorbs the oil into its pores.

Crude oil is a heterogeneous mixture of many hydrocarbon molecules, which are separated from each other during the refining process. The greater the number of carbon atoms in an organic molecule, the higher will be that molecule's boiling point and viscosity. Organic molecules with less than five carbon atoms are gaseous. Molecules with 6–9 carbon atoms are liquid and used for light fuels, such as kerosene and gasoline. Molecules with the largest number of carbon atoms are solids and semi-solids, such as waxes and tars.

Use of Fossil Fuels

Coal is the most abundant non-renewable energy source and is used for about 27% of the world's energy needs. It is mined in either a surface mine or a shaft mine. Coal is bulky to transport, and combustion of it yields oxides of sulfur (remaining in the coal as a result of decomposed proteins) that, when combined with water, produces acid rain. Also, rain runoff flowing through open mines produces acidic mine drainage.

Nuclear Power

Radioactivity

Radioactivity is caused when unstable nuclei undergo a nuclear reaction and give off high-energy electromagnetic radiation. Unstable nuclei are created when the ratio of neutrons to protons is not appropriate. At this point the nuclear forces between nuclear particles cannot overcome the electrical repulsion between protons, a nuclear reaction occurs, and energy is emitted.

Energy is produced during a nuclear reaction because an extremely small portion of the mass of the nuclear particles is converted directly into energy ($E = mc^2$). The resulting energy is usually of very short wavelength and high frequency—far outside the visible electromagnetic spectrum. High-energy electromagnetic radiation can cause severe damage to biological tissue.

Nuclear Reactions

In Chapter Two, chemical reactions were described as a change in the way atoms were arranged together. For example, one atom of carbon combines with two atoms of oxygen to form one molecule of carbon dioxide. Both reactants and products have one atom of carbon and two atoms of oxygen, just rearranged.

Nuclear reactions are similar to chemical reactions because matter is conserved at a superficial level. They are different in that nuclear reactions do not rearrange matter like a chemical reaction; a nuclear reaction actually changes the identity of matter from one type of element to another by adjusting the number of either neutrons or protons in the nucleus. Five types of nuclear reactions are important in AP Environmental Science.

1. **Neutron capture**: In this reaction, a nucleus absorbs free neutrons to become an isotope—often unstable and radioactive—of the original nucleus. This is the type of nuclear reaction whereby free neutrons from the Chernobyl explosion converted non-radioactive nuclei into unstable, radioactive nuclei.

    e.g., $4\ ^1n\ +\ ^{127}I \longrightarrow\ ^{131}I$

2. **Alpha decay**: This naturally occurring type of decay emits an alpha particle (helium nucleus) as the unstable parent atom attempts to become more stable. As a result, the atomic number of the element decreases by two, and the total mass number decreases by four.

    e.g., $^{238}U \longrightarrow\ ^{234}Th\ +\ ^4He$ (alpha particle)

3. **Beta (β) decay**: This naturally occurring decay emits an antineutrino from the nucleus, which later decomposes into an electron. The result is that a neutron in the nucleus turns into a proton and the atomic number of the element increases by one.

    e.g., $^{234}Th \longrightarrow\ ^{234}Pr\ +\ ^0e$

4. **Fission**: Fission is the breakup of large, unstable nuclei into many smaller nuclei. It is usually catalyzed by bombarding the large nuclei with neutrons. This is the reaction that takes place in nuclear power plants and in atomic weapons—such as those dropped on Japan to end World War II. The neutrons that are given off in a fission reaction, in turn, catalyze other large nuclei to split, which can result in an explosive chain reaction unless the neutrons are absorbed. Fission reactions in nuclear power plants are moderated by control rods, which slow the fission reactions by absorbing neutrons.

    e.g., $^1n\ +\ ^{235}U \longrightarrow\ ^{97}Kr\ +\ ^{141}Ba\ +\ 3\ ^1n$

5. **Fusion**: Fusion is the combination of two small nuclei into one larger nuclei. This type of nuclear reaction only occurs at very high

temperatures, such as those that occur on stars—including our sun—and with thermonuclear weapons. To date, we have not figured out how to control this reaction well enough to generate electricity.

e.g., $^2H + {}^2H \rightarrow {}^4He$

Nuclear Reactors

1. Anatomy of a Reactor

   While nuclear reactors differ in size and levels of containment, all nuclear reactors are similar in that they:

   - use fuel packed together in a fuel assembly

   - use control rods—or moderators—to absorb neutrons and control the rate of fission inside the reactor core

   - heat a material that is in contact with the fuel, which also cools the reactor core

   - use a heat-exchanging material to carry the kinetic energy of the originally heated material to the turbine

   - use a coolant to prepare the heat-exchanging medium for reheating

2. Types of Nuclear Reactors

   - Boiling water reactors (BWRs) allow water to come in direct contact with the fuel assembly. This radioactive water then leaves the containment structure and drives the turbine directly. This is probably the dirtiest type of nuclear reactor and carries the highest chance of accident.

   - Pressurized water reactors (PWRs) allow water to circulate through the core of the reactor to absorb heat and cool the fuel rods. When this water reaches a high enough pressure, it heats a secondary circuit of water, which evaporates to steam and drives the turbine. This is the type of reactor that is most often used in commercial nuclear power plants.

   - Heavy water reactors (HWRs) use heavy water—water whose hydrogen has an extra neutron—as both a cooling agent and a moderator.

   - Graphite reactors use graphite—or carbon—as both a moderator and cooling agent. Then they blow gas through the reactor to carry the heat to steam generators, which in turn drives the turbines. Chernobyl was a graphite reactor.

- High temperature, gas-cooled reactors (HTGCRs) coat the fuel pellets with a ceramic, then blow helium through the pellets as a coolant. A melt-down is not possible unless operators put too many pellets together. While this seemed to be a promising design because fuel could be added and removed while the reactor was operating, no such reactors have successfully remained operational.

Health Risks Due to Radioactivity

Radioactivity is energy emitted from nuclear reactions. Like any other types of energy, if it is high enough, radiation can damage humans. Our bodies are constantly subjected to background radioactivity due to cosmic rays, the sun, building materials, and other sources. For the most part, our cellular machinery has evolved to be able to handle doses of radioactivity at these levels. However, irresponsible use of nuclear power or explosives in the world can expose humans to much higher levels of radiation. To better understand the risk of biological damage due to radiation, one must understand how doses of radioactivity are measured.

1. Dose Measurements

    The metric units of energy are **joules**. The units of energy measurement for radiation are **rads** (radiation absorbed dose). One rad equals 10^{-2} joules. The amount of biological damage is related to both energy, measured in rads, and the type of subatomic particle colliding with the body. Therefore, **rems** (roentgen equivalent for man) are a more accurate unit when considering biological risk. One rem equals the number of rads multiplied by a constant that is related to the type of particle causing the radiation. The biological effects of a radioactive dose depend on

    a. the type of particle involved,

    b. the overall energy of the radiation,

    c. the length of time of exposure,

    d. the tissues that have been exposed, and

    e. the chemical properties of the radioactive element.

2. Biological Damage

    Biological damage can occur in two general ways: there can be damage to somatic cells, or there can be damage to the genetic machinery inside the cell.

    Somatic, or general body cells, can be physically damaged due to the energy of an acute exposure (burns, for example). Even if the somatic cell damage has not caused immediate death or is not visible, less evident body cells may be damaged enough to cause death in a few days or weeks.

With lower level chronic exposure, there may be **genetic** changes within the cell that cause some form of cancer, or have teratogenic effects. Below is a list of various doses, causes for the doses, and possible effects. This chart represents an approximation because every body responds differently to radiation.

Table 10.2 Cause and effects of various radioactive doses. The highest doses would occur through nuclear accidents, war, or chronic exposure to nuclear waste. Mrem = millirem = 1/1,000 rem

Dose (mrem)	Cause	Effect
3 mrem	5-hour flight	None
7 mrem/yr	Building materials	None
50 mrem/yr	Cosmic radiation	None
50 mrem	Diagnostic x-ray	None
200 mrem/yr	Typical annual dose	None
700 mrem	Brain scan	None
1,000 mrem/yr	Safety threshold	None
10,000 mrem/yr		Cancer risk
25,000–50,000 mrem		Decrease in white blood cells
350,000–500,000 mrem		Risk immediate death, half die in 30 days

Pollution Risks of Nuclear Power

Unlike power plants that use fossil fuels, producing electricity from nuclear-generated steam does not produce air pollution. However, every step of the mining, refining, processing, use, and storage of nuclear fuel contains the possibility of introducing radioactivity into the environment.

Ideally, once nuclear fuel is used up in the environment, it can be re-processed and used again. Practically, this only happens in a few countries. In the United States, nuclear fuel is not re-used, but stored for many years at the site of use. We are still trying to figure out how to store nuclear fuel over many years so that future generations are protected.

Again, while nuclear power plants do not generate air pollution, they do need vast amounts of water to cool the nuclear core and/or the steam that the core produces. For this reason, nuclear power plants tend to be located near rivers, and much of the river is diverted through the plant to provide the necessary cooling. In some locations, local authorities allow the temperature of rivers to be increased by 5°C, which is enough to decrease the amount of dissolved oxygen in the river and can put the temperature of the river outside the tolerance range of key organisms. Consequently, this type of thermal pollution has the ability to be just as devastating to the environment as the emission of more toxic forms of pollution.

Renewable Energy Sources

Renewable energy sources can be regenerated within our life time at a rate that exceeds its use. Energy sources that are considered renewable include solar, wind, biomass, geothermal, hydro, and tidal sources.

Solar Heating

Passive solar heating uses design of orientation, special materials, and space to maximize the retention and flow of solar energy. Passive solar heating does not involve moving parts or an input of energy.

Some passive designs include the following:

1. A Trombe wall: a massive wall built behind a window exposed to the sun that will retain and re-emit solar heat

2. Extended eves that block the summer sun

3. A greenhouse with vents into a living space

Active Solar Heating

"Active" refers to the input of energy in order to gain full benefit from the solar heating design. For example, water that is heated by the sun on the roof of a house may be considered passive, but if an electrical pump is used to circulate water through the system, it would be considered active solar heating. Adding a pump or fan to a passive system makes it an active system. For example, a fan that circulates air past a Trombe wall creates an active solar heating system.

Photovoltaic Cells

Photovoltaic cells convert solar radiation directly into electricity. Photovoltaic cells are composed of layers of semiconductor wafers that give off electrons via the photoelectric effect when photons strike the wafers. As a result, the moving electrons flow along a conductor and provide a current of electricity.

Photovoltaic cells were initially used during the space program to provide ongoing power for satellites. With improved semiconductor designs, photovoltaic efficiency has also improved.

Fuel cells are actually a method for energy storage, rather than generation. Photovoltaic cells collect solar energy and convert it directly to electricity, which is directed to a membrane that separates hydrogen and oxygen from water molecules. At a later time when the energy is needed, the membrane can be changed to combine the hydrogen and oxygen back together to form water, and the electricity initially put into the process is returned with a high level of efficiency (very little energy is lost to heat) and no pollution is created.

Wind

About 2% of the solar energy that strikes the Earth heats the air and creates wind, which carries the ability to do mechanical work. Windmills are from 35–60% efficient in converting mechanical energy into electrical energy.

Wind energy had a large impact on the American westward expansion. Windmills provided a critical source of self-sufficient energy to pump water out of aquifers for cattle, crops, and towns. Today, giant wind farms use large aerodynamically designed rotors to produce electricity but thousands of the small windmills are still used in ranches and rural areas.

Pros:

1. Wind energy is pollution-free.

2. Wind is renewable indefinitely.

3. The land under wind farms is more easily used for other purposes—grazing cattle, for example—than the land under solar collectors or reclaimed coal mines.

Cons:

1. Like solar power, energy obtained from wind must be stored using batteries or fuel cells.

2. Wind is unreliable in most areas. There are only a few locations where the wind is constant enough to be a dependable source of energy.

3. Windmills and rotors take up space and can be unsightly. Windfarms are often in remote areas and might not be noticed, but they hinder the sense of open space or wilderness.

Biomass

Plants convert the energy of solar photons into chemical energy stored inside the bonds of organic molecules. About half the energy that plants absorb undergoes this conversion. When those chemical bonds are broken—such as when wood is burned on a fire—most of the energy is released.

Besides burning plant matter or animal waste, another way to capture solar energy is to ferment organic material. Yeasts can metabolize sugars to produce alcohol, which can then be stored or burned to regain the chemical energy stored in the bonds.

Bacterial digestion of carbon wastes—such as animal dung—can produce methane, or natural gas, which can be burned to provide heat or run a generator. However, in the Third World countries that are inclined to use this process, the dung may be better used as a source of nutrients for crops.

Geothermal

Geothermal energy from mantle heating is possible in a few places in the world. Steam or hot water that is warmed by the Earth's mantle is drawn out of deep wells and used to drive turbines to generate electricity, or pumped directly through buildings to provide heat. Once the kinetic energy of the steam or hot water has been used, the water is further cooled and returned to the ground to be re-heated.

Geothermal steam can be depleted without warning, which effectively makes it a nonrenewable source of energy. Geothermal steam usually contains dissolved salts that corrode and encrust pipes and generating equipment; some of the chemicals used to clean the pipes and turbines are very toxic. Because of its lack of availability, potential for depletion, and the dissolved ions, obtaining energy from geothermal steam is only practical in those very few areas where "clean" steam—very nearly distilled water—is available.

Geothermal energy from heat pumps, unlike geothermal steam heat, can be used anywhere in the world. Geothermal heat pumps use the same principles that operate a refrigerator to pump heat into or out of the Earth using a series of pipes and a heat-carrying fluid. The Earth provides a nearly limitless heat sink that absorbs heat during hot days and provides heat during cold days. Geothermal heat pumps are being used increasingly in homes and large city buildings; the larger the building, the greater must be the exposed surface of pipe underground. Geothermal energy using heat pumps is a renewable energy source, and is very promising.

Hydroelectric

Pros:

1. Producing electricity from falling water is a highly efficient, nearly pollution-free method of producing electricity. About 85% of the gravitational potential energy contained in water by virtue of its high position can be converted into electricity when the water passes through a turbine at the end of its fall.

2. Because no chemical or nuclear reaction is used to heat steam to drive the turbine, no toxic by-products are produced that could potentially spoil the environment.

3. Hydroelectric power is a renewable resource. After the water passes through a dam, solar energy evaporates the water and moves it uphill so that it may go through the cycle again.

Cons:

1. The water impounded behind a dam devastates the previously existing habitat.
2. Silt flowing into the impounded water piles up behind a dam to reduce its effectiveness.
3. The river is blocked so that species of fish and other organisms may not pass through.
4. The slow flow of water behind the dam can create oxygen-depleted or pathogenic aquatic systems. For example, schistosomiasis is a disease that is caused by parasitic flatworms that live within snails, which live in the waters behind dams in tropical areas.
5. The impounded water can also displace people who lived along the former river bank. For example, the Three Gorges project in China will displace about one million people, who will need to compete for resources with existing residents in some new place.
6. When a large body of water is impounded behind a dam in a warm climate, a considerable amount of water is lost to evaporation. In some countries, this water might be more useful for drinking or producing crops, rather than being stored so that a fraction of it may pass through a turbine to provide electricity.

Tidal

Tidal movement is a unique source of hydroelectric power and shares the same pros and cons of that source. It is the only type of power that ultimately comes from a non-stellar source—in this case, the movement of the moon. Like wind and solar power, the electricity generated from tidal movement is intermittent; energy must be stored in order to serve a constant need. An additional drawback for capturing tidal energy is that saltwater must flood normally brackish estuaries, which alters that biome considerably. Also, retention of saltwater for too long in inland estuaries may promote saltwater intrusion of freshwater aquifers. Finally, because retaining water from tides only gives the water a few feet of gravitational potential energy, there must be a very large area that is flooded in order to obtain a small amount of energy. For example, water that falls 700 feet at Hoover Dam provides 40 times more power than the water that falls 17 feet in the Bay of Fundy—and most tidal drops are much less than this.

Patterns of Human Energy Use

A Brief History of Human Energy Use

Hunters-Gatherers

Bands of people from prehistoric times have used fire from wood fuel as a primary source of energy. Using this type of energy required bands of people to move from place to place and allow the land to replenish itself before the band returned. However, increased population growth placed too many people on the land and risked depleting this style of energy use.

The Neolithic Revolution

With the advent of farming and animal domestication about 10,000 years ago, energy derived from a wood-burning fire was supplemented with muscle energy from animals (to pull carts and plows). Farming allowed a population of people to concentrate its sources of food and grow crops to fuel domesticated animals, but the source of energy was still limited to the energy captured by crops and trees.

The Industrial Revolution

Four major advances caused the Industrial Revolution: improved agricultural practices, the invention of the steam engine, the ability to make steel, and the increased use of coal as an energy source. Each of these advances played off each other to allow for greater population densities in cities, improved manufacturing and technology, and stimulated economic growth.

Improved agricultural methods allowed for increased population densities, which was supported by the use of coal instead of wood for heat in more populated areas. The availability of coal facilitated the discovery of how to make steel from iron, and the invention of the steam engine helped to mechanize modern production and pump water from deep coal mines—further improving the supplies of coal. Major businesses grew up around each of these developments, which allowed people to support their families with salaries from factories rather than raise crops and livestock. Greater wealth allowed each member of society to use a greater amount of energy; per capita energy use increased by a factor of eight.

The Automobile Society

The use of a new energy source—refined crude oil—allowed automobile owners to live farther from work and market, but consume ever greater amounts of energy per capita in the course of daily life. Owning cars allowed families to live in suburbs and commute longer to work. The advent of the suburbs corresponded with an increase in labor-saving devices—such as a garage-door opener, and luxuries—such as air conditioning. For example, some metropolitan areas nearly double summer energy production simply to cool air. Few people owned cars or cooled air 100 years ago, but it is difficult to function in a highly developed society today without a car—and without using the energy needed to operate a car. Likewise, few people in developed countries expect to live and work in a warm climate without the luxury of cooled air.

Developed vs. Undeveloped Countries

Different countries in the world are at different stages of development. With each stage of development, per capital energy use increases dramatically. The hunter-gatherer only uses fire to cook food and stay warm in the winter. The farmer uses that amount of energy, and also the energy needed to plow fields. The industrialized worker needs energy to spin cotton, manufacture goods, and build bridges. The automobile owner needs all those types of energy, plus energy to operate a car.

Currently one-fifth of the world's population lives in a developed country, but that 20% of the population uses more than 80% of the world's energy supply. An American uses more energy in a day than a person in an undeveloped country uses in a year.

Current Residential and Commercial Energy Use

In a developed country today, energy comes from many sources in the following approximate proportions.

Table 10.3 Sources of Energy

Source	Percentage of Energy Provided
Oil	36%
Coal	26%
Gas	23%
Nuclear, solar, wind, hydro	9%
Biomass, including wood	6%

Of these energy sources, natural gas is the most efficient. Less than 10% of its energy is lost in processing and transport, and it needs very little refining. Also, because it contains more hydrogen per carbon atom, it produces less carbon dioxide—and therefore contributes less to global warming—than other carbon-based fossil fuels.

Table 10.4 Uses of Energy

Use	Precentage of Energy Used
Residential and light commercial	35%
Industrial	35%
Transportation	25%
Other	5%

AP Environmental Science

Energy Conservation

The following represent a few recommendations that the average consumer can follow to conserve our energy resources, and reduce the environmental impact of obtaining the energy needed to operate our society.

1. Live closer to markets and work
2. Use vehicles that use less energy for construction and for daily operation
3. Use products and eat foods that require less energy to manufacture
4. Reduce, reuse, recycle
5. Live in homes and work in buildings that require less energy to operate. Such buildings would be engineered to take advantage of the sun for heating, use geothermal resources for heating and cooling, and minimize the use of fossil fuels.
6. Engage in recreational activities that require less energy

Case Summaries

Ohio River System

Principle mentioned in this case:

- **Environmental damage related to power production**

In the 1930s, a series of locks and dams allowed tug-driven barges to navigate far up the Ohio River. Because of the easy access to coal transported by barges, and cooling water from the river, electric utilities built nearly 50 coal-fired power plants to provide power for Chicago and much of the Eastern Seaboard. Additionally, the river navigation system allowed for easy transport for chemical, cement, crude oil, and grain industries also near the river. Inexpensive power and transportation provided the essential infrastructure to fuel population and economic growth in the area. However, the acid rain and heavy metal pollution from the high density of coal-fired power plants is a severe environmental health liability for the area. Considerable pressure is exerted on industries—particularly the power industry—to reduce emissions.

Dam-Breaching: A Post-Hydroelectric Era?

Principles mentioned in this case:

- **Reduction in hydroelectric power facilities**
- **Restoration of rivers**

Since the early 1900s, the Army Corps of Engineers and some major engineering firms have been building dams to produce inexpensive hydroelectric power. However, states are beginning to feel that the benefits of a free-flowing river outweigh the convenience and cost-savings of hydroelectric power, plentiful local water supply, and lake-style recreation. For years, the U.S. Fish and Wildlife Service recommended removal of the Edwards Dam on the Kennebec River in Maine. Finally, the Federal Energy Regulatory Commission allowed the dam to be opened and permit migration of key fish species upstream. Removal of other dams in the northwest on the Elwa and Snake Rivers are also being considered.

Oil Transport from Alaska

Principles mentioned in this case:

- **Environmental damage related to energy production**
- **Bioremediation**
- **Effect of oil production on wildlife**

Large oil reserves were discovered on the northern coast of Alaska in 1967. Locked in by ice for the majority of the year, some method other than oil tanker had to be devised to transport the oil to warmer ports and on to refineries. In the early 1970s, a 1,300-kilometer-long pipeline was constructed between Prudhoe Bay to Valdez. The 1.2 meter diameter pipeline had to be elevated above the tundra so that the oil—super-heated to facilitate rapid flow—would not melt the permafrost, and would not be a barrier for animal migration.

In 1989, a supertanker named *Exxon Valdez* ran aground in Prince William Sound, releasing many millions of gallons of oil and devastating the marine ecosystem for thousands of miles of coastline. Although Exxon poured large sums of money into the attempted remediation, the most successful cleanup was performed by naturally occurring bacteria that digested the oil.

Persian Gulf Wars: Oil and Foreign Policy

Principle mentioned in this case:

- **Political instability as a result of reliance on fossil fuels for an energy source**

Iraq, led by Saddam Hussein, invaded Kuwait in August 1990 and laid claim to the rich oil fields of this small country. The United States immediately protected next-door Saudi Arabia from the same fate, and then attacked Iraqi troops to regain Kuwait.

The conditions of surrender in the conflict demanded that Iraq not build up a long-range missile arsenal—which had been used against troops and Israel during the conflict, and to not build or stockpile weapons of mass destruction (Iraq had already used chemical weapons on Kurdish towns in northern Iraq). United Nations inspectors were to make sure that such an arsenal was not developed. Intermittent non-compliance from Iraq eventually led to the second Gulf War, which began in the spring of 2003.

The Arctic National Wildlife Refuge: Oil and Biodiversity

Principle mentioned in this case:

- **Biodiversity**

Two different legislative acts in 1960 and 1980 set aside about 40 million acres on the northern coast of Alaska for wilderness protection, called the Arctic National Wildlife Refuge (ANWR). This area is a sensitive tundra biome that is rich with wildlife and provides rangeland for the second largest caribou herd in the world. However, increased energy demand and depleted domestic oil reserves (and subsequent increased reliance on foreign oil) puts pressure on Congress to authorize drilling for oil in the refuge.

The James Bay Power Project: Return to Hydroelectric Power

Principles mentioned in this case:

- **Hydroelectric power from the tides**
- **Estuary ecosystem**
- **External cost paid by indigenous peoples**

The James Bay Power Project is a plan for hydroelectric power that would impound vast amounts of water in northeastern Quebec. The plan would impede all of the free-flowing rivers in about one-fifth of this large province, thereby destroying the habitat for animals that depend on those rivers—and threatening the indigenous cultures that depend on those animals for food. Possible environmental damage includes siltation, nutrient pollution, habitat destruction, thermal pollution, and salinization. Each of these consequences would eliminate many critical species, as well as destroy the way of life for local Inuit and Cree populations. After the effects of the first phase of the project were observed, the possibility of further environmental damage has halted the project, but the need for power in Canada and the northeastern United States—which would be the largest user of James Bay power—continues to grow.

Three-Mile Island: A Hint of Nuclear Catastrophe

Principle mentioned in this case:

- **Nuclear accident**

Three Mile Island is a nuclear power plant located outside of Harrisburg, Pennsylvania, on the Susquehanna River. In March 1979, a faulty pump and gauge led to a decision to override the emergency cooling system at an electricity-producing nuclear reactor, and the temperature of the reactor increased and caused a partial core meltdown. Some radioactive steam was released, and many people evacuated local neighborhoods. The reactor was never restarted after the accident, and its fuel was removed entirely in 1990. This incident eroded American confidence in the safety of operating nuclear-powered electrical generation plants—let alone the controversy in finding a place for nuclear waste, and very few new nuclear power plants were built in the United States after this.

Yucca Mountain: An Attempt to Complete the Nuclear Fuel Cycle

Principles mentioned in this case:

- **Nuclear fuel cycle**
- **Storage of nuclear waste**

The nuclear fuel "cycle" is not really a completed cycle because much of the radioactive waste created is not reprocessed and, in fact, has no final resting place. High- and medium-level waste is often stored on-site where it was used, or shipped by rail to sites such as the Idaho National Engineering and Environmental Laboratory, near Idaho Falls, Idaho. Other similar sites exist in South Carolina, New York, and Washington.

While some other countries reprocess spent nuclear fuel rather than store it in temporary locations, no country has yet developed a permanent storage facility for high-level radioactive waste.

One proposal is to put the waste in borosilicate glass containers, which would then be encased in steel canisters and stored above ground until it could be safely buried.

One possible permanent storage site is Yucca Mountain, which was chosen because it is geologically stable and located in an isolated area of Nevada, near a nuclear test site. It is also very dry; the water table is about 2,000 feet below the mountain and, therefore, unlikely to be polluted. Although some exploratory tunneling has occurred, actual construction of the site has not in the face of local protests. Construction is scheduled to begin in 2005.

Chernobyl Nuclear Catastrophe

Principles mentioned in this case:

- **Severe nuclear disaster**
- **How countries respond to nuclear disaster**

On April 25, 1986, at Power Station Four in the small city of Chernobyl in Ukraine, technicians began a test to see how much power would be produced from a turbine that was still spinning after the steam had been turned off. This test initiated a sequence of events that led to the rapid overheating of the reactor, which deformed the core so that the control rods could not be further inserted. As a result, the core melted and caused an explosion.

The explosion released vast amounts of neutrons that, through neutron bombardment, caused nonradioactive elements to become radioactive. In addition to the immediate fatalities of workers and severe radiation exposure to nearby populations, air currents carried radioactive particles high into the atmosphere where normal convective downdrafts created "hotspots" in various locations in Europe.

Wherever radioactivity increased in communities around Europe, officials responded by giving children large doses of nonradioactive iodine. Neutron bombardment of iodine in the environment had created radioactive iodine (I-131), which would be metabolized and imbedded in young thyroid glands, where it could cause thyroid cancer. Flooding the body with nonradioactive iodine decreased the chance of the thyroid assimilating the radioactive isotope, and therefore decreased the chances of thyroid cancer.

Overall there were about 30 immediate fatalities, and hundreds of people were to die over the subsequent months due to radiation-related complications. Over 100,000 people were evacuated from nearby towns and villages, some of which have since become ghost towns and have never been re-inhabited.

It took ten days and heroic sacrifices to flood the core with neutron-absorbing boron and decrease the outpouring of radioactive material. By November 1986, the Ukrainian government had covered the reactor core with cement, but further containment is planned for the future.

Soviet Nuclear Legacy

Principle mentioned in this case:

- **Ongoing effects of irresponsible use of nuclear power**

While the Chernobyl accident was an immediate embarrassment to the struggling Soviet leadership, four decades of the Cold War arms buildup and economic shortcuts left environmental calamities that eventually became known after the Soviet Union fell.

More than 100 nuclear bombs had been detonated to move earth for mining or construction. Over 400 nuclear explosions occurred in weapons testing. Radioactive waste dumps are found throughout the region, including over 600 dump sites near Moscow. A now-infamous example is the dumping of waste into the Techa River, causing widespread, unmonitored contamination near Chelybinsk. The health of the environment and the people of the former Soviet Union will be affected for many years to come because of irresponsible use of nuclear devices.

Environment and Society

Chapter 11

Environment and Society

Economic Forces

Market Forces

Ecological health is closely related to economics. In the original Greek, **eco-** refers to *the home*, and **economics** refers to *taking care of the home*; **ecology** refers to *how the home operates*. In the modern world, taking care of the home seems heavily driven by the financial bottom line, but it would seem that to take good care of the home—or our environment—it would make sense to understand how it operates, and vice versa. Indeed, much of the pollution and land misuse in the environment has its origins in people wanting to cut financial corners to maximize profit, or companies simply providing a product or service for "the market." What are the market forces that drive people and companies to care more about money in the bank than breathing clean air?

Supply and Demand

Supply is the amount of product that is available for sale. **Demand** is the amount of product desired by buyers. As the demand goes up, the supply goes down and the price buyers are willing to pay goes up. If there is more supply than demand, sellers are willing to let go of the product at a lower price. In general, there is an inverse relationship between supply and demand; as one goes up, the other goes down, and the price shifts with the demand. These simple relationships between supply, demand, and price form the foundation for **classical economics**.

However, there are other forces that determine prices in a market. While demand from consumers in a climate of low supply can drive up the price, a high price due to low supply can also whittle away at demand. For example, if computers are priced low enough, more people will be willing to buy one and incorporate it into their lifestyle. However, if each computer carried with it an environmental clean-up tax to help remediate each of the 29 Superfund toxic waste sites in Santa Clara County, California—many of which have been created from the silicon chip industry—then perhaps more people would think twice about purchasing a computer. In this example, a localized ecosystem has paid the price for technological and economic development that is experienced by many. In paying that price, the supply for computers can meet demand, but the buyer has not really paid for the overall cost of developing and building the computer. So the rules of supply and demand are a rough indicator of how market forces determine price, but it is an incomplete model of all the costs associated with a product (see **Internal and External Costs** later in this chapter).

Economic growth based on simple supply-demand rules results in environmental degradation because the environmental cost is not easily factored into the real price of a product. A society might combat the unfair payment of environmental costs by a small number of people by spreading the costs over all the people, such as with a tax. Another way to manipulate the supply-demand equation is to use laws or tax breaks to artificially increase demand.

Using Policy to Create Demand

Two policy tools are used to create consumer demand and therefore stimulate production at a price that will generate a profit. First, the state or federal government can create tax breaks for using environmentally friendly products. For example, in the 1970s, President Carter introduced tax breaks for homeowners who incorporated alternative energy technologies in their homes. A private sector variation of this tool is seen in the use of rebates by power companies. Many electric power companies are finding that it is less expensive to offer rebates to consumers who conserve energy, or purchase energy-saving devices, than it is to find new sources of electrical power. Therefore, demand for these energy-saving devices increases because the rebate adjusts the overall price paid by the consumer.

A second policy tool that artificially increases consumer demand is the enactment of state and federal laws. For example, in the 1960s, it would not have been economically feasible to study the environmental consequences of a building project. However, when the federal government enacted the National Environmental Policy Act (NEPA, 1969), then all major projects required an Environmental Impact Statement. With the drop of a gavel, now all major projects require a scientific study to determine the environmental consequences of the project. Building technology, food production, energy production, mining, and many other industries contain many examples of using laws to increase demand for products (see **Environmental Laws** later in this chapter).

Another way to shift the balance between supply and demand is to use technology to decrease the pressure of low supply.

Effect of Technology on Supply and Demand

Technology is the use of science to improve transportation, communication, manufacturing, and the use of materials. When these aspects of our society improve, there is a shift in supply, demand, and prices. For example, consider the price of a shovel in the Klondike during the Alaska gold rush. Before trains could bring in commodities in bulk, a shovel had to be brought in by a mule or by hand. Since a shovel was a very useful tool in finding gold and creating new wealth, the low supply and high demand caused the prices of shovels to be high. However, improved transportation brought in many more shovels and increased the supply, so the price of a shovel dropped.

In the early stages of personal computing, a computer cost many thousands of dollars and was limited in its ability. However, improved technology allowed the mass production of many units that were increasingly more powerful, so the price of computers began to drop. As this process has continued, people now expect the market of computers to be deflationary—every year you can get a stronger computer for less money. Increased technology works to decrease cost by increasing supply and/or increasing efficiency. Technological improvements in manufacturing, materials, transportation, and communication tend to make the biggest difference between a **frontier** (or pioneer) **economy** and a more productive, more efficient **developed economy**.

Another major contributor to the value of a product—or the ability to create products in the future—is the amount of *capital* that has been invested in developing, manufacturing, and distributing the product.

Capital

Capital is a resource—or group of resources—that can be used to create more wealth—usually financial. There are different types of capital that can be turned into wealth:

1. **Natural capital** includes natural resources such as land, minerals, water, air, or energy. Natural resources can be further broken into either renewable or nonrenewable resources, and other divisions of resources based on whether or not their recovery is economical and technically realistic.

    a. **Renewable resources** can be replenished within a human lifetime, such as wood or sunlight.

    b. **Nonrenewable resources** cannot be replaced and are usually geological resources, such as mined minerals, metals, or fossil fuels. Intangible resources have nonmarket value, such as a beautiful valley or a scientifically interesting tree.

    c. **Proven resources** are thoroughly mapped and can be realistically recovered, both economically and technologically.

    d. **Known resources** are generally located but not entirely mapped, but it is not currently economically realistic to recover them, although it may be so in the future.

    e. **Recoverable resources** are available with current technology, but it is not economically feasible to do so.

2. **Human capital**, sometimes called intellectual capital, includes the cumulative education and experience contained in a group of people.

3. **Manufactured capital** includes the time and materials that have gone into a completed product that continues to serve a purpose, such as tools, roads, sewer systems, completed buildings, etc.

4. **Social capital** includes shared values among groups of people, such as trust, morale, and community organization.

Accounting for the Flow of Capital

While it is easy to place value on some financial assets, such as goods and services provided by a government or a corporation, it is difficult to assess the value of all assets—such as clean groundwater or biological diversity. In accounting for the volume of business accomplished, most economists refer to either the Gross National Product or the Gross Domestic Product, which are based on financial transactions. This is not unlike defining biological productivity using a number of different measures.

Gross National Product (GNP) is the most common way to assess a country's economic output. GNP can be calculated from the amount of goods and services purchased by households in a year. GNP can also be calculated by the total cost of production of all goods and services.

Gross Domestic Product (GDP) includes only the goods and services produced within national borders. This attributes the production of goods and services of multinational corporations to the country that plays host to the actual production. Many economists prefer using GDP rather than GNP as a gauge of a country's production because it more accurately describes where the wealth resides.

However, neither GNP nor GDP distinguish between economic activities that are beneficial and economic activities that are harmful and may diminish economic activity in the long run. For example, a small country may temporarily increase its GDP by temporarily increasing farm production using nonsustainable methods, but that country may experience a future drop in GDP because the farmers have depleted soil nutrients or allowed soil to erode. GDP and GNP do not take into account natural, human, or social capital that may play a vital role in the development of future wealth. Likewise, a major oil spill would prevent oil from being able to successfully make it to consumers, but the spill would actually increase the GDP or GNP because the company has still paid to get the oil out of the ground, and is further paying for people to clean up the spill. More money has been spent by the corporation and so the GDP and GNP increases, but there is a significant loss in natural capital and the consumers have not gained the benefit of using the product.

One alternative to GNP is the **Index of Sustainable Welfare** (ISW), which takes into account resource depletion, unpaid labor, and environmental damage.

Even though a country's GNP is increasing, it may decrease the ISW. The United Nations Development Program (UNDP) tracks the development of countries using the **Human Development Index** (HDI), which assesses quality and length of life, education, and standard of living in a country. Some economists feel that the ISW and HDI combine to give a better picture of the natural, human, and social capital, rather than simply measuring the money that changes hands. For example, the GNP of the United States more than doubled between 1950 and 1986, but the ISW remained virtually unchanged.

Cost-benefit Analysis

A **cost-benefit analysis** (CBA) is used to evaluate public projects to assess the capital costs that must be paid in order to gain benefits—or value—for a large group of people. Quantitative amounts are assigned to the capital expended and the value gained. Many people undergo a simple CBA when comparing benefits and drawbacks in making personal decisions. The use of CBA began with the Federal Navigation Act of 1936, when the Army Corps of Engineers was attempting to assess the benefits of improving navigation of the nation's waterways. Since then, CBA continues to be used to assess value of a project, particularly government projects.

The difficulty of using CBA is that there is a large amount of uncertainty and arbitrariness in assigning dollar values to an intangible resource. The danger lies in agencies using CBA to justify using resources by under-assigning value to intangible resources. For example, what is the dollar value of a breath of fresh air, or the monetary worth of not being able to swim in a lake anymore?

In assigning costs in a CBA, there are different types of costs to keep in mind: fixed costs, marginal costs, intangible costs, internal costs, and external costs.

Costs

Fixed costs are the costs paid to make a product or provide a service that does not change as production increases. For example, a manufacturer pays an accountant to audit its records. Even if the manufacturer doubles production, the cost of the audit will probably be the same; this is a good example of a fixed cost. As more units of a product are produced, then the fixed cost can be spread over a larger number of units. When companies merge into larger companies, there is usually an immediate savings of certain fixed costs.

Variable costs are the costs that increase as the number of products produced increases, such as for raw materials to manufacture a product. Perhaps the cost does not increase in a linear manner. For example, it may be cheaper to purchase the raw materials for 10 units than for 100 units, but it might not be ten times cheaper.

Marginal cost is the cost of making one additional unit of a product or service, or the total costs (fixed and variable) per item when one more unit is produced. The marginal cost of production will decrease as more units are produced. While a company would naturally strive to minimize the marginal cost per unit, profits would be hindered if too many units were produced and the oversupply decreased the price per unit.

Margin of diminishing returns are the additional benefits gained by the buyer by procuring one more unit of the product or service. For example, if eating one bowl of ice cream is a pleasurable experience, how does that pleasure per bowl diminish by eating two bowls, or three? By eating more ice cream, the "return"—or benefit gained—per bowl diminishes. As the benefit diminishes, so does the demand. You might pay $40 for an oil change if your car really needs it, but you would be less likely to pay $80 for two oil changes if your car only needs one.

Thomas Malthus applied the idea of diminishing returns to populations. He felt that workers would respond to rising wages by having more laborers go to work (increasing the supply of laborers). This would both decrease the demand for more laborers (and decrease the wages), and also increase the population of future laborers (which further increases the supply of laborers and decreases the wages). As a result, classical economists assume that continued economic growth of the company is needed to accommodate the increased workers, or workers would end up working for lower and lower wages. However, unlimited economic growth as measured by total wages paid (and reflected in the GNP) does not take into account the environmental consequences of continued growth of both production and population. Some economists would say that accounting for environmental consequences increases the effects of diminishing margins of return. Would that second bowl of ice cream taste good at all if the consumer was mindful of the environmental effects of engaging in gluttony?

Internal and External Costs

Internal costs are those costs that were actually experienced to manufacture a product. For example, a computer would need the raw materials, the labor costs to construct the parts and put the parts together, the buildings to build and store computers before they are sold, the transportation to take the computer to the consumer, and the administrative costs to be sure that all of these things occurred at the right times and in the right amounts. For the most part, the consumer pays for the internal costs plus an amount that represents profit to the owners or shareholders of a company.

However, there are often costs, called **external costs**, to people or society that are not experienced by the company and are not passed on to the consumer. External costs are felt by someone, or some aspect of society, but since they are not paid by the people who turn the resources into the product, establish the price for the product, and make the profit, they do not enter into the pricing of the product, or the decision by the consumer to purchase the product.

For example, people use a given amount of electrical energy based on the price they pay for the energy. If the price of the energy goes up, people conserve more energy so they don't have to pay so much. The movie *Erin Brockovich* dramatized an actual event in which a California electrical utility cut corners in using and disposing of toxic chromium compounds that it used to clean the pipes in a power plant. The hexavalent chromium then entered the groundwater and made several local residents extremely ill. The movie catalogs Erin Brockovich's effort to lead the community in a lawsuit to receive payment from the utility for damages and suffering. The health costs paid by the residents in the community are an example of external costs. Litigation is one way to *internalize* external costs.

Establishing laws and creating taxes are also ways to internalize external costs. For example, passing of the Surface Mining Control and Reclamation Act (SMCRA, 1977) now assures that the price for newly mined coal resources includes the cost of reclaiming the land after mining is complete. In order to begin mining operations, a mining company must first establish a trust fund that finances the cleanup of the area after mining operations have ceased. The cost of environmental clean-up is factored into analyzing whether or not it is cost-beneficial to undergo the mining operation in the first place, and the total cost is passed onto the companies who purchase the minerals from the mine, and ultimately to the consumer.

To internalize external costs means that the consumer is paying for the full cost of a product or service—or **true cost**, rather than having someone who is not gaining the benefit of the product or service unjustly bearing the cost. Because of this, internalizing external costs is also called **full-cost analysis,** or **true-cost pricing**.

Different Economic Systems

Classical Economics

Adam Smith's book *Inquiry into the Nature and Causes of the Wealth of Nations* (1776) is the basis of classical economic thought. In it, Smith identifies the economic force of the cumulative decisions of individuals—following the rules of supply and demand—as creating an "invisible hand" that promotes the interests of society by creating demand for resources. The resulting balance between supply and demand, with competition between both buyers and sellers, keeps prices and quality in a reasonable range, maximizes personal liberty, and yields the greatest good.

Smith's thinking lies in stark contrast with Garret Hardin's *Tragedy of the Commons* (1968), which argues that resources held in common tend to be overused and destroyed if left to the cumulative decisions of individuals. Hardin argues that the increase in marginal cost incurred by an individual does not take into account the external cost of degrading the environment. Therefore, the classical market established by supply and demand does not necessarily yield the greatest good.

With the advances in scientific and mathematical thinking toward the end of the nineteenth century, classical economists invoked rigorous mathematical modeling to establish a "value free" approach to predict markets. This quantitative form of economics is sometimes referred to as Neoclassical Economics, but it is no different from classical economics in that supply and demand form the fundamental forces in determining price, and continued economic growth is essential to maintain full employment. Natural resources are viewed as annual costs rather than ongoing assets.

Political Economics

As the scientific revolution took off near the end of the nineteenth century, social engineers attempted to construct a philosophy based on value systems and equal relationships among social classes. Karl Marx's focus on economic production in a manner that attempted to maximize equity between the classes formed the basis of communist revolutions in the former Soviet Union and China. However, his idealistic view of the relationship between values and pricing did not succeed as it became embodied by a large, centralized government that did not place equal value on all people, or on environmental resources.

Ecological Economics

Ecological economics attempts to take into account the importance of natural, human, and social capital in cost-benefit analysis. John Stuart Mill begins to identify the importance of natural, human, and social capital in *Principles of Political Economy*, in which he says that when we meet our materialistic needs, we still have much growth in our social and aesthetic lives to consume our energies, rather than simply continuing to use resources. More recently, E.F. Schumacher in his book, *Small is Beautiful: Economics as if People Mattered*, established a way of thinking about resources that venerated small, earth-friendly solutions. To Schumacher, the hope for our planet is to shift our values from large, energy-consuming projects to small-scale solutions that take into account external as well as internal costs.

Survival Economy

Many countries do not have the luxury of either a vibrant GNP—even at the expense of the environment—or a high ISW. Such countries are constantly on the brink of social, economic, or environmental disaster and have a very difficult time just feeding their people, let alone helping them and further generations to live long, healthy lives. Such an economy is called a **survival economy**, although it barely enables survival. Most of the country's basic needs are derived directly from nature with little thought to degradation or depletion. About half of the world's population is in this type of desperate survival economy, which includes much of China, Africa, India, and Southeast Asia. These countries await a demographic transition (see Chapter Five) in order to be able to allow the leaders and the populace to plan further into the future.

Environmental Ethics

The ethics of a group of people are based on that group's **morals** (sense of right and wrong) and the **value** the group places on items or actions. Morals are related to values in that it may be okay to destroy something of little or no value, but if the value of that object increases, it becomes wrong to destroy it. Conflict arises when people with different morals and values attempt to solve problems together, or when people ignore the implicit moral imperative of the society.

Adult humans are—or should be—*moral agents*, who are capable of determining right from wrong and acting upon that determination and who take responsibility for their actions and for the well-being of those weaker than themselves. A *moral subject* is under the responsibility of a moral agent, just as a child would be under a parent. Whether or not people recognize and step up to their role as a moral agent for others seems to be the largest question.

Even though most societies generally feel that its weaker members deserve treatment that is equal to the stronger members, that has not always been the case. Consider the lower view of women, children, or minorities in previous times, or the use of slaves, for example. The increasing circle of moral subjects is called *moral extensionism*. However, whether or not a moral agent recognizes responsibility toward a moral subject depends on the value placed on the subject.

Nonmarket Values

Instrumental Value

Instrumental value of a commodity or capital good refers to value placed on the item as a result of its usefulness. For example, a cow might have more instrumental value to humans than a bear because the cow can provide milk over time. Even the value of a beautiful view or a sunset has some instrumental value because it can inspire us to be healthier and more productive.

Intrinsic Value

Intrinsic value of a commodity or capital good refers to value placed on the item as a result of its very existence. Intrinsic value is sometimes also called **inherent value**, or the value that is inherited automatically by an organism or an object. For example, in the example of the sunset from the previous paragraph, the sunset has instrumental value if someone sees it and then goes and experiences greater productivity as a result; if it still contains value if nobody sees it, then it has intrinsic value.

People draw the line at different points about whether or not something contains intrinsic value. Take, for example, the moral imperative to not kill. Where does one draw the line for killing other living things? We can't live without killing at least plants in order to survive, so we would be killing ourselves if we didn't kill something. Some people draw the line of killing on the plant-animal border by ascribing greater intrinsic value to animals: it is morally acceptable to kill plants, but not animals. Some people would ascribe greater intrinsic value to sentient animals—or those animals that can think and experience sensations. To these people, it would be okay to kill insects that attack crops, but not okay to eat cattle. Still other people will ascribe greater value to organisms that are more intelligent; to many people who eat meat, the thought of eating a dog or a dolphin seems out of the question.

Aesthetic Value

Aesthetic value of a commodity or capital good refers to value placed on an item because it is pleasing or beautiful. To many, aesthetic value is a type of instrumental value—particularly if aesthetic value is closely linked to the perception of quality of life.

Cultural Value

Cultural value of a commodity or capital good refers to value placed on an item because it is a part of the history of a group of people. For example, an old cemetery that is in the path of development may not have instrumental or aesthetic value, but those who remain in the community may wish to visit deceased relatives, or not wish to see a sacred place be destroyed. Likewise, an old church or historical site, or even a favorite restaurant, may contain value to a group of local people that would not contain value to a new person moving into the area.

Scientific and Educational Value

Scientific or **educational value** is conferred upon a resource if it contains lessons—whether already understood or as yet to be discovered. For example, Antarctica contains enormous scientific value, even though the continent may not contain any value as a profit-generating tool.

Social Justice Value

A piece of land, a resource, or an action may contain value based on fairness or equity toward a group of people. For example, dumping toxic wastes from a large city in a small, low-income area—possibly occupied by minority residents—means that those residents—by virtue of the fact that their political and economic voice is not loud enough—pay a price in health and wellness for the consumption habits of the wealthier people in the city. In this sense, "the greatest good for the greatest number of people" is not valid if it causes marginalized peoples to unjustly pay the price for the poor stewardship of others. One form of social justice is the elimination of environmental hazards based on race, or eliminating **environmental racism**. The use of Third World countries as depositories of toxic wastes is considered **toxic colonialism**.

Diverse World Views and the Environment

Different people in varied cultures have different attitudes toward the relationship between humans and the environment. Among people in a single culture, individuals may believe any of the following:

1. Nature is beautiful and inspirational, in balance with itself.
2. Nature is chaotic, with living things seeking to survive at all costs.
3. Nature is dangerous, and humans must cushion themselves from its hazards.
4. Nature is a commodity to be used as raw materials.
5. Nature provides pleasure and entertainment.

Different schools of philosophies contain varied amounts of the above views toward nature. Below are summaries of some of the major philosophical schools and their attitudes toward nature.

Universalists believe that fundamental and unchanging laws govern the universe, nature, human interactions, and ethics. Rules of right and wrong (morals) hold true in all situations. The rules of the universe that are discovered through reason and empiricism are equally unchanging. If a universalist asserted that an animal contained intrinsic value, then that value would hold in all situations.

Relativists believe that moral decisions and ethical values depend on the context of the situation, rather than unchanging universal laws. All ethics are contextual, and there are no truths that hold up in all situations. Even scientific laws have special circumstances where they shift; for example, Newton's Laws hold true unless an object moves at a velocity that approaches the speed of light, at which point other rules seem to come into play. The relativist might say that another living thing should be protected in one context, but not in another.

Nihilists believe that the world makes no sense, all is arbitrary, and there is no real meaning or purpose in life other than a random, instinctive struggle for survival. The nihilist might use this reasoning to justify using resources in order to survive now, without thinking about what future generations might need. They will have their own struggle for existence. Survival is the greatest good, and there is no reason to exercise a sense of right or wrong unless it helps one to survive.

Utilitarians attempt to find the greatest good for the greatest number of people. Of course, people may not agree about the definition of "good." To some, "good" might mean pleasure and self-indulgence; to others, "good" might mean survival; to yet another group, "good" might mean an atmosphere that promotes kindness.

Anthropocentric philosophers put paramount importance and value on humans. Humans are the only beings with intrinsic value; nature carries only instrumental value.

Biocentric philosophers put equal value on all living creatures. The highest value is placed on biodiversity—as measured by sustaining the size of populations and the diversity of species. However, humans are seen as typically irresponsible moral agents.

Ecocentric viewpoints are carried by many ecologists who argue that individuals may experience pain, but the highest value is to allow the group to continue to undergo a strong process of evolution or adaptation. The whole group and a healthy relationship to the entire system are considered more important than the individual's well-being.

Ecofeminism argues that most other viewpoints stem from a patriarchal, sexist worldview that operates through domination and exploitation of resources and marginalized peoples at the whim of (typically male) individual egos. Ecofeminists are nonviolent, collaborative, pluralistic, and relational; they see themselves as nurturers of people and the environment.

Neo-Malthusians see the world as driven by scarcity and competition with a disastrous eventual outcome of misery, illness, and starvation because nobody will act to reverse the current trend.

Technological optimists, in stark contrast to neo-Malthusians, feel that technology and ingenuity will continue to help societies conquer environmental problems and steadily increase the carrying capacity of the Earth.

Environmental History, Laws, and Regulations

History of Thinking in the Environmental Movement

Plato (fourth century, B.C.) identified the sequence of events that resulted from the deforestation of Greece.

Pierre Poivre, in his role as French governor of an island in the Indian Ocean in the late 1700s, identified the environmental and social consequences of deforestation and wildlife depletion on the island.

Henry David Thoreau wrote *Walden* (1854) and other works of naturalist philosophy; he appreciated the environment for its intrinsic value, and for the value in building inspiration in people.

John Audubon was a skilled artist and naturalist who is most famous for his work *The Birds of America*.

George Perkins Marsh wrote *Man and Nature: Physical Geography as Modified by Human Action* (1864), which stated that every human action disturbs some aspect of nature. In his travels, Marsh observed the effects of overgrazing by goats on steep hillsides and the inefficient use of resources in the American West. Marsh heavily influenced Gifford Pinchot and Theodore Roosevelt.

John Muir organized the Sierra Club and convinced the federal government to begin Yosemite National Park in 1890 and create 21 million acres of forest preserves. Muir, a biocentric conservationist, also authored the book *Yosemite*. Muir challenged Pinchot's utilitarian conservationism as being too anthropocentric.

Theodore Roosevelt, as the twenty-sixth president of the United States, set aside large amounts of public land by establishing national parks, national forests, and wildlife refuges.

Gifford Pinchot was a utilitarian conservationist who coined the phrase still used for the U.S. Forest Service and National Forests public lands (which he helped found), "the greatest good for the greatest number for the longest time."

Upton Sinclair wrote *The Jungle* (1906) to describe the unhealthy and unjust working conditions in a Chicago meatpacking plant. Increased public consciousness of poor working conditions led to the formation of unions in America and to legislation to improve working conditions and ensure food quality.

Aldo Leopold wrote *A Sand County Almanac* (1949), which introduces the idea of a biological community as interdependent populations. Leopold helped found the Wilderness Society and pioneered new techniques in game management.

Ansel Adams used his photographic genius to acquaint people with the American landscape—particularly that of Yosemite National Park and Kings Canyon in California.

Rachel Carson, in her book *Silent Spring* (1962), outlined the effects of biomagnification and bioamplification of DDT and other toxic substances. *Silent Spring* brought to the public consciousness the insidious nature of environmental pollution and degradation; many consider Carson to be the key catalyst of escalation of the environmental movement in the late 1960s and early 1970s.

Paul Ehrlich, in his seminal book *The Population Bomb* (1968), identified the profound effect of unattenuated world population growth. A biology professor and noted entomologist at Stanford University, Dr. Ehrlich also founded in 1968 the nonprofit organization Zero Population Growth (ZPG), which works to increase public awareness of the need to decrease population growth.

Garrett Hardin is an ecologist who is best known for his essay "Tragedy of the Commons" (1986) in which he cites the ineffectiveness of cumulative personal decisions in managing common resources.

Government Organization

The following chart outlines the cabinet-level agencies under the Executive Branch of the United States Government.

AP Environmental Science

Table 11.1 Federal Government Branches and Agencies

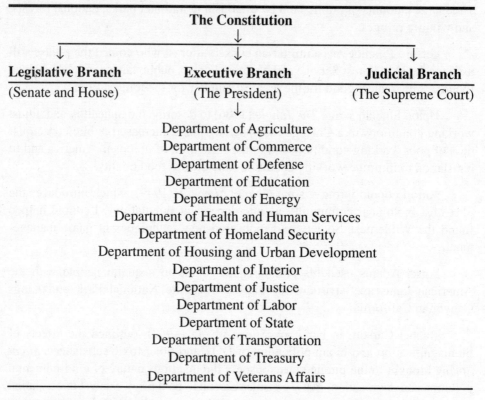

Government Agencies

Cabinet-level agencies within the Executive Branch contain these agencies that address environmental and health issues.

Department of Agriculture

- Farm Service Agency

- National Forest Service (NFS)

    The NFS manages national forests and some wilderness areas. National forests are used for many reasons, which include recreation, habitat protection, logging, and mining. Wilderness areas are protected from any human intervention, including the building of roads or leases for harvesting natural resources.

- National Resource Conservation Agency

Department of Commerce

- National Oceanic and Atmospheric Administration (NOAA)

    NOAA monitors ocean and atmospheric environmental conditions on a global scale. The National Weather Service is a sub-organization of NOAA.

Department of Defense

- Army Corps of Engineers

    The Army Corps of Engineers has, in the past, provided the engineering support for large projects, such as building locks and dams to increase the navigation of inland waters, coastal erosion prevention, and flood control. There may be a defense-related objective in the event of war, but in the meantime, some economic gain is experienced by the society.

Department of Energy

- Office of Environmental Management

    This agency oversees many environmental issues within the Department of Energy, which examines the feasibility of obtaining energy resources from nuclear power, oil exploration, and solar sources.

Department of Health and Human Services

- Food and Drug Administration (FDA)

    The FDA is responsible for setting and enforcing standards for food, food additives, drugs, and cosmetics. It is the major approval agency for medicines and foods on the public market.

- National Institutes of Health (NIH)

    The NIH monitors and investigates health trends in the population of the United States.

- Centers for Disease Control and Prevention (CDC)

    CDC tracks emerging and existing diseases and threats to environmental health.

Department of Homeland Security

- Federal Emergency Management Agency (FEMA)

    FEMA coordinates emergency responses in situations of crisis—including environmental crisis, such as a major oil spill or hurricane.

Department of the Interior

- Bureau of Land Management (BLM)

    BLM manages many of the non-park federal lands, particularly in the western United States, with a total of about one-eighth of the land area of the country. This includes the leases for natural resources (timber, mineral, wildlife, energy, recreation sites, wilderness areas, and historical sites).

- Fish and Wildlife Service

  "Game and Fish," as many outdoor sports enthusiasts call this agency, regulates hunting and fishing licenses.

- National Park Service

  NPS manages and oversees the national parks and many of the wilderness areas in the United States.

- Office of Surface Mining, Reclamation, and Enforcement

  This agency oversees and enforces the Surface Mining Control and Reclamation Act to be sure that states set up proper procedures that assure that mining companies reclaim land when coal mining operations are complete.

- United States Geological Survey (USGS)

  USGS is responsible for surveying and mapping the terrain and mineral resources in the United States.

Department of Labor

- Occupational Safety and Health Administration (OSHA)

  OSHA enforces health and safety laws in the workplace, particularly the Occupational Safety and Health Act (also called OSHA).

Cabinet-Level Independent Agencies

- Environmental Protection Agency (EPA)

  The EPA was established by NEPA in 1970 to coordinate, investigate, and enforce environmental issues and laws.

Other Independent Agencies and Commissions

- Nuclear Regulatory Commission (NRC)

  The NRC regulates commercial nuclear power plants and the waste generated from these plants.

- Tennessee Valley Authority (TVA)

  Oversees the series of dams, locks, and electrical generation projects along the rivers in the Tennessee Valley.

Environmental Laws

In each of the following sections, APES students are encouraged to know about the most important law in the field and its basic tenets. Other laws are listed as a reference.

General

Must-Know Laws:

National Environmental Policy Act (NEPA, 1969)

- Established the Council on Environmental Quality and the Environmental Protection Agency (EPA)
- Directs EPA to take into account environmental consequences when making or recommending decisions
- Requires Environmental Impact Statements (EIS) for major federal construction projects

Occupational Safety and Health Act (OSHA, 1970)

- Provides standards for health and safety in the workplace
- Provides workers with the ability to hold employers accountable

Air

Must-Know Law:

Clean Air Act (1970)

- Established air-quality standards for primary and secondary pollutants
- Established "criteria pollutants" as the most threatening to human health
- Requires states to develop clean air plans, which include the emission testing of automobiles

Other Laws:

Asbestos Hazard and Emergency Response Act (1986) Requires schools to be inspected for and mitigate asbestos

Montreal Protocol (1987, amended 1990 and 1992) sets a timetable for phasing out the use of ozone-depleting substances

Kyoto Protocol (1997) Agreement among 150 nations to reduce greenhouse gases. The United States does not participate in the agreement.

Water

Must-Know Law:

Clean Water Act (1972)

- Set into motion as a result of Santa Barbara oil spill in 1969
- Sets goal of creating "fishable, swimmable" waterways by setting water quality standards

Other Laws:

Federal Water Pollution Control Acts (1948) Allows the Surgeon General to assess and address pollution in interstate waters

Water Resources Planning Act (1965) Assesses water resources in each region of the United States

Safe Drinking Water Act (1974) Sets standards for municipal water systems that treat water; also protects groundwater resources

Ocean Dumping Ban Act (1988) Prohibits ocean dumping of sewage and industrial waste

Energy

Must-Know Law:

U.S. Public Utility Regulatory Policy Act (PURPA, 1978)

- Requires utilities to purchase power from small power producers at market value

Land

Must-Know Law:

Surface Mining Reclamation and Control Act (SMCRA, 1977)

- Requires mining companies to put funds in escrow to assure land reclamation after coal mining operations have ceased
- Creates fund to reclaim abandoned coal mines
- Creates "state primacy" to give states control of mining reclamation

Other Laws:

National Park Service Act (1916) Established the National Park Service

Taylor Grazing Act (1934) Established grazing districts in public lands, except for national forests and national parks

Soil Conservation Act (1935) Established the Soil Conservation Service in wake of the Dust Bowl

Wilderness Act (1964) Allows establishment of wilderness areas

Federal Land Policy and Management Act (1976) Outlines use of public lands, vests BLM with authority to manage most non-park public land, and ends homesteading initiated by the Homestead Act of 1862

Toxic Substances

Must-Know Laws:

Federal Insecticide, Fungicide, and Rodenticide Act (FIFRA, 1947)

- Regulates manufacture and use of pesticides
- Pesticides must be registered, approved, and labeled

Federal Food, Drug, and Cosmetic Act (FFDCA, 1906)

- EPA sets tolerance levels for the amount or toxic residues that may lawfully remain on food, drug, and cosmetic products marketed in the United States
- Delaney Amendment (1958) requires absolute lack of hazard for food and drugs, further amended (1996) to define "acceptable risk" as one case of cancer in a million exposures

Comprehensive Environmental Response, Compensation and Liability Act (CERCLA, 1980, amended 1994)

- Established Superfund for emergency response and remediation of toxic sites

Other Laws:

Federal Hazardous Substances Act (1960) Requires cautionary labels in household products

Resource Conservation and Recovery Act (RCRA, 1976, amended 1984) Regulates hazardous waste storage and disposal by tracking them from "cradle-to-grave"—or throughout entire period that the wastes may potentially be harmful: from procurement, through their use, and including the time after disposal

Biodiversity

Must-Know Laws:

Species Conservation Act (1966)

- Secretary of the Interior charged with compiling a list of endangered species
- Provided funds to purchase land to add to other federal lands to form the National Wildlife Refuge System
- Banned importation of species that are endangered globally

Endangered Species Act (1973)

- Creates list of endangered species
- Protects habitat of endangered species
- Directs FWS to prepare recovery plans for endangered species

Lacey Act (1990)

- Allows the Department of the Interior to restore populations of scarce or extinct animals, or introduce scarce populations to new habitats

Environmental Policy and Management

Managing the health of a large number of people and the environment, both in the present and into the future, requires an ability to understand issues and solve problems from an intuitively wise and intellectually interdisciplinary manner. A good scientific understanding of the issue is necessary, of course, but so is an understanding of cultural values, economic and ethical trade-offs, and a sense of detecting well-being that goes beyond the limits of profitability, existing laws, and personal ethics. Because our awareness of environmental health and our own wellness continues to expand and require an increasingly complex and integrated understanding to be able to make right decisions, clearly there are many ways to go about managing any type of environmental problem.

Assessing a risk is often the first step in solving an environmental problem. However, an individual's perception of risk may differ from the current scientific understanding of the risk. Nevertheless, once a risk has been established, how does a society reduce the risk for the future? The field of environmental policy would include the laws enacted, any taxes levied, economic climate and cultural values established, available education, and other efforts made within government, corporations, and institutions to minimize environmental damage.

Risk Assessment

Risk Perception

Perception of various types of environmental or health risk during daily life can be altered by a number of factors, some of which have little to do with a factual level of risk that is really experienced. While in the past, some of these factors have worked to underplay the actual level of environmental risk, it may work in the other direction—to overstate the level of actual risk and prompt people to overreact.

1. Bias Based on Personal Experiences

    A prior personal trauma will skew the perception of risk for the same event in the future. For example, if a person was the victim of an armed robbery, their sense of risk of a future armed robbery will be much higher in the future than it was previously, and may be higher than common sense would warrant.

2. Weighing Probabilities

    When getting ready to go to school in the morning, a person might hear a radio report say that there is a 40% chance of rain that day. In that person's mind, they begin to use probability to assess the risk of getting unpleasantly wet at some point during the day. As a result of that assessment, the person will make decisions to take an umbrella, raincoat, or wet weather shoes, or simply not prepare to be rained on.

    The problem with using probabilities as an accurate determination of risk is that an individual will experience more or less of the risk factors associated with an event than the general population. For example, if one in 200 teenagers contracts HIV, that suggests that all teenagers have a one-in-200, or 0.5%, chance of getting HIV. However, individuals may increase, decrease, or virtually eliminate that probability based on their lifestyle. If a teen is not sexually active, does not engage in intravenous drug use, does not get a tattoo with an unclean needle, and does not work around bodily fluids, then the actual risk is far less than 0.5%—it is virtually 0%. Probabilities establish the risk to an overall population, but it may not be an accurate way for an individual to assess personal risk.

3. Irrational Sense of Invulnerability

    Some people have a difficult time believing that they can be involved in any type of calamity or disaster. Such a person's perception of risk is almost non-existent when, in fact, it may be very high. A group of people may choose to not evacuate in the face of a large hurricane because they feel "those things just happen to other people."

4. Bias Based on Media Reports

Sometimes we assess risk based on a media report (radio, television, magazine, or newspaper) that intentionally or unintentionally suggests that a risk is larger or smaller than it really is. Sometimes just the tone of an announcer's voice may send a signal of urgency that prompts people to perceive the risk to be higher than it really is.

5. Fear of the Unknown

During the Three-Mile Island nuclear disaster, many thousands of people were evacuated from nearby communities because the threat was unseen, and news reports were uncertain. Sometimes the uncertainty itself creates a sense of fear that heightens the perception of an environmental risk.

Risk Assessment to Determine Policy or Action

1. Government Agencies that Assess Risk

    Several government agencies are charged with assessing risk experienced by consumers. The following represent a few examples:

    a. The Food and Drug Administration (FDA) assesses the risks of food, food additives, and medicines before they are released for market.

    b. The Consumer Product Safety Division helps to identify risks to consumers from nonfood, nonmedicinal products.

    c. The Environmental Protection Agency (EPA) determines risk by determining acceptable standards of pollutants in the air and water.

    d. The Centers for Disease Control (CDC) assesses risk due to disease and social factors.

    e. The Occupational Safety and Health Administration assesses risk in the workplace.

2. Private and Corporate Risk Assessment

    NEPA requires large construction projects to generate an environmental impact statement (EIS) to determine possible impact of the project on human health, wildlife, and the surrounding environment.

    The need to comply with existing laws forces organizations to plan for being environmentally accountable. For example, a mining operation must put into escrow an amount of money that will finance the reclamation of the site after mining operations have ceased. Before these laws came into being, there was very little accountability for environmental health.

The threat of a lawsuit also forces accountability onto private sector organizations. While it may not be economically desirable in the short term for a utility to clean up a pool of solvent used to maintain pipes in a power plant, the possibility of a very large lawsuit and poor public relations—both of which will adversely affect stock prices and the performance evaluation of corporate officers—make it cheaper to do the right thing in the first place. The threat of a lawsuit is one tool to internalize previously external costs.

To minimize the possibility of corporate officers breaking laws that result in prison time, large payouts to litigants, or poor public relations, many privately held organizations do what they can to assess risk—or investigate with "due diligence"—before a project begins. That often includes hiring an independent consultant to investigate possible outcomes, since poor decisions can sometimes be made within an organization for the sake of personal gain, or because of the sway of politics and personalities that occur within any organization. The independent consultant can then carry the weight of reporting potentially bad news, so that a better decision is made on behalf of the whole organization rather than simply advancing someone's personal career.

Environmental Management

In making decisions about limiting the impact on the environment, lawmakers, corporations, and private organizations can employ any of the following concepts.

Conservation

Conservation refers to simply not using and protecting resources that might otherwise be expended with a less responsible pattern of use. For example, utility companies are finding that it is less expensive to educate consumers about using less power than it is to find new sources of power for energy-ignorant consumers. In this case, the policy that makes the most economic and environmental sense is conservation, which can be achieved by educating the consumers and using incentives to decrease consumption.

Preservation

Preservation refers to providing an ample sink of resources so that they might be enjoyed by others in the future. It is different from conservation in that preservation protects actual resources, whereas conservation simply decreases use of resources—whether or not they have been identified, isolated, or protected for others. For example, the Park Service Act of 1916 sought to preserve natural features, unique populations, and historical objects "by such means as will leave them unimpaired for the enjoyment of future generations." Similar acts to preserve nature are the 1964 Wilderness Act, which established wilderness areas, and the establishment of wildlife refuges, which began under Theodore Roosevelt in 1901. Currently, about one percent of the United States is preserved as a wildlife refuge. The most recent and massive addition to this block of land is the 1980 Alaska National Interest Land Act, which nearly tripled the total acreage preserved as a wildlife refuge.

Restoration

Restoration refers to bringing a damaged ecosystem back to its unspoiled, natural condition. For example, the Nature Conservancy is restoring a 40,000-acre prairie in Kansas by using periodic fires and by introducing bison. The fires tend to kill exotic plants that have not adapted to fire and help natural prairie grasses reseed themselves. The bison serve as seed dispersal agents as they eat the grass in one area and eliminate the nutrient-rich, seed-laden waste as they roam. The gentle disturbance from bison hooves also provide opportunities for other organisms to build the soil and help establish a more complete, complex prairie biome. The hope is that the area will be completely restored to its prior prairie grandeur.

Remediation

Remediation refers to using chemical, biological, or physical methods to remove chemically active (either hazardous or toxic) pollutants. Chemical treatment may include neutralization of acids or oxidants. Examples of biological treatment include bacterial digestion of oil or nutrients and using water plants to remove nutrients from wastewater. An example of a physical method includes the vaporization of hydrocarbons in an underground plume from a broken oil transport pipeline in a process known as *air stripping*. In another example, CERCLA was passed to establish a "superfund" that would begin cleaning up toxic waste dump sites across the country. Taxation of a large number of people provided the funds for these massive cleanups. In this case, the government took an active role in remediation.

Reclamation

Reclamation used to refer to using large water projects to bring water to otherwise unarable land. However, now this term usually refers to the rehabilitation of land, to return a massively scarred, denuded, or devastated area to a condition that is environmentally useful and socially or politically acceptable. For example, when a mining operation wishes to create an open-pit mine, they are required by the Surface Mining Control and Reclamation Act (enforced by the Office of Surface Mining Reclamation) to put into escrow enough money to pay for total reclamation of the land after mining operations are complete. Reclamation might include reburying mine tailings so that surface runoff does not carry soluble toxic ions into waterways, refilling open pits to allow plants to grow and wildlife to thrive, and digging out "high walls"—or the cliff-like edge of an open-pit mine—so that it does not present a hazard to hikers and that the terrain may take on a more natural topography.

Mitigation

Mitigation is a general term that refers to finding a solution to a problem. In the context of environmental degradation, mitigation usually refers to establishing another ecosystem elsewhere of comparable health and magnitude in exchange for damage done as a result of developing a nearby area. For example, the Fish and Wildlife Conservation Act of 1980 prompted states to develop plans to conserve wildlife resources. One such state plan occurred in California where a developer had to establish a remote preserve to protect the habitat of a rare frog before the state and county officials would grant a permit to undergo construction that would decrease the habitat of the frog in another area. Such mitigation hopefully accomplishes a greater benefit for the frog and, by providing housing that is necessary, for a growing population.

Case Summaries

A Civil Action

Principles mentioned in this case:

- **Internalizing external costs**
- **Using taxation and establishing laws to clean up a site for which no direct responsibility may be attributed**

As early as the eighteenth and nineteenth centuries, Woburn, Massachusetts, was an industrial city, with toxic effluent from local tanneries polluting the groundwater. In 1958 the city drilled two new wells, which they ended up using to accommodate the growing population even though the city engineer warned that the water in the wells was not fit to drink. In the early 1970s, children in the area began to contract cancer. The families of those children sued W.R. Grace & Company and Beatrice Foods for damages caused by improper disposal of toxic wastes near the wells.

However, the case settled out of court because it could not be proven with certainty that Grace and Beatrice had provided the actual pollutants that had caused the illnesses. Even though it was clear that the plaintiffs had illegally buried drums of toxic waste on the property, it was also clear that the groundwater had been polluted even before that violation.

Such litigation is an example of external costs becoming internalized. The $8 million paid by Grace now becomes an expense that is ultimately taken out of the profit to shareholders, or passed on to consumers.

Also, since the site ended up becoming one the most expensive Superfund cleanup sites in Massachusetts, it is an example of the greater populace paying a tax to clean up a massive spill, the responsibility for which is difficult to ascribe to any one person or organization.

This case was the subject for Jonathan Harr's 1995 novel, *A Civil Action*, which was later made into a movie of the same title.

The Northern Spotted Owl

Principles mentioned in this case:

- **Clash of biocentric and utilitarian world views**
- **Legislation shifts economic demand**
- **Laws of supply and demand**
- **Internalizing external costs**

The Endangered Species Act establishes a list of endangered species and protects their habitat from destruction. The protection of the endangered northern spotted owl depended upon preserving old-growth forests in the Pacific Northwest. Financial analysts estimated that preserving the habitat to save about 2,000 owls would cost billions in lost revenue and decrease jobs in Washington State and Oregon. While the old-growth forests were protected, the perceived supply of lumber plummeted, which increased the cost of lumber and home-building.

This conflict was between loggers, who viewed timber as an economic resource and a way to feed their families, and environmentalists, who created legislation to protect a rare species. The utilitarian world view clashed head-on with the biocentric world view. Many would argue that the increase in the cost of lumber was the real fair market value in a world that desperately needs to protect endangered species, and the resulting economic depression in the area was simply a backlash of an over-dependence on the natural capital that is eventually depleted. Others would argue that it is silly to have such a drastic effect on the financial well-being of families just to protect some birds that nobody notices anyway.

Practice Exam 1

Practice
Exam 1

AP Environmental Science

Practice Exam 1
This exam is also on CD-ROM in our TestWare®.

Section I

TIME: 90 minutes
100 multiple-choice questions

(Answer sheets appear in the back of this book.)

> **DIRECTIONS:** Each of the questions or incomplete statements below is followed by five suggested answers or completions. Select the best answer for each question and then fill in the corresponding oval on the answer sheet.

Questions 1–4 refer to the following answers. Select the one lettered choice that best fits each statement.

(A) $CaCl_2$

(B) Hg^+

(C) $C_6H_{12}O_6$

(D) H^+

(E) CO_3^{2-}

1. An ionic compound

2. An organic molecule

3. A polyatomic ion

4. An acidic proton

Questions 5–8 refer to the following answers. Select the one lettered choice that best fits each statement.

    (A) Kinetic energy

    (B) First Law of Thermodynamics

    (C) Second Law of Thermodynamics

    (D) Conservation of mass

    (E) Nutrient cycle

5. Exchanges of energy always yield useless heat

6. Chemical reactions have the same amount of reactants as products

7. Energy is neither created nor destroyed

8. Energy is absorbed by an object

Questions 9–12 refer to the following answers. Select the one lettered choice that best fits each statement.

    (A) Marine pelagic

    (B) Freshwater aquatic

    (C) Tundra

    (D) Desert

    (E) Savannah

9. Short growing season in warm, dry environment, requiring plants to retain water

10. Similar to grassland biome, but drier

11. Occurs at either high latitude or high altitude

12. Coral reef ecosystem is part of this larger biome

Questions 13–15 refer to the following map. Select the one lettered choice that best fits each statement.

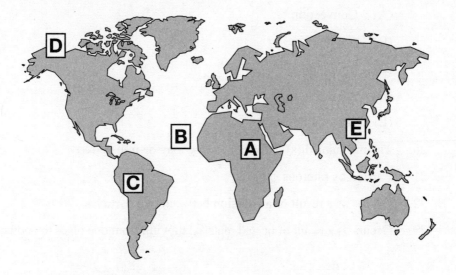

13. Vast tundra provides a home for caribou.

14. Incubating ground for North American hurricanes

15. Likely location for savannah biome

Questions 16–19 refer to the following answers. Select the one lettered choice that best fits each statement.

(A) Nitrates

(B) Nitrites

(C) Ammonia

(D) Atmospheric nitrogen

(E) Nitrogen dioxide

16. Produced in the atmosphere by lightning

17. Fixation in the nodules of legumes produces this form of nitrogen.

18. Full oxidation during secondary sewage treatment produces these ions.

19. When mixed with water vapor, this forms acid rain.

AP Environmental Science

Questions 20–23 refer to the following answers. Select the one lettered choice that best fits each statement.

(A) Convection

(B) Conduction

(C) Radiation

(D) All of the above

(E) None of the above

20. The mechanism(s) by which energy may be transferred

21. Carried by photons

22. Occurs as a result of a collision between two particles

23. Occurs as a result of heated material moving from one place to another

Questions 24–27 refer to the following answers. Select the one lettered choice that best fits each statement.

(A) Igneous rocks

(B) Metamorphic rocks

(C) Sedimentary rocks

(D) All of the above

(E) None of the above

24. This type of rock is susceptible to chemical and mechanical weathering.

25. This type of rock is created by the cooling of magma within the lithosphere.

26. This type of rock is produced when small particles, transported by erosion, settle on top of each other and become compressed.

27. Marble is an example of this type of rock.

Questions 28–31 refer to the following answers. Select the one lettered choice that best fits each statement.

(A) O horizon
(B) A horizon
(C) E horizon
(D) B horizon
(E) C horizon

28. Primarily composed of larger fragments of weathered parent material that lies directly above the parent material

29. Topsoil

30. Organically rich surface material that contains partially decomposed plants, animals, bacteria, and fungi

31. Subsoil layer that is rich with inorganic minerals

Questions 32–36 refer to the following answers. Select the one lettered choice that best fits each statement.

(A) El Niño/Southern oscillation
(B) Global warming
(C) Ozone depletion
(D) Milankovitch cycle
(E) Leapfrog effect

32. Results in pollutants migrating to polar regions

33. Results in periodic weather disturbances in North America

34. Chronic change due to greater absorption of long-wavelength infrared light that is re-radiated from the surface of the Earth

35. Caused by human pollution and results in greater levels of skin cancer in New Zealand

36. A "wobble" in the Earth's orbit that may affect long-term climate fluctuations

Questions 37–39 refer to the following answers. Select the one lettered choice that best fits each statement.

(A) Amount of biomass produced through photosynthesis

(B) Amount of herbivore biomass

(C) Total amount of biomass in a biological community

(D) Entropy

(E) Two of the above

37. Primary productivity

38. Secondary productivity

39. Total productivity

Questions 40–43 refer to the following answers. Select the one lettered choice that best fits each statement.

(A) Mutualism

(B) Parasitism

(C) Commensalism

(D) Interspecific competition

(E) Intraspecific competition

40. Demonstrated by even spacing of penguins on the shore

41. Demonstrated by the two organisms involved in forming lichens

42. Demonstrated by nectar-producing flowers

43. Demonstrated by broad-leafed plants in a rainforest

Questions 44–47 refer to the following answers. Select the one lettered choice that best fits each statement.

    (A)   Fertility

    (B)   Fecundity

    (C)   Demographic transition

    (D)   Birth rate

    (E)   Total fertility rate

44. The number of children born to an average woman during her entire reproductive life

45. The number of births in a year per thousand people

46. Represents a drop in both death and birth rates

47. The ability of an individual to reproduce

48. Which of the following international agreements called for the unilateral decrease by all nations in the production of greenhouse gases?

    (A)   Ontario Agreement

    (B)   Kyoto Protocol

    (C)   Sao Paulo Conference

    (D)   Toronto Agreement

    (E)   UN Global Warming Initiative

49. Which of the following trophic levels would contain the greatest amount of biomass in a given ecosystem?

    (A)   Bacteria

    (B)   Photosynthetic algae

    (C)   Herbivores

    (D)   Primary carnivores

    (E)   Top carnivores

AP Environmental Science

50. Genetic redistribution and recombination in organisms is achieved through

    (A) cloning
    (B) the founder effect
    (C) adaptive radiation
    (D) sexual reproduction
    (E) divergent evolution

51. This environmental crisis involved the production of water-reactive gas that killed thousands in the middle of the night.

    (A) Love Canal
    (B) Bhopal Crisis
    (C) *Exxon Valdez*
    (D) DDT
    (E) Chernobyl

52. Limestone is used to remove oxides of sulfur and nitrogen in this technique:

    (A) Electrostatic precipitator
    (B) Fluidized bed combustion
    (C) Catalytic converter
    (D) Staged burner
    (E) Bag filters

53. This type of water pollution causes pipes to be encrusted and decreases the effectiveness of soap.

    (A) Minimata disease
    (B) Eutrophication
    (C) Fecal coliform
    (D) Sediment
    (E) Calcium ions

54. Easter Island is an island on which human overpopulation depleted the island's resources and converted it from a forested paradise to a barren island that could barely support a small population. It appears that the population exceeded the island's

    (A) doubling time
    (B) stationary phase
    (C) environmental resistance
    (D) biotic potential
    (E) carrying capacity

55. Which of the following is an example of *active* solar heating in a residential setting?

    (A) Extended eves that block the hot summer sun, but allow the winter sun to enter the house
    (B) Photovoltaic cells that convert sunlight directly to electricity
    (C) A system that pumps water to the roof to be heated by the sun, and then stores it in a water tank to be used later
    (D) Using heavy insulation in the attic
    (E) Wrapping the hot water heater and pipes

56. Which of the following factors accompanies urbanization?

    (A) Increased price of property in rural areas
    (B) Increased standard of living in rural areas
    (C) Increased specialization of jobs in urban areas
    (D) Decreased crime in urban areas
    (E) Decreased amount of energy used per capita

AP Environmental Science

57. Which of the following is an energy-saving technique that can be accomplished anywhere, regardless of sun exposure?

    (A) Geothermal heating and cooling
    (B) Electrical production from photovoltaic cells
    (C) Passive solar heating through building design
    (D) South-facing windows
    (E) Batch-heating of water

58. Most of the property damage from hurricanes occurs from

    (A) storm surges
    (B) winds
    (C) tornadoes
    (D) looting
    (E) crashing waves

59. Which pair of phrases best describes the effect that O_3 molecules have on the environment?

    (A) Greenhouse effect; lung tissue irritant
    (B) Ozone depletion; Greenhouse effect
    (C) Blocks UV rays; lung tissue irritant
    (D) Ozone depletion; carcinogen
    (E) Greenhouse effect; causes nerve damage

60. Most ozone depletion occurs under conditions that create a catalytic effect. One of the essential conditions is the presence of an ice crystal. These ice crystals can be found in clouds that occur in the

(A) troposphere

(B) mesosphere

(C) stratosphere

(D) ionosphere

(E) biosphere

61. The air pressures in the equatorial latitudes tend to be higher than the pressures at the same altitude about 30 degrees north or south because

(A) the area 30 degrees north and south of the equator tends to have more seasons, which cause those areas to have more storms and lower air pressures

(B) the area 30 degrees north and south tends to be deserts, which are hotter and tend to have lower pressures than equatorial regions

(C) the equatorial region gets more sun, which causes the air to expand and to move north and south along the surface of the earth

(D) the equatorial region gets more sun, which expands the air, lowers the air pressure, and causes the air to rise

(E) the equatorial region gets more sun, which causes the air to heat up and increase pressure; equatorial regions actually have higher pressures than areas that are 30 degrees north and south

62. Which of the following gases pose the highest risk of lung cancer?

(A) Carbon monoxide

(B) Ozone

(C) Radon

(D) Oxides of sulfur

(E) Oxides of nitrogen

AP Environmental Science

63. The crude birth rate for a population is 32 births per 1,000 people, and the crude death rate is 14 deaths per 1,000 people. Calculate the annual percent growth of the population.

    (A) 3.2%

    (B) 18%

    (C) 1.8%

    (D) 2.8%

    (E) 14%

64. A major oil tanker has run aground and spilled millions of barrels of oil. Even though the oil company spends many millions of dollars on the clean-up effort, and there is less profit for the oil company, the GNP actually increases as a result because

    (A) other oil companies will make more money as a result of out competing the oil company whose tanker ran aground

    (B) the increased price that results from a decrease in supply will cover the loss that was paid to clean up the spill

    (C) the clean-up companies made more money than the oil company lost

    (D) additional money is spent on the clean-up that would not ordinarily be spent

    (E) the GNP doesn't actually increase in this scenario

65. A company that produces interior paint is faced with a toxic waste problem. Which of the following solutions require the least overall energy and will result in the safest outcome for the environment?

    (A) Recycling or reusing the waste

    (B) Incinerating the waste

    (C) Injecting the waste underground

    (D) Storing the waste in a salt mine

    (E) Storing the waste in a landfill

66. Which of the following would NOT be considered a strategy used by a population that demonstrated *logistic* growth?

    (A) Give careful care to the young

    (B) Evolve into a relatively large organism

    (C) Take more time to reach sexual maturity

    (D) Produce a large number of offspring

    (E) Be a carnivore

67. Large water projects that slow the flow of rivers may show many economic benefits, but such projects in warm climates can also lead to

    (A) schistosomiasis

    (B) malaria

    (C) valley fever

    (D) HIV

    (E) river blindness

68. Which of the following describes the process by which one community outcompetes another community for a habitat?

    (A) Natural selection

    (B) Adaptive radiation

    (C) Convergent evolution

    (D) Primary succession

    (E) Secondary succession

69. Which of the following units are equivalent to the units, *ppm*?

    (A) Molarity

    (B) mg/kg

    (C) mg/L

    (D) g/L

    (E) Mg

70. Ecosystem complexity is most easily measured by

    (A) the number of species in the ecosystem

    (B) the number of individuals in the ecosystem

    (C) the number of species at each trophic level in the ecosystem

    (D) the total number of producers in an ecosystem

    (E) the total number of top carnivores in an ecosystem

71. Which sequence of responses toward solid waste leads to the least amount of energy used?

    (A) Reduce, reuse, recycle

    (B) Recycle, reduce, reuse

    (C) Reuse, recycle, reduce

    (D) Reuse, reduce, recycle

    (E) Reduce, recycle, reuse

72. An area that was formerly a toxic waste site but is now safe enough to use for residential property is considered

    (A) a brownfield

    (B) a mistake

    (C) a leachfield

    (D) restored

    (E) a superfund site

73. A government agency is considering the costs and benefits of a project that would produce pollution in an area. A *true cost analysis* would include

   (A) analyzing only the measurable monetary costs of materials to produce the project

   (B) analyzing only the medical costs of people affected by the pollution

   (C) comparing the profit and expenses of the project

   (D) the cost of materials, in addition to administrative costs of the project

   (E) analyzing the costs of medical care and loss of happiness and productivity of the people who would be harmed by the project, in addition to the cost of materials

74. In Greek, the "*eco*" portion of both *eco*nomics and *eco*logy refers to

   (A) "health"

   (B) "the home"

   (C) "the study of"

   (D) "nature"

   (E) "wealth"

75. The wealth that can be generated as a function of land resources is referred to as

   (A) social capital

   (B) manufactured capital

   (C) human capital

   (D) natural capital

   (E) total capital

76. How much of the energy needed to produce a single aluminum can is saved by recycling the can?

    (A) 5%
    (B) 10%
    (C) 20%
    (D) 60%
    (E) 95%

77. Which of the following is NOT a trait that would put a species at risk of being endangered now or in the near future?

    (A) The species is migratory and depends on the availability of healthy ecosystems in more than one area.
    (B) The species competes with humans for food or habitat.
    (C) The species has a high reproductive rate.
    (D) The species is higher on a food chain.
    (E) The species has low genetic diversity.

78. Which of the following organisms would be most at risk for accumulating fat soluble pollutants?

    (A) Fish
    (B) Birds
    (C) Small herbivores
    (D) Insects
    (E) Algae

79. Which of the following would be a good measure of primary productivity?

    (A) The mass of insect excrement

    (B) The mass of green leaves produced on one acre of land

    (C) The volume of oxygen used through respiration

    (D) The number of animals born or hatched in a given season

    (E) The number of species that can live in one area

80. A species has been decimated due to overhunting. To save the species, a nonprofit organization creates a private reserve and helps a single breeding pair build a new population that will later be released into the wild. Which of the following best describes the chances that this species will survive?

    (A) As long as enough individuals are produced in the reserve before being set free, the species can survive.

    (B) As long as the species can survive in captivity, there will always be a chance that the species can be reintroduced in the wild and survive.

    (C) Even though the species can survive in captivity or on a reserve, the same conditions that led to being threatened will prevent it from being successfully reintroduced in the wild.

    (D) Limited genetic diversity will create a population that is not viable, and the species will not survive.

    (E) As long as the population is protected from hunting, the species can survive.

81. This federal law protects habitats for endangered species.

    (A) NEPA

    (B) FFDCA

    (C) ESA

    (D) SMCRA

    (E) CERCLA

82. This disease results from a lack of iodine in one's diet.

    (A) Kwashiorkor

    (B) Anemia

    (C) Beriberi

    (D) Scurvy

    (E) Goiter

83. Which of the following represents the nutrient-rich portion of a lake that is closest to the bottom?

    (A) Limnion

    (B) Hypolimnion

    (C) Mesolimnion

    (D) Littoral zone

    (E) Pelagic zone

84. All of the following impacts are typically true of current home construction techniques EXCEPT:

    (A) Heavy dependence on energy

    (B) Exposure of humans to toxic substances

    (C) Large production of waste

    (D) Insufficient use of solar energy

    (E) Sustainability

85. A temperature inversion is different from a normal weather pattern because

    (A) hot, warm winds blow pollution out to sea

    (B) pollutants get trapped in warm air

    (C) warm air is trapped under a blanket of cool air

    (D) cool air is trapped under a blanket of warm air

    (E) a dome of warm air holds in pollutants and diverts weather

86. Which of the following water compartments contains the greatest amount of fresh water?

    (A) Ice and snow
    (B) Rivers and streams
    (C) Lakes and reservoirs
    (D) Groundwater
    (E) Atmospheric water vapor

87. Which of the following occurs when freshwater is too quickly removed from a well near the coast, and saltwater begins to come out of the well?

    (A) Infiltration
    (B) Transpiration
    (C) Intrusion
    (D) Compaction
    (E) Depression

88. A solution has a hydrogen ion concentration of 1.0×10^{-3} moles per liter. The pH of this solution is

    (A) 1
    (B) 3
    (C) 7
    (D) 11
    (E) 14

89. Once radiant energy has penetrated the clouds and has been absorbed by the Earth, which of the following occurs?

    (A) All the energy is eventually re-radiated as long-wavelength infrared light.

    (B) All the energy is re-radiated as short-wavelength UV light.

    (C) The energy is entirely absorbed by the oceans and the trees.

    (D) The energy is entirely absorbed by the atmosphere to create convection cycles.

    (E) The energy is entirely trapped by greenhouse gases.

90. The equator is warmer than the poles because

    (A) the equatorial regions experience hot downdrafts in atmospheric convection cycles

    (B) the equatorial regions are closer to the sun

    (C) the oblique angle of the sun spreads the energy out over a larger area in polar regions

    (D) there is more ice in the polar regions

    (E) the sun spends less time shining on polar areas

91. The number of species that a single ecosystem can support is most related to

    (A) the overall amount of resources available

    (B) the amount of competition between species

    (C) the amount of sunshine available to the ecosystem

    (D) the amount of water available to the ecosystem

    (E) the abundance of top carnivores in the system

92. How many years will it take for a population to quadruple in size if it demonstrates an 8% annual rate of growth?

    (A) 3 years

    (B) 4.5 years

    (C) 9 years

    (D) 18 years

    (E) 36 years

93. In the wastewater treatment process, aluminum sulfate is used to help precipitate suspended toxins in a process called

    (A) flocculation

    (B) absorption

    (C) stripping

    (D) incineration

    (E) bacterial remediation

94. Which of the following had the greatest impact on reducing population in the U.S.?

    (A) Demographic transition

    (B) The Great Depression

    (C) Post–World War II era

    (D) The flu of 1918

    (E) Post–World War I era

95. Which of the following has the effect of increasing toxicity to human populations?

    (A) Chemical synergism

    (B) Metabolic degradation

    (C) Excretion

    (D) DNA repair enzymes

    (E) Glomerular filtration

96. Which of the following series of numbers demonstrates arithmetic growth?

    (A) 2, 4, 6, 8

    (B) 2, 4, 8, 16

    (C) 4, 16, 126, 432

    (D) 3, 6, 12, 24, 48

    (E) All of the above are arithmetic series

97. Carbon filters improve water quality by

    (A) disinfecting pathogens

    (B) clinging to small particles and improving taste

    (C) removing large foreign objects

    (D) absorbing calcium ions that make water "hard"

    (E) adding carbon atoms that provide important nutrients

98. All of the following practices help conserve freshwater EXCEPT

    (A) not ordering a newspaper

    (B) buying food in bulk that uses less packaging

    (C) putting silt dams around construction sites

    (D) choosing drought-resistant plants for landscaping projects

    (E) All of the above help conserve water

99. Which of the following is NOT a unit of energy?

    (A) Joule

    (B) Watt

    (C) BTU

    (D) Kg × m² / sec²

    (E) All of the above are units of energy.

100. Which is the correct sequence in which coal is made from organic material?

    (A) Peat, anthracite, lignite, bituminous

    (B) Peat, lignite, bituminous, anthracite

    (C) Peat, bituminous, anthracite, lignite

    (D) Peat, anthracite, bituminous, lignite

    (E) Anthracite, lignite, bituminous, peat

AP Environmental Science

Practice Exam 1

Section II

Time: 90 minutes

> **DIRECTIONS:** You have 90 minutes to answer all four of the following questions. It is suggested that you spend approximately half your time on the first question and divide the remaining time equally among the next three. In answering these questions, you should emphasize the line of reasoning that generated your results; it is not enough to list the results of your analysis. Include correctly labeled diagrams, if useful or required, in explaining your answers. A correctly labeled diagram must have all axes and curves labeled and must show directional changes.

1. A major nuclear power plant generates 15,000 MW-hr of electricity each day.

    a. How many tons of coal would be needed to produce the equivalent amount of energy per day, if one ton of coal produces 2,000 kW-hr of energy?

    b. Describe the type of nuclear reaction that takes place inside a nuclear power plant.

    c. Describe the role of a control rod in the operation of a nuclear power plant. What is a likely environmental impact if control rods are removed?

    d. Describe the environmental consequences of using nuclear power.

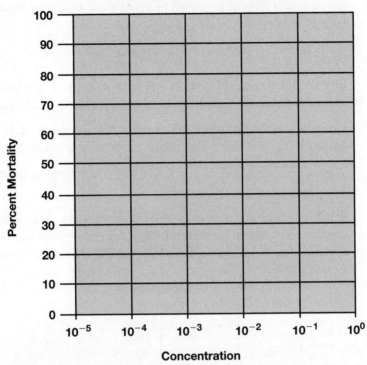

2. Use the following data table to answer the questions.

Percent of daphnia dead	Molar concentration of toxin A
0	10^{-5}
10	10^{-4}
23	10^{-3}
38	10^{-2}
62	10^{-1}
100	10^{0}

a. Construct a dose-response curve and calculate the LD_{50} for the daphnia with respect to toxin A.

b. Another toxin, toxin B, has an LD of 3.0×10^{-3} moles per liter with respect to daphnia. Which toxin, A or B, is most toxic to daphnia? Why can or can't this data give you some idea of the toxicity of these compounds to humans?

c. Toxin B is water soluble, and toxin A is a fat-soluble hydrocarbon. Which toxin has the greatest potential for biomagnifica-tion in the environment? Explain your answer.

AP Environmental Science

d. Describe the concept of tolerance limits. Which law would require the understanding of tolerance limits of these toxins on human populations?

3. a. Identify the major differences between a tropical rainforest biome and a desert biome in terms of temperature, moisture, and latitude.

   b. Describe two major plant adaptations that occur in each of these biomes to accommodate the unique conditions found in that biome.

   c. Why is it important to sustain a diversity of ecosystems in addition to a diversity of individual species?

   d. What scarce resource in each of these biomes results in strong interspecific competition for that resource? For each biome, also identify a resource that is not scarce, for which there is little or no competition.

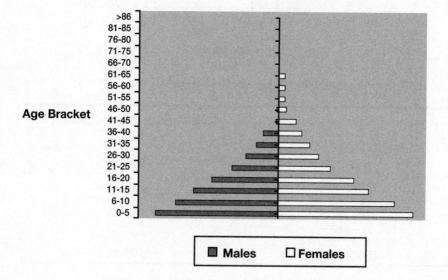

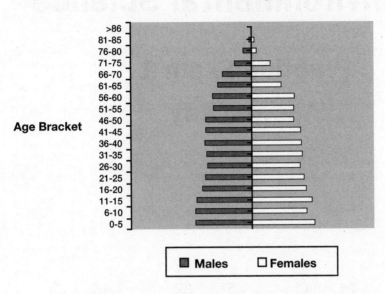

4. a. Use the above age-structure histograms to compare the following traits between Country A and Country B.

   i. The average age of the population

   ii. The probable life expectancy

   iii. The likelihood for children to reach the age of adulthood

   iv. The level of fertility

   b. Describe the typical sequence of changes in birth rate and death rate within a country as it undergoes a demographic transition.

   c. Describe the cultural changes that need to take place in a country in order for it to undergo a demographic transition.

AP Environmental Science

Practice Exam 1
Answer Key

Section I

1.	(A)	26.	(C)	51.	(B)	76.	(E)
2.	(C)	27.	(B)	52.	(B)	77.	(C)
3.	(E)	28.	(E)	53.	(E)	78.	(B)
4.	(D)	29.	(B)	54.	(E)	79.	(B)
5.	(C)	30.	(A)	55.	(C)	80.	(D)
6.	(D)	31.	(D)	56.	(C)	81.	(C)
7.	(B)	32.	(E)	57.	(A)	82.	(E)
8.	(A)	33.	(A)	58.	(A)	83.	(B)
9.	(D)	34.	(B)	59.	(C)	84.	(E)
10.	(E)	35.	(C)	60.	(C)	85.	(D)
11.	(C)	36.	(D)	61.	(D)	86.	(A)
12.	(A)	37.	(A)	62.	(C)	87.	(C)
13.	(D)	38.	(B)	63.	(C)	88.	(B)
14.	(B)	39.	(C)	64.	(D)	89.	(A)
15.	(A)	40.	(E)	65.	(A)	90.	(C)
16.	(A)	41.	(A)	66.	(D)	91.	(A)
17.	(C)	42.	(A)	67.	(A)	92.	(D)
18.	(A)	43.	(D)	68.	(E)	93.	(A)
19.	(E)	44.	(A)	69.	(B)	94.	(D)
20.	(D)	45.	(D)	70.	(C)	95.	(A)
21.	(D)	46.	(C)	71.	(A)	96.	(A)
22.	(B)	47.	(B)	72.	(D)	97.	(B)
23.	(A)	48.	(B)	73.	(E)	98.	(E)
24.	(D)	49.	(B)	74.	(B)	99.	(B)
25.	(A)	50.	(D)	75.	(D)	100.	(B)

Detailed Explanations of Answers

Practice Exam 1

Section I

1. **(A)** An ionic compound is made of a metal and a nonmetal that are held together with an ionic attraction. Only (A), calcium chloride, involves a metal and a nonmetal.

2. **(C)** An organic molecule is a carbon-based collection of nonmetal atoms, held together with covalent bonds.

3. **(E)** A polyatomic ion is a group of elements that are held together with covalent bonds, and the whole molecule carries a non-neutral charge. Only the carbonate ion (E) meets these criteria.

4. **(D)** An acid is defined as a proton donor. The simplest of all acids is the hydrogen ion (D).

5. **(C)** The Second Law of Thermodynamics states that all processes work to produce waste heat, called entropy. Entropy is also considered a measure of disorder because waste heat is distributed widely and randomly to the surrounding environment.

6. **(D)** A chemical reaction is a reflection of the Law of Conservation of Mass. The number and type of elements—and therefore total mass—on each side of the equation must balance.

7. **(B)** The First Law of Thermodynamics states that energy is neither created nor destroyed; it only changes form during chemical changes.

8. **(A)** To have energy be absorbed by an object, a collision must occur in which kinetic energy is transferred from one object to another.

9. **(D)** Two biomes experience very short growing seasons: deserts and tundra. Because of the low temperatures in a tundra, most of the water is bound up as ice crystals outside of the plants during most of the year. Deserts have less

water in any form—liquid or solid—so plants must retain water internally. If that were to be the case in a tundra biome, the water inside the plants would freeze and cause damage to the plants.

10. **(E)** Savannah biomes are very much like grassland biomes with the possible exception that a savannah is drier and often interspersed with drought-tolerant trees.

11. **(C)** There are two types of tundra: arctic and alpine. Arctic tundra occurs at high latitude; alpine tundra exists at high altitude.

12. **(A)** The marine pelagic system refers to the open ocean environment, which would include coral reef ecosystems. No other answer choices refer to a marine (salt water) ecosystem.

13. **(D)** The north slope of Alaska is predominantly tundra and provides the necessary habitat for caribou.

14. **(B)** Hurricanes that typically hit the southeastern United States begin to form in the late summer in the equatorial Atlantic, off the coast of Africa, where high levels of water evaporation store the energy needed to drive the storm. The prevailing ocean currents and the Coriolis effect direct the storm across the Atlantic and toward North America.

15. **(A)** Savannah biomes tend to occur in the dry sections of Africa, often adjacent to deserts or latitudes where warm, descending air promotes a hot, dry climate.

16. **(A)** Lightning superheats air—which is 80% nitrogen gas—and oxidizes the nitrogen to the fullest extent. To put this another way, oxidation of nitrogen means that oxygen atoms are added to the nitrogen atom. Oxidizing the nitrogen atom results in adding as many oxygen atoms to the nitrogen atom as is possible.

17. **(C)** Nitrogen fixation occurs when nitrifying bacteria in the nodules of legumes reduce atmospheric nitrogen gas into ammonia. Sometimes farmers take advantage of this by rotating crops with legumes so that these bacteria "fix" nitrogen into the soil after it has been depleted by more nitrogen-demanding crops.

18. **(A)** Once again, "full oxidation" results in adding as many oxygen atoms to nitrogen atoms as is possible, resulting in nitrate ions. Sewage treatment relies on bacterial digestion to oxidize the nitrogen in human waste from ammonia to nitrites, and then from nitrites to nitrates. Secondary treatment ceases with the release of nitrate ions into the environment—which can cause oxygen-demanding waste and accelerate eutrophication in streams, rivers, and lakes.

Practice Exam 1 – Section I: Detailed Explanations of Answers

Tertiary treatment of waste includes a chemical or biological step to use up the nitrate ions and prevent nutrient pollution.

19. **(E)** Water vapor adds to nitrogen dioxide to form both nitrous and nitric acid.

20. **(D)** Energy can be transferred by convection (warm material moving from one place to another), conduction (energy transferred from one particle to another via a collision), or radiation (photons moving through space).

21. **(D)** Radiation is the type of energy transfer that relies upon photons carrying energy from one place to another as a ray of light.

22. **(B)** In all collisions, energy is conserved. For example, when one football player hits another, energy is transferred from one player to another in the form of sound or movement. On a microscopic level, an incoming particle collides with another and gives the receiving particle energy that it did not have before. This is what is occurring in conduction.

23. **(A)** Convection is the type of energy transfer that occurs when a material that contains energy moves from one place to another. For example, radiation may heat the air in one place in a room; but convection drafts will allow the warmer air to pass to a different place in the room.

24. **(D)** All rocks can undergo chemical or mechanical weathering. Chemical weathering occurs when oxidizing agents or acids react with and dissolve a portion of a rock. Chemical weathering also makes a rock more vulnerable to mechanical weathering, which takes place as a result of the force of wind or water breaking off small pieces of a rock.

25. **(A)** Igneous rock is the type of rock that is formed when magma cools.

26. **(C)** Sedimentary rock is formed when small particles of rock layer upon other small particles of rock and eventually are hardened by compression.

27. **(B)** Marble is sandstone that has undergone extreme pressure so that the particles in the sandstone fuse together more tightly. This is an example of metamorphic rock.

28. **(E)** The C horizon contains weathered parent material.

29. **(B)** The A horizon in a soil profile contains top soil, a blend of organic material and the sand, silt, or clay that is also available in the surrounding environment. The top soil is loosely held and allows water to seep into it and roots to push through.

30. **(A)** The O horizon is the organically rich surface material that contains partially decomposed plants, animals, bacteria, and fungi. Some areas will have more O layers than others; some areas have no O layer at all (a desert, for example).

31. **(D)** The B horizon makes up the hard-packed subsoil. In many cases, the B horizon is so densely packed that it might resemble a sedimentary rock.

32. **(E)** The leapfrog effect is responsible for transporting pollutants from warmer climates to arctic regions. The leapfrog effect depends on the fact that pollutants will more easily vaporize at warmer temperatures than they do at cooler temperatures which, combined with the convection cells being able to transport material from one spot on the globe to another, results in the overall deposition of toxins at high latitudes.

33. **(A)** El Niño/Southern Oscillation (ENSO) results in periodic weather disturbances in North America because it brings warm, moisture-laden air toward the western coast of South America. The Coriolis effect sweeps that air toward the north, where it comes in contact with the cooler circumpolar vortex. The collision of the warm, moist air with the cooler air mass results in wetter weather.

34. **(B)** The greenhouse effect results when certain gases (greenhouse gases, such as carbon dioxide and methane) absorb the weaker, long-wavelength light that has been re-radiated by the Earth. The Earth re-radiates this energy because, upon absorbing the incoming energy from the sun, it is warmed. The increase in energy creates thermal motion, which itself radiates the low-frequency radiation back out into space. It is this re-radiated energy that is captured by greenhouse gases.

35. **(C)** Ozone depletion is caused by anthropogenic CFCs and other gases migrating to the stratosphere, then catalyzing a reaction that converts ozone into oxygen gas. Ordinarily, ozone in the stratosphere provides a protective blanket around the Earth that filters out much of the damaging UV light. Without the ozone layer, the damaging UV light penetrates the atmosphere and reaches the surface. The first ozone hole was discovered as a cycling lobe over Antarctica. Periodically, a lobe of ozone hole would pass over nearby latitudes—including New Zealand.

36. **(D)** The Milankovitch cycle refers to a wobble in the axis of the Earth's orbit—like a top spinning quickly, but also experiencing a periodic shift in the direction of its axis of rotation. A shift in the Earth's axis would result in some variation of seasons, since it is the angle of the Earth that determines the location and magnitude of the seasons. The Milankovitch cycles are thought to be responsible for major climatic shifts (including ice ages) that occur every 100,000 years or so.

Practice Exam 1 – Section I: Detailed Explanations of Answers

37. **(A)** Primary productivity can be measured in a number of ways, as long as it refers to the conversion of solar energy into biomass through photosynthesis.

38. **(B)** Secondary productivity can also be measured in a number of ways as long as it refers to the level of productivity that uses biomass produced through photosynthetic activity. That would include the total biomass of herbivores.

39. **(C)** Total productivity of an ecosystem refers to the productivity of all trophic levels in an ecosystem. One way to measure productivity is by measuring biomass, so measuring the total biomass is one way to measure total biological productivity.

40. **(E)** Penguins space themselves evenly when a population inhabits a shoreline because they are establishing territory with which to protect and raise chicks. With little territory, they may not be able to successfully raise the next generation, so they compete among themselves (intraspecific competition) for space that will help assure the survival of the young.

41. **(A)** Although lichens may look like moss, they are not really plants at all. Actually, they are a group of algal cells in a mutualistic relationship with a type of fungus.

42. **(A)** Flowers produce nectar to attract bees and hummingbirds, so that those animal species can aid the plant in spreading its pollen. The animal benefits from a rich source of calories. It is a mutualistic relationship.

43. **(D)** Broad leaves in a rainforest are an adaptation to compete with other species for sunlight, which is a limited resource in the dark zone underneath the forest canopy. Since the competition for sunlight is between different species, the broad leaves represent a form of interspecific competition.

44. **(A)** Fertility refers to the number of children that a woman gives birth to during her lifetime.

45. **(D)** The number of births in a year per thousand people is defined as the birth rate, or crude birth rate, for that population.

46. **(C)** A demographic transition represents a drop in both death and birth rates, and results in slower population growth. Usually a demographic transition accompanies the development of a culture or country.

47. **(B)** Fecundity is the ability to reproduce. Unfortunately, idiomatic use of the term *fertility* has come to also express the ability of someone to reproduce, but in fact fecundity is the correct term.

48. **(B)** The Kyoto Protocol was an agreement between 150 nations to reduce the production of gases that cause the greenhouse effect.

49. **(B)** Photosynthetic algae are the producers in the options listed, and would produce the greatest amount of biomass. The other options depend on one producer or another, and would not represent as much biomass as a producer.

50. **(D)** Sexual reproduction serves the purpose of combining genes from male and female, in a manner that increases the mixing of traits and increases the chance of survival for an organism.

51. **(B)** Bhopal is a city in India in which a chemical explosion at a Union Carbide plant created a toxic, deadly cloud that passed through the slums of the city.

52. **(B)** Limestone provides the calcium (Ca) needed to combine with sulfur dioxide to form calcium sulfite, calcium sulfate, or gypsum in a fluidized bed combustion. This scheme captures sulfur dioxide from possible emission as air pollution and converts it to a solid waste.

53. **(E)** Water with calcium ions dissolved in it is considered "hard water" and causes calcium encrustation of pipes and appliances. The calcium can be removed by passing the hard water through a water softener, which exchanges the dissolved calcium ions with dissolved sodium ions. There is less calcium in the water and the appliances are saved, but it is not a good source of drinking water for people who are on a low sodium diet.

54. **(E)** The desecration of Easter Island was an example of a population exceeding the carrying capacity of the island. Once the carrying capacity of the island was exceeded, the population depleted needed natural resources and then died back. The current population lives within a new, lower carrying capacity.

55. **(C)** A solar heating device that requires a pump needs energy in order to function and is considered "active" solar heating. "Active" solar heating requires an input of electricity and involves moving parts.

56. **(C)** Increased urbanization will increase crime and pollution, decrease economic attractiveness of rural jobs and property, and increase the specialization of jobs in the urban setting.

57. **(A)** Geothermal energy production is one sustainable method that can be used regardless of sun exposure. Newer geothermal heat exchangers can be used anywhere, even where no geothermal warm water is produced. All the other options depend on adequate sun exposure.

Practice Exam 1 – Section I: Detailed Explanations of Answers

58. **(A)** Most hurricane damage occurs from the storm surge—the swell of water that is pushed ahead of a hurricane and can cause water to come many miles inland in low-lying coastal areas.

59. **(C)** Ozone molecules block UV light in the stratosphere, but cause irritation to sensitive lung cells when they are produced in the troposphere and inhaled by humans.

60. **(C)** Ozone depletion occurs in the stratosphere, at very high altitudes.

61. **(D)** The direct exposure of equatorial latitudes to solar rays causes equatorial air to heat and rise. Due to the rising air, equatorial areas experience lower pressures. The warmed air rises to high altitudes and moves north or south. It descends when it collides with cooler air, and creates high pressure desert zones.

62. **(C)** Radon gas is the only carcinogen on this list; it is the second leading cause of lung cancer.

63. **(C)** The difference between the crude birth rate and the crude death rate represents the population growth rate per 1,000 people. This value can be converted to a percentage by multiplying by one-tenth, thereby making the growth per hundred (percent) rather than growth per thousand. Those calculations yield the answer 1.8%.

64. **(D)** GNP is based on overall money spent, not the amount of profit. An oil spill would actually increase the amount of money spent getting the oil out of the ground and increase GNP, even though there was less profit on the oil that was sold.

65. **(A)** This is similar to the "reduce-reuse-recycle" axiom; less overall energy will be spent—and pollution created—by reducing or reusing the toxic compounds than by storing or incinerating them. Both storage and incineration cause different environmental problems.

66. **(D)** Producing a large number of offspring is a growth strategy by r-strategist, or logistic, populations.

67. **(A)** Schistosomiasis is a parasite that lives in humans and snails. The slow-moving water allows the snail vectors to breed and spread the disease.

68. **(E)** Secondary succession occurs when one biological community—or combination of flora and fauna—outcompetes another for a given location. For example, when a beaver dies, the pond created by the beaver eventually fills in. Terrestrial grasses replace aquatic species, which are in turn replaced by small shrubs, and eventually the climax community.

69. **(B)** Of these answers, only one has a million of the denominator in the numerator. There are one million mg in a kg.

70. **(C)** Complexity is a measure of the number of overall biological relationships in a community, which can be measured by the number of each species in different niches at each trophic level.

71. **(A)** Reduce, reuse, recycle is the sequence that leads to the least overall use of energy, and the production of the least amount of solid waste. First, the waste producer should attempt to reduce waste production; then if that can't be done, reuse the waste that is produced; then when those two options are exhausted, recycle the waste that is eventually produced.

72. **(D)** An area that was formerly a toxic waste site but has remediated enough to live on is considered restored. A brownfield is an area of dubious toxicity, and represents a high liability to future owners.

73. **(E)** True cost analysis takes into account all costs to human, social, natural, and intellectual capital. That would include not only the cost of health care for anyone affected, but also the loss of wages, well-being, and joy.

74. **(B)** "Eco" refers to the home. *Eco*nomics is the "maintaining of the home"; *eco*logy is "the study of the home."

75. **(D)** Land resources represent natural capital. Natural capital can be further divided into known and proven resources.

76. **(E)** Recycling an aluminum can saves 95% of the energy that would ordinarily be needed to produce an aluminum can from scratch. Another way to phrase that would be that it takes the same amount of energy to produce 20 recycled cans as it does to produce one unrecycled can.

77. **(C)** High reproductive rates would be a trait of an r-strategist, which would have a much lower chance of being endangered. The other traits—migratory, competes with humans, high on the food chain, and low genetic diversity—all increase the risk of being endangered.

78. **(B)** Birds are higher on the biomass pyramid and are therefore more prone to biomagnification of fat-soluble toxins.

79. **(B)** Primary productivity refers to the productivity of photosynthesis, which could be measured by leaf mass. Insect excrement and oxygen demand are other measures of productivity, but not photosynthetic productivity.

Practice Exam 1 – Section I: Detailed Explanations of Answers

80. **(D)** Species preservation that comes from a single breeding pair has limited success because of the drastic drop in genetic diversity. Because of the founder effect, it is highly doubtful that a single breeding pair would contain the complement of traits needed to continue to adapt to environmental conditions over a long period of time. The resulting population from a single breeding pair will probably not form a "minimum viable population."

81. **(C)** The Endangered Species Act (ESA) identifies species as rare or endangered and protects their habitat.

82. **(E)** Goiter occurs when the thyroid gland cannot produce enough thyroid hormone to meet the body's needs and compensates by enlarging. This can be a result of a lack of iodine, as iodine is vital to the formation of thyroid hormone.

83. **(B)** The hypolimnion is the portion of the lake that is closest to the bottom. The limnion is the portion exposed to the sun; the mesolimnion is the portion in the middle that usually contains the thermocline; and the littoral zone is that portion closest to the shore. The pelagic zone exists in the open marine environment.

84. **(E)** Current practices in home construction depend heavily on energy, use considerable packaging, and introduce many toxic substances into the new home.

85. **(D)** A temperature inversion exists when a layer of warm air traps cool air near the ground. Inversions are created in a number of different ways. One way inversions are created is when cool evening air slips down over a mountain or hillside and displaces the warm air next to the Earth. Then the cool air remains trapped by surrounding terrain and capped by the blanket of warm air above.

86. **(A)** Ice and snow contain by far the largest amount of fresh water (2% of total global water versus 0.28% for groundwater—which is next closest in the ranking); however, the water trapped in ice and snow is not available for use. Groundwater provides the largest usable source of freshwater.

87. **(C)** Saltwater intrusion occurs when the freshwater side of the groundwater has been heavily used and is being replaced by encroaching salt water.

88. **(B)** pH is defined as the negative of the logarithm of the molar concentration of hydrogen ions. Put a different way, the pH of a solution is the negative of the exponent of the H^+ concentration. A pH of 3 corresponds to a hydrogen ion (acid) concentration of 10^{-3} moles acid per liter of solution. pH 4 corresponds to 10^{-4}; pH 5 corresponds to 10^{-5}, and so forth.

89. **(A)** The Earth re-radiates energy because, upon absorbing the incoming energy from the sun, it is warmed. The increase in energy creates thermal motion, which itself radiates the low-frequency radiation back out into space. It is this re-radiated energy that is captured by greenhouse gases.

90. **(C)** The poles are cooler than the equator because the sun strikes the poles at a steeper angle, like a shadow covering more ground when the sun is low. As a result the incident energy from the sunlight is spread over a larger surface area than it is at the equator, and less energy is absorbed per unit area.

91. **(A)** Resource portioning will continue to catalyze speciation and divergent evolution—and hence new species—as long as there are available resources. Competition alone is not enough to prompt speciation; resources must also be available. The presence of sun and/or water alone is not enough; competition and the right nutrients must also be available for speciation to be promoted (consider a desert, for example).

92. **(D)** A close approximation called the "rule of 72" may be used to relate percent growth with doubling time, where doubling time × percent growth = 72. With an 8% annual growth, the doubling time would be 72/8, or 9 years. Therefore, in order for the population to double again, or quadruple, it would have to undergo two doubling times, or 18 years.

93. **(A)** Flocculation is a process that uses the insoluble trait of aluminum sulfate to attract other insoluble ions, as well as suspended particles. When the suspended solids are drawn toward the aluminum sulfate, they may then be filtered off to result with a cleaner product.

94. **(D)** The post-war eras only increased population levels. The demographic transition and the depression slowed growth, but did not reverse it. Only the great flu of 1918 decreased the population.

95. **(A)** Chemical synergism refers to the situation where the combined effect of two chemicals together is greater than the sum of the two chemicals experienced separately. This increases the toxic effect of the chemicals; all other options listed work to decrease the toxic effect of chemicals.

96. **(A)** 2, 4, 6, 8, now this is something to appreciate. Each of these numbers increases by a steady, fixed amount (+2), which is a trait of arithmetic growth. In the other options, the numbers were increasing by amounts that were, themselves, also increasing—a trait of geometric and logarithmic growth.

97. **(B)** Carbon filters work to improve the taste of water by using the free electrons in the carbon atom to cling onto impurities that may cause the water to taste bad.

Practice Exam 1 – Section I: Detailed Explanations of Answers

98. **(E)** Freshwater is a precious resource that is in dwindling supply. In fact, all these options use freshwater. A single copy of the Sunday *New York Times* requires about 280 gallons of freshwater to go from tree to doorstep. Likewise, the production of packaging requires much more water than we would imagine.

99. **(B)** Watts are actually a unit of power, not energy. Power × time (kW-hour, for example) are units of energy because power equals the amount of energy exerted in a given amount of time. Joules are the metric units of energy; BTUs are the British units of energy. Kg × m^2/sec^2 are the fundamental units of joules, and are therefore also units of energy.

100. **(B)** Coal is produced by the steady, long-term compression of organic material (usually from plants). Peat is the first stage, moving to soft lignite coal, then bituminous, ending with highly compressed and very hard anthracite coal.

AP Environmental Science

Detailed Explanations of Answers

Practice Exam 1

Section II

1. (10 points)

    a. (3 points; +1 for answer, +1 for set up, +1 for correct units)

    $$\frac{15,000 \text{ MW-hr}}{\text{day}} \times \frac{1,000 \text{ kW}}{\text{MW}} \times \frac{1 \text{ ton coal}}{2,000 \text{ kW-hr}} = 7,500 \text{ tons coal}$$

    b. (2 points)

    Fission is the predominant nuclear reaction that provides energy inside a nuclear power plant.

    c. (2 points)

    Control rods absorb neutrons, so that the fission reactions do not get out of control. Fission reactions release many neutrons, which then go and catalyze further fission reactions. Absorbing neutrons is a way to be sure that too many fission reactions are not going to melt down the core of the reactor.

    d. (3 points; +1 for each point made, up to three)

    - Mining of the ore creates radioactive piles of tailings, scars the land, creates opportunity for exposure for miners, creates airborne particulate matter.

    - Milling creates waste out of 99% of the material that comes out of the mine. Machinery in the milling process becomes radioactive.

    - Processing creates some amount of waste.

    - Energy production at the reactor is somewhat clean, as long as the reactor is well operated and no accidents happen. However,

Practice Exam 1 – Section II: Detailed Explanations of Answers

the actual splitting of atoms in the core by neutrons creates further radioactive elements, which will eventually need to be disposed of.

- Cement production to build a reactor creates particulate matter and scars the landscape to obtain sand and gravel.

- Once any portion of the process, most particularly the fuel cells, are spent, it becomes either high-level or low-level nuclear waste. This needs to be stored in a safe place for many thousands of years. Creating such a site for storage requires a massive construction project that needs cement—which scars the land to obtain sand and gravel—and a natural feature (a mountain) that can be dedicated to storing the waste.

- Transportation between each of the steps in the fuel cycle requires energy, which creates waste that depends on the energy used. For example, much of the materials will travel by train. Trains run on diesel fuel, which is obtained from crude oil, which has other impacts on the environment.

2. (10 points)

   a. (2 points)

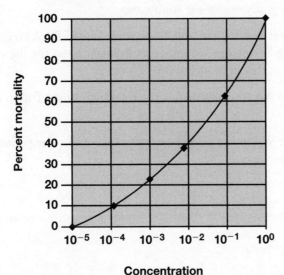

   b. (3 points)

   +1 point: Toxin B is more toxic to daphnia. (The lower LD_{50} means that less toxin is needed to kill 50% of the population.)

   +2 points possible; +1 point for each

AP Environmental Science

- Data is able to give a general comparison of toxicity
- Because of the differences in physiology between daphnia and humans, the data cannot be relied upon to absolutely determine toxicity levels for humans. Every species demonstrates different tolerance levels to toxins.

    c.   (3 points; +1 for explanation of biomagnification, +1 for mentioning accumulation, +1 for identifying the fat soluble toxin)

The fat soluble toxin, toxin A, presents the greatest potential for biomagnification in the environment. Fat soluble toxins can accumulate in cells, then be retained until that cell is digested by a predator. When many cells are consumed by the predator, it will also retain the fat soluble toxin and end up with a higher dose than was experienced by its prey. In this way, fat soluble toxins accumulate within individuals, and magnify in tissues as they pass from one trophic level to the next.

    d.   (2 points)

+1 for:

Tolerance limits establish the comfort zone and the range of survivability that any organism has to a particular toxin or abiotic factor.

+1 for mentioning any one of the following:

- The Federal Food, Drug and Cosmetics Act (FFDCA) requires knowledge of tolerance limits that humans have for the most important toxins in food and drugs.
- The Clean Air Act establishes the tolerance limits for air pollutants.
- The Clean Water Act establishes tolerance limits for pollutants in our waterways.

3. (10 points)

    a.   (3 points)

+1

Tropical rainforests are warm and moist, with little temperature fluctuation between day and night and during the different seasons.

+1

Deserts are hot and dry, with wider extremes between day and night temperatures.

+1

Tropical rainforests occur near the equator, where sunlight is direct and ascending air allows high humidity. Deserts tend to occur at about 30º latitude, where descending air is hot and arid.

b. (2 points)

+1 for any one of the following desert adaptations, all of which focus on water retention

- Small leaves minimize water lost to transpiration.
- Waxy leaves retain water.
- Thorns protect the plant from animals that would eat the plant to obtain the stored water.

+1 for any one of the following rainforest adaptations, most of which have to do with the lack of available sunlight underneath the canopy

- Broad leaves outcompete other plants for limited sunlight.
- Broad, shallow roots allow trees to grow tall (to compete for sunlight) in shallow, infertile top soil. (The top soil is infertile because the intense competition for organic material results in the fact that it tends to be used immediately by some organism once it is available; it doesn't lay on the ground and decay, thereby contributing to soil formation.)

c. (2 points)

+1

Species and genetic diversity are not possible without sustaining the habitats in which those species thrive.

+1 for any one of the following reasons that biodiversity in general is important:

- Biodiversity is important for sustaining a wide variety of foods.
- Drugs and medicines come from exotic species in a wide range of ecosystems.
- There are ecological benefits to sustaining biodiversity; the global ecosystems are connected as one depends on the good health of the others.
- We enjoy aesthetic, scientific, educational, and cultural benefits from knowing about the different ecosystems and their relationship to surrounding cultures.

- Some would also say that biodiversity has an intrinsic value to exist for its own sake.

d. (3 points)

Scarce resources

- Soil nutrients for the rainforest
- Water for the desert

Resources that are not scarce

- Water for the rainforest
- Sunlight for the desert

4. (10 points)

a. (4 points, +1 for each of the following)

i. The average age of population A is far less than the average age of population B.

ii. The life expectancy of population A is less than the life expectancy of population B.

iii. There is fewer percentage of children in population A than in population B.

iv. Fertility is a measure of the number of births by women in the population, and is best described by the crude birth rate (births per 1,000 people per year). With this definition, population A is more fertile because a smaller number of adult women have a larger number of children.

b. (4 points; +1 for each stage)

Stage One: Both birth and death rates are high and fluctuate with the following possible reasons.

I. Little access to birth control

II. High infant mortality

III. Children are used as a "social security" to provide wealth for the parent generation.

IV. Cultural values encourage having children (pronatalist pressure) or discourage birth control.

> V. High death rates exist, especially among children. This may be due to poor medical support, infanticide, disease, famine, or poor hygiene.

Stage Two: Birth rates remain high, but death rates begin to fall rapidly, which causes rapid population growth. This may be caused by any of the following.

> I. Improved medical care, sanitation, or water quality
>
> II. Food production and distribution improves or increases
>
> III. Some decrease in infant mortality

Stage Three: Birth rates fall while death rates continue to fall. The total population growth rate becomes arithmetic (constant), rather than exponential. This may occur for any of the following reasons.

> I. Contraception is used more often.
>
> II. Parents sense less need to have large families and reset goals for family size.
>
> III. Wealth—or the perception of wealth—per family increases.
>
> IV. The attitude toward women changes in the population. A career path becomes an encouraged and acceptable alternative to finding self-fulfillment in having children.
>
> V. Education

Stage Four: Both birth and death rates remain low, resulting in a steady population size. The age structure diagram is rectangular and block-like. If the birth rate continues to drop below death rates, the age structure diagram will take on an "inverted pyramid" and the population size would actually begin to decrease. This is already happening in some European countries.

> c. (+1 for each of the following, up to 2 points maximum)
>
> - Increased wealth decreases the need to have children provide for parents in old age, or use children for labor
>
> - Increased technology improves food production, food distribution, and medical care. This in turn decreases death rates, which in turn decreases birth rates.
>
> - Increased opportunities and education for women create other avenues for fulfillment other than childbearing.

Practice Exam 2

Practice
Exam 2

AP Environmental Science

Practice Exam 2
This exam is also on CD-ROM in our TestWare®.

Section I

TIME: 90 minutes
100 multiple-choice questions

(Answer sheets appear in the back of this book.)

> **DIRECTIONS:** Each of the questions or incomplete statements below is followed by five suggested answers or completions. Select the best answer for each question and then fill in the corresponding oval on the answer sheet.

Questions 1–4 refer to the following answers. Select the one lettered choice that best fits each statement.

    (A)   Electrons

    (B)   Protons

    (C)   Neutrons

    (D)   Isotopes

    (E)   Ions

1. A positively charged nuclear particle

2. A negatively charge subatomic particle

3. An atom or group of atoms with an imbalance of electrons and protons

4. Two atoms with the same number of protons but a different number of neutrons

Questions 5–7 refer to the following answers. Select the one lettered choice that best fits each statement.

    (A)   Cells

    (B)   Communities

    (C)   Populations

    (D)   Species

    (E)   Ecosystems

5. The largest organization of matter listed here

6. Contains groups of different species

7. ATP is a major currency of energy inside this structure.

Questions 8–11 refer to the following answers. Select the one lettered choice that best fits each statement.

    (A)   Boreal forest

    (B)   Temperate rainforest

    (C)   Deciduous forest

    (D)   Tropical rainforest

    (E)   Kelp forest

8. Found in coastal waters

9. Northernmost forests near tundra

10. Highest level of precipitation of any biome

11. Climax community contains hardwood trees

Questions 12–15 refer to the following map. Select the one lettered choice that best fits each statement.

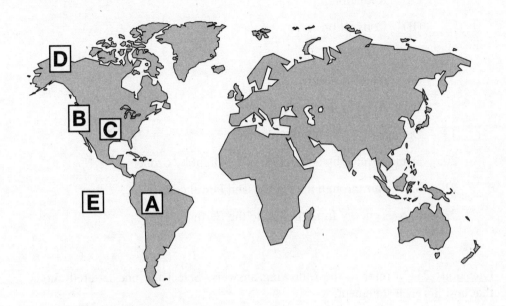

12. Area of the world that contains Taiga forests

13. Ogallala aquifer

14. Rain shadow exists on the eastern side of these mountains.

15. Tropical rainforests exist here.

Questions 16–18 refer to the following answers. Select the one lettered choice that best fits each statement.

(A) Nitrates

(B) Nitrites

(C) Ammonia

(D) Atmospheric nitrogen

(E) Nitrogen dioxide

16. Form of nitrogen excreted as waste by mammals

17. Responsible for decreasing dissolved oxygen in streams

18. Results from high temperature combustion of air

AP Environmental Science

Questions 19–22 refer to the following answers. Select the one lettered choice that best fits each statement.

    (A)   Radiation

    (B)   Convection

    (C)   Conduction

    (D)   All of the above

    (E)   None of the above

19. Demonstrated by mantle plumes

20. Demonstrated by absorption of UV light by ozone

21. Moves air through the Hadley and Ferrel cells

22. Carries energy from the sun to the Earth

Questions 23–26 refer to the following answers. Select the one lettered choice that best fits each statement.

    (A)   Divergent boundary

    (B)   Convergent boundary with oceanic and continental plates

    (C)   Convergent boundary with two oceanic plates

    (D)   Convergent boundary with two continental plates

    (E)   Transform boundary

23. The San Andreas Fault between the Pacific Plate and the North American Plate in California is an example.

24. The Himalayas were formed from this type of boundary.

25. This type of boundary created the Marianas Trench.

26. The mid-Atlantic rift forms this type of boundary.

Questions 27–30 refer to the following answers. Select the one lettered choice that best fits each statement.

(A) Clay
(B) Silt
(C) Sand
(D) Humus
(E) Loam

27. Portion of soil that contains fine particles that experience an ionic attraction to one another to make it impermeable to water
28. A mixture of equal amounts of sand, clay, silt, and humus
29. Organic material composed of partially decayed plants and animals
30. Coarse inorganic particles used in glass production

Questions 31–33 refer to the following answers. Select the one lettered choice that best fits each statement.

(A) Founder effect
(B) Minimum viable population
(C) Fragmentation
(D) Competition
(E) At least two of the above

31. Reduces genetic biodiversity
32. Prevented by establishing a large core of strict wilderness protection, ringed by steadily decreasing levels of protection
33. Caused when niches of two species overlap

Questions 34–37 refer to the following answers. Select the one lettered choice that best fits each statement.

    (A) Wilderness area

    (B) National forest

    (C) National park

    (D) BLM land

    (E) Wildlife refuge

34. Does not allow any mechanical device, tool, or vehicle

35. Allows multiple use to allow for the "greatest good for the greatest number of people"

36. Allows developers to lease land for mining

37. Preserves natural and historic features for future generations

Questions 38–41 refer to the following graph. Select the one lettered choice that best fits each statement.

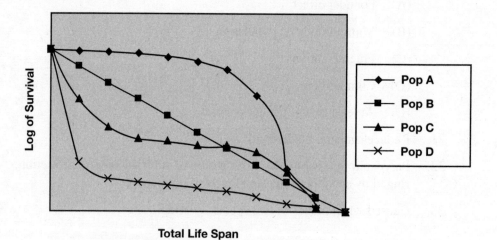

38. Which population demonstrates a culture that has undergone a demographic transition?

39. Which population demonstrates a constant rate of death throughout the life span of members of the population?

40. Which population shows the highest infant mortality?

41. Which population demonstrates the lowest infant mortality?

Questions 42–46 refer to the following answers. Select the one lettered choice that best fits each statement.

(A) Secchi disc

(B) Calcium ion concentration

(C) pH

(D) Conductivity

(E) Dissolved oxygen

42. Measure of acidity

43. Measure of water hardness

44. Measure of turbidity

45. Measure of salinity

46. Decreased by nutrient pollution

Questions 47–51 refer to the following answers. Select the one lettered choice that best fits each statement.

(A) Carbon monoxide

(B) Carbon dioxide

(C) Nitrogen dioxide

(D) Methane

(E) Radon

47. Produced by combustion of air

48. Is further oxidized by catalytic converter

49. An important indoor air pollutant

50. Produced by bacterial digestion of hydrocarbons in solid waste

51. An important precursor to acid rain

AP Environmental Science

52. This environmental crisis was made infamous by Carson's *Silent Spring*.

    (A) Love Canal

    (B) Bhopal

    (C) *Exxon Valdez*

    (D) DDT

    (E) Chernobyl

53. In this air pollution control device, fuel is first combusted in a high temperature, low oxygen environment.

    (A) Electrostatic precipitator

    (B) Fluidized bed combustion

    (C) Catalytic converter

    (D) Staged burner

    (E) Bag filters

54. This is caused by the biomagnification of mercury in fish.

    (A) Minimata disease

    (B) Eutrophication

    (C) Fecal coliform

    (D) Sediment

    (E) Calcium ions

55. Which of the following cycles is most responsible for transferring energy from one part of the Earth to another?

 (A) Carbon cycle

 (B) Nitrogen cycle

 (C) Phosphorous cycle

 (D) Sulfur cycle

 (E) Water cycle

56. This federal law requires government contractors to file an environmental impact statement.

 (A) NEPA

 (B) FFDCA

 (C) ESA

 (D) SMCRA

 (E) CERCLA

57. Which of the following pairs of variables is most instrumental in determining the identity of a biome?

 (A) Temperature and available moisture

 (B) Temperature and altitude

 (C) Altitude and latitude

 (D) Temperature and latitude

 (E) Available moisture and pH

Questions 58–60 refer to the following graph.

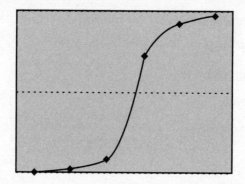

58. If the above is a dose-response curve, what might be the units of the *x* and *y* axes, respectively, and what might be calculated from the chart?

    (A) Number of organisms; time; population growth

    (B) Number of organisms; concentration; strength of toxin

    (C) Dose; percent of organisms dead; safety range of drug or toxin

    (D) Concentration; number of organisms dead; LD_{50}

    (E) Dose; population size; disease frequency

59. If the above is a population growth curve, what type of population would this depict?

    (A) A population that has undergone overshoot and die-back

    (B) A population that demonstrates a Malthusian growth strategy

    (C) A population that demonstrates a logistic growth strategy

    (D) A population with extrinsic regulation of growth

    (E) A population that tends to produce many offspring and not invest many resources in each individual

60. If the curve represents a population growth curve, which of the following principles is NOT demonstrated by this particular curve?

    (A) Overshoot

    (B) Log phase

    (C) Replication, or lag phase

    (D) The effect of environmental resistance

    (E) Carrying capacity

61. Which of the following types of toxic compounds is most often responsible for causing the growth of cancerous cells?

    (A) Teratogen

    (B) Mutagen

    (C) Irritant

    (D) Allergen

    (E) Oxidation agent

62. From which living kingdom come bacterial pathogens?

    (A) Animalia

    (B) Plantae

    (C) Protista

    (D) Monera

    (E) Fungi

63. The major source of energy for deep-sea producers is

    (A) photosynthesis

    (B) chemosynthesis

    (C) either photosynthesis or chemosynthesis

    (D) decomposing debris that falls from above

    (E) warmth from solar energy

64. The government agency that oversees the management of national forests is the

    (A) Department of the Interior
    (B) Department of Agriculture
    (C) Department of Health and Safety
    (D) Department of Justice
    (E) Environmental Protection Agency

65. A poor country that is having a hard time providing health services to its population, and where the standard of living is quite low, is faced with an offer by a developed country to receive payment in order to harbor solid waste in landfills. Which of the following will probably drive the poor country's decision to accept the waste?

    (A) Environmental economics
    (B) Marxian economics
    (C) Survival economics
    (D) Socialist economics
    (E) Capitalism

66. Which of the following does NOT usually accompany a thunderstorm?

    (A) Updrafts
    (B) Low pressure zone
    (C) Cumulonimbus clouds
    (D) Warm front
    (E) Precipitation

67. Which of the following accompanies an El Niño event?

    (A) Movement of water from west to east across the equatorial Pacific

    (B) Movement of water from east to west across the equatorial Pacific

    (C) Movement of warm water from north to south in the central Pacific

    (D) An increase in precipitation in Indonesian rainforests

    (E) An increase in ozone depletion

68. Ozone depleting gases are particularly damaging because

    (A) ozone depleting molecules are also toxic

    (B) so many different types of molecules deplete ozone

    (C) they act as catalysts; a few molecules can do a lot of damage

    (D) they don't work at lower altitudes

    (E) most of them also cause the greenhouse effect

69. Which of the following types of molecules is NOT included in commonly used pesticides?

    (A) Carbamates

    (B) Organophosphates

    (C) Heavy metals

    (D) Chlorinated hydrocarbons

    (E) Dioxins

70. A "debt-for-nature-swap" refers to

    (A) charging corporations money for damaging the environment

    (B) providing low interest loans to developers who also protect endangered habitats

    (C) charging taxes to polluters, which then go to pay off the public debt

    (D) forgiving Third World debt in exchange for preserving endangered habitat

    (E) forcing polluting countries to loan money to tribal leaders in order to protect endangered habitat

AP Environmental Science

71. The mobility of a toxin moving through the environment is determined by

    (A) the toxin's attraction to water and soil particles

    (B) the retention of the toxin by living organisms

    (C) the toxicity of the toxic compound

    (D) the ability of the toxin to biomagnify

    (E) At least two of the above

72. Which of the following is NOT a form of energy?

    (A) Heat

    (B) Sunlight

    (C) Work performed

    (D) Electric power

    (E) All of the above are different forms of energy.

73. How much energy is used by a 100-watt light bulb that shines for 100 seconds?

    (A) 1.0 joule

    (B) 1.0 watt

    (C) 1,000 joules

    (D) 10,000 joules

    (E) 10,000 watts

74. Genetically modified organisms (GMOs) represent an environmental concern because

    (A) they give plants a bacterial infection

    (B) less pesticide will be used on GMO crops

    (C) farmers must rebuy seed each year

    (D) GMOs infect weeds with superior survival traits

    (E) GMO crops provide too much profit for large corporations

75. Which of the following is not a practice associated with sustainable agriculture?

    (A) Low-till farming

    (B) Crop rotation

    (C) Biological control of pests

    (D) Perennial polyculture

    (E) Monoculture

76. To eat beef rather than grain, a consumer is using _____ as much total energy to gain the calories they need.

    (A) sixteen times

    (B) fifty times

    (C) twice

    (D) one-fourth

    (E) one-sixteenth

77. Most of the oxygen in the early atmosphere was originally created by

    (A) fungi

    (B) plants

    (C) bacteria

    (D) volcanoes

    (E) None of the above

78. Which of the following gases occupies the greatest percentage of the current atmosphere on Earth?

    (A) Nitrogen gas

    (B) Oxygen gas

    (C) Carbon dioxide gas

    (D) Ozone

    (E) Water vapor

79. A stream has a pH of 6.0. This stream is
    (A) strongly acidic
    (B) weakly acidic
    (C) neutral
    (D) weakly basic
    (E) strongly basic

80. A population demonstrates Malthusian growth and tends to overshoot a carrying capacity. Which of the following are probably traits of this population?
    (A) Individuals are probably small and do not care for their young.
    (B) Individuals are probably large K-strategists.
    (C) The species is probably a climax species in an ecosystem.
    (D) Individuals care for their young and have few offspring.
    (E) Individuals are large; each individual uses a lot of resources.

81. Which of the following are density-dependent factors that affect population size?
    (A) Food availability
    (B) Disease
    (C) Weather
    (D) Two or more of the above
    (E) None of the above

82. Which of the following is another term for Malthusian growth?
    (A) K-strategist
    (B) r-strategist
    (C) Logistic growth
    (D) Irruptive growth
    (E) Two or more of the above

83. The greatest factor that contributes to whether or not a toxin will magnify in the environment is

    (A) solubility
    (B) persistence
    (C) both solubility and persistence
    (D) neither solubility nor persistence
    (E) persistence and one other variable

84. Toxins can be broken down in the atmosphere by

    (A) oxidation
    (B) photolysis
    (C) reaction with water
    (D) Toxins are never broken down in the atmosphere
    (E) Two or more of the above

85. Microbes best remediate toxins in which medium?

    (A) Air and soil
    (B) Water and soil
    (C) Air and water
    (D) Only water
    (E) Only soil

86. Which of the following types of waste occupies the greatest percentage of the residential waste stream?

    (A) Food
    (B) Glass
    (C) Paper
    (D) Metal
    (E) Plastic

87. Which of the following represents the largest drawback of solid waste incineration?

    (A) Less material to recycle

    (B) Production of dioxins

    (C) Finding a place to put the ash

    (D) Water pollution

    (E) Running out of dump space

88. Which of the following elements does NOT need to be added to make a healthy compost pile?

    (A) Carbon

    (B) Oxygen

    (C) Phosphorous

    (D) Nitrogen

    (E) Water

89. Which of the following is NOT a pathogen?

    (A) Botulism bacteria

    (B) Cold virus

    (C) Malaria mosquito

    (D) Athlete's foot fungus

    (E) Protist that causes giardia

90. Which of the following choices refers to the shaded portion of the following graph?

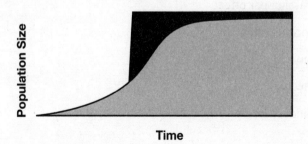

    (A) The biotic potential
    (B) The carrying capacity
    (C) The difference in population size between actual population and carrying capacity
    (D) Logistic growth
    (E) The number of organisms that could live under ideal conditions

91. Channelization of a waterway improves _____ but it reduces _____.

    (A) navigation… hydroelectric power production capability
    (B) biodiversity… hydroelectric power production capability
    (C) navigation… abundance of wetlands
    (D) biodiversity… siltation
    (E) There are no drawbacks to channelization.

92. Which of the following is an example of water pollution that could potentially experience biomagnification?

    (A) Salinization
    (B) Selenium
    (C) Hard water
    (D) Nutrients
    (E) Fat soluble pesticides

93. When arctic caribou begin to retain industrial toxins in their fat cells, the toxins probably reached the caribou because of

    (A) the grasshopper effect
    (B) the Coriolis effect
    (C) el Niño
    (D) plate tectonics
    (E) local industries

94. The California Water Project and the Mono Lake water diversion program is most like the Aral Sea incident in Asia because

    (A) wetlands were formed from diverted water
    (B) channelization allowed for better navigation
    (C) both projects involve large hydroelectric projects
    (D) both projects involve conserving water for later generations
    (E) diverted water left a more highly saline lake

95. Which of the following would NOT be a good candidate for composting?

    (A) Hair
    (B) Peanut butter
    (C) Grass clippings that are free of fertilizers and pesticides
    (D) Dead leaves
    (E) Horse manure

96. In the "Tragedy of the Commons," the real tragedy involves

    (A) the loss of profit
    (B) the loss of cattle
    (C) the loss of a common pasture
    (D) the loss of personal choice
    (E) the loss of technology

97. Recycling aluminum helps the environment most because

    (A) only five percent of the energy is required
    (B) it creates fewer bauxite mines
    (C) it decreases the production of poisonous gases during processing
    (D) it costs less to soft drink bottling companies
    (E) it makes landfills smaller

98. Heap leach extraction is the process that is used to refine which of the following from ore?

    (A) Uranium
    (B) Gold
    (C) Aluminum
    (D) Halite
    (E) Gravel

99. What percentage of refined uranium typically results from the milling process?

    (A) Less than 1%
    (B) 5–10%
    (C) About 20%
    (D) About 50%
    (E) Over 80%

100. Environmentally sustainable economics and ethics are both marked by

    (A) political activism
    (B) the idea that resources should never be used for profit if they are obtained at the cost of endangered habitat
    (C) survival economics
    (D) consumers of resources paying for indirect costs, rather than having those costs shouldered unjustly by marginalized people
    (E) consumerism

AP Environmental Science

Practice Exam 2

Section II

Time: 90 minutes

DIRECTIONS: You have 90 minutes to answer all four of the following questions. It is suggested that you spend approximately half your time on the first question and divide the remaining time equally among the next three. In answering these questions, you should emphasize the line of reasoning that generated your results; it is not enough to list the results of your analysis. Include correctly labeled diagrams, if useful or required, in explaining your answers. A correctly labeled diagram must have all axes and curves labeled and must show directional changes.

---

1. A major coal-fired electrical power plant uses 4,500 tons of coal each day. Each pound of coal can produce 5,000 BTUs of electrical energy (3,400 BTUs are equivalent to 1.0 kW-hr of energy).

   a. How many kW-hr of electrical energy are produced by the plant each day?

   b. Beginning with the coal, identify the energy transformations that take place in order to produce electricity in the power plant.

   c. Assuming that each energy transformation is 80% efficient, calculate the total efficiency of the power plant.

   d. Two-staged burners are used to prevent the formation of NO_x in the power plant. The formation of about 0.75 lb. of NO_x per million BTUs is prevented by the burners.

i. Describe how two-staged burners prevent the formation of NO_x.

ii. How many pounds of NO_x are prevented per day by the operation of the two-staged burners?

2. Use the data table to answer the following questions.

Measurement	Location A	Location B
DO (ppm)	10.0	3.0
Temperature (ºC)	18.0	27.4
pH	7.5	6.0
Nitrates (ppm)	2.0	6.0
Diversity (#)	16.0	7.0
Abundance (#)	525.0	24.0

a. Which point on the stream, A or B, seems to demonstrate the greatest ecological health? Use the data to give a solid rationale for your answer.

b. Which of the above measurements represent "biotic" factors in the stream environment?

c. What device would be used to measure dissolved solids? What device would be used to measure undissolved solids?

d. Make a hypothesis about what may have caused the change in stream health between point A and point B. Support your hypothesis with data from the chart.

3. a. Identify four human activities that put rare species at risk of being endangered.

    b. Identify four traits of species that predispose them to becoming endangered.

    c. Identify an international treaty or federal law that helps protect endangered species; describe how the treaty or law works to accomplish its goal.

    d. In a brief paragraph, compare the effectiveness of a captive breeding program with a habitat protection program in being able to protect an endangered species.

4. Answer the following questions in response to this excerpt, which comes from the newsletter published by a utility company in the northwest.

    ### Watt Addresses Community About Dam Breaching Proposal

    Earlier this month, Ms. Ima Watt, CEO of the utility Upper Snake River Electric (USRE), addressed a public hearing about breaching three Snake River dams.

    Ms. Watt stated that she is very interested in meeting the needs and wishes of the consumers in the Snake River region. "Our decision should be based upon good science. We have all benefited from inexpensive power along the Snake River. We need to be sure that we make the right decision," states Watt.

    Electricity is also sold to Seattle-based industries—particularly the aluminum smelting industry. "Dam breaching," says Watt, "will decrease the revenue from the power that is sold to the aluminum companies, and it will increase farming costs among the potato growers in the region, which ultimately increases the price of food."

    Watt also expressed a concern about maintaining the salmon population in the upper Snake River. "Many people depend on the revenue from the fishing business, and don't we all love to go fishing with our families now and then?" Already, fish ladders allow salmon to live and breed in Snake River reservoirs.

    While many spoke to the need to allow the river to flow freely, Watt pointed out that lowering the level of the reservoirs as a result of breaching a dam would expose the silt along the shoreline, which would create a new air pollution problem.

"USRE is all about helping to protect the environment that we all share," affirmed Watt.

"When one analyzes all the costs and benefits, it just doesn't make sense [to breach the dams]," concluded Watt. Watt received an ovation upon completing her remarks.

a. Ms. Watt felt that the decision to breach the dams should be made based on "good science." Identify three reasons Watt used to not breach the dams.

b. Describe two ecological benefits gained from dam breaching that are not mentioned by the article.

c. While Ms. Watt implied that revenues in the region may decline as a result of dam breaching, describe how breaching the dam might actually contribute to increasing the GDP or GNP.

d. Watt implied that the utility had undergone a cost-benefit analysis regarding the breaching of dams. Identify three external costs for hydroelectric power production along the Snake River.

AP Environmental Science

Practice Exam 2

Answer Key

Section I

1.	(B)	26.	(A)	51.	(C)	76.	(A)
2.	(A)	27.	(A)	52.	(D)	77.	(C)
3.	(E)	28.	(E)	53.	(D)	78.	(A)
4.	(D)	29.	(D)	54.	(A)	79.	(B)
5.	(E)	30.	(C)	55.	(E)	80.	(A)
6.	(B)	31.	(E)	56.	(A)	81.	(D)
7.	(A)	32.	(C)	57.	(A)	82.	(D)
8.	(E)	33.	(D)	58.	(D)	83.	(A)
9.	(A)	34.	(A)	59.	(C)	84.	(E)
10.	(D)	35.	(B)	60.	(A)	85.	(B)
11.	(C)	36.	(D)	61.	(B)	86.	(C)
12.	(D)	37.	(C)	62.	(D)	87.	(B)
13.	(C)	38.	(A)	63.	(B)	88.	(C)
14.	(B)	39.	(B)	64.	(B)	89.	(C)
15.	(A)	40.	(D)	65.	(C)	90.	(A)
16.	(C)	41.	(A)	66.	(D)	91.	(C)
17.	(A)	42.	(C)	67.	(A)	92.	(E)
18.	(E)	43.	(B)	68.	(C)	93.	(A)
19.	(B)	44.	(A)	69.	(E)	94.	(E)
20.	(C)	45.	(D)	70.	(D)	95.	(B)
21.	(B)	46.	(E)	71.	(A)	96.	(D)
22.	(A)	47.	(C)	72.	(D)	97.	(A)
23.	(E)	48.	(A)	73.	(D)	98.	(B)
24.	(D)	49.	(E)	74.	(D)	99.	(A)
25.	(C)	50.	(D)	75.	(E)	100.	(D)

Detailed Explanations of Answers

Practice Exam 2

Section I

1. **(B)** Two major types of particles make up the nucleus of an atom: protons and neutrons. Of those two, only the proton carries an electrical charge—which is positive.

2. **(A)** The negatively charged particle in the atom does not exist in the nucleus, like the proton and neutron; it exists in clouds outside the nucleus and is called the electron.

3. **(E)** If there is an imbalance of electrons and protons, that means that there is an imbalance of charges, and the atom or group of atoms makes up an ion.

4. **(D)** Isotopes have the same number of protons, but a different number of neutrons. For example, carbon has six protons. By definition, anything with six protons in the nucleus is carbon; the number of protons determines the identity and the chemical reactivity of the element. However, it is possible to have carbon with six neutrons (carbon-12) or carbon with eight neutrons (carbon-14). Carbon-12 is an isotope of carbon-14.

5. **(E)** The organization of matter ranges from subatomic particles to galaxies, but the largest level of organization on this list is an ecosystem. Ecosystems contain all of the other levels of organization, but not vice versa.

6. **(B)** Groups of different species—or different populations—are called communities.

7. **(A)** ATP is adenosine triphosphate, which is an important currency of energy for the cell.

8. **(E)** Kelp forests use brown algae rather than trees to adorn the undersea landscape in coastal waters.

9. **(A)** Boreal forests, or Tiaga forests, represent the northern most forest type. Boreal forests tend to border on both arctic and alpine tundra.

10. **(D)** Tropical rainforests receive the highest volume of rainfall of any biome.

11. **(C)** Another name for hardwood trees is deciduous.

12. **(D)** The Tiaga, or boreal forests, represent the northernmost type of forest.

13. **(C)** The Ogallala aquifer lies underneath about 11 Midwestern states. It is a major source of water for residential, industrial, and agricultural use. It took millions of years to allow the water to accumulate, but we are removing it much faster than it is being replaced.

14. **(B)** The rain shadow exists on the leeward side of mountain ranges. The eastern slopes of the Cascades, Rockies, and Sierras are all much drier and warmer than the windward western slopes. In order to rise over the mountains, the air must decrease in pressure, which decreases the amount of moisture that can remain in the vapor phase. As a result, the air releases moisture as rain or snow on the windward side. After the air crosses over the summit, it descends and compresses, causing the now-drier air to become warmer.

15. **(A)** Tropical rainforests occur in equatorial zones in South America and Indonesia, in particular.

16. **(C)** Animals secrete nitrogen in the form of ammonia, which then becomes oxidized by nitrifying bacteria in the nitrogen cycle.

17. **(A)** Nitrates contribute to nutrient pollution, which increases algae production, which is then decomposed by decomposers that undergo respiration. Organisms undergoing respiration require oxygen, so the end result of adding nutrients is a drop in dissolved oxygen in the stream.

18. **(E)** One of the major hurdles to global clean air is solving the dilemma of having a combustion reaction that does not reach high enough temperatures to convert atmospheric nitrogen into nitrogen dioxide. The nitrogen dioxide then combines with water to form acid rain, or can be converted to photochemical oxidants by the sun.

19. **(B)** Mantle plumes are a form of convection, as magma that is heated by nuclear reactions at the center of the Earth rises and eventually encounters the cooler crust.

Practice Exam 2 – Section I: Detailed Explanations of Answers

20. **(C)** When UV light is absorbed by ozone, the ozone molecule absorbs—or conducts—the light energy so that it does not pass through to the Earth.

21. **(B)** The vertical cycles of air that are created by atmospheric convection include the Hadley and Ferrel cells.

22. **(A)** The energy of the sun comes through space via electromagnetic radiation, carried by photons of light.

23. **(E)** The San Andreas fault is the spot where the Pacific plate moves north along the edge of the North American plate. This lateral movement of one plate past another is called a transform boundary.

24. **(D)** The Himalayas were formed when the Indian subcontinent and the Asian continental plates collided. The lighter Indian plate slid underneath the Asian plate and caused the elevated Tibetan plateau. The edges of the plates bunched up to form the Himalayan mountains.

25. **(C)** The Marianas Trench was created when two oceanic plates collided and a subduction zone was created.

26. **(A)** The mid-Atlantic rift is where new plate material is being formed, or a constructive boundary. This is also called a divergent boundary because the plates are moving away from each other as new plate material is formed.

27. **(A)** Clay is made of very fine particles. It is impermeable to water partially because of the mobile electrons on the finely divided particles. This ionic nature makes the clay sticky and holds it together.

28. **(E)** Loam is a mixture of sand, clay, silt, and humus. Different types of loam exist with different relative percentages of each of these components.

29. **(D)** Humus is organic material that contains partially decayed plants and animals.

30. **(C)** Sand is the most coarse soil particle listed.

31. **(E)** Both the Founder effect and fragmentation decrease genetic biodiversity. The Founder effect occurs when a single breeding pair becomes isolated and starts a new population. However, the gene frequencies in that population are limited by the genes that were available from the ancestral pair. As a result, the population has lower genetic biodiversity. Fragmentation also occurs when breeding pairs become isolated from one another. As with the Founder effect, the genetic diversity of future generations will then be limited by the ancestral pair.

32. **(C)** Fragmentation of populations occurs when the habitat is segmented by development and roads. However, if the habitat is left as a whole and not segmented, each mature adult has access to all the breeding pairs in the population, and genetic diversity is maximized.

33. **(D)** Niches are defined as both the habitat a species occupies and the role the species takes in the ecosystem. For example, a bear and a wolf may live in the same area, but since they do not eat the same foods they do not compete for food resources. The degree to which the niche of two different species overlaps determines the amount of interspecific competition the two species undergo.

34. **(A)** Wilderness areas do not allow any mechanical device or vehicle. Recreational enthusiasts challenge this point by taking snowmobiles into wilderness areas, but this disrupts the habitat and restricts the effectiveness of creating the preserve in the first place.

35. **(B)** National forests, administered by the U.S. Forest Service, were originally inspired by Gifford Pinchot, who felt that the forests should be preserved for the greatest good for the greatest number of people over the longest period of time. Pinchot is considered a utilitarian conservationist who believes that "greatest good" may include economic uses, such as logging and mining. Biocentric conservationists would argue that there is intrinsic value for preserving wildlife, even if there are steep economic costs.

36. **(D)** The Bureau of Land Management (BLM), which is managed by the Department of the Interior, leases land for mining, logging, and other commercial operations. BLM land is intended to be used with moderation, not just for preservation.

37. **(C)** The National Park Service is charged with protecting and preserving national historic landmarks and natural features of aesthetic and geological importance.

38. **(A)** Population A has undergone a demographic transition because it experiences the highest survivability of children, young adults, and mature adults. It would correspond to an age-structure diagram that is more "block-like" and less pyramidal. That type of age-structure diagram and this survivability chart indicate a population that tends to live longer lives, with lower birth and death rates.

39. **(B)** Population B demonstrates a constant rate of death throughout the life span of the members in the population. There are no plateaus in the chart; a plateau would suggest continued survival through that age bracket.

40. **(D)** Population D shows the highest infant mortality because it shows the steepest decline in survivability early in life.

Practice Exam 2 – Section I: Detailed Explanations of Answers

41. **(A)** Population A—the same population that has undergone a demographic transition—demonstrates the lowest infant mortality rate. Again, the plateau in the chart indicates that the population experiences continued survival at that age bracket.

42. **(C)** pH is a measure of acidity. Since pH is a logarithmic scale, each pH level corresponds to a tenfold increase in acid concentration. A pH of 3 corresponds to a hydrogen ion (acid) concentration of 10^{-3} moles acid per liter of solution. The pH is equivalent to the negative of the exponent of the concentration. pH 4 corresponds to 10^{-4}; pH 5 corresponds to 10^{-5}, and so forth.

43. **(B)** Water hardness refers to the concentration of calcium ions. Water can be made "softer" by replacing the calcium ions with sodium ions. Hard water causes calcium encrustation in pipes and appliances and prohibits soap from working properly, and so is somewhat undesirable. The sodium in the water causes hypertension in some people.

44. **(A)** Turbidity is caused by undissolved solids that are suspended in the water. Turbidity is measured with a secchi disc—a black and white disc—by dropping it into the water and determining the depth at which the black and white pattern on the disc is no longer visible.

45. **(D)** Salinity is caused by solids that are dissolved in the water. When ionic solids dissolve, the charged particles separate and enable the solution to conduct electricity. Therefore, the greater the electrical conductivity of the solution, the greater the salinity, or dissolved solids.

46. **(E)** Nutrient pollution increases algal growth, which increases decomposer growth, which decreases the amount of oxygen dissolved in the water.

47. **(C)** Nitrogen dioxide is produced when air—which is mostly nitrogen and oxygen gases—is combusted. The nitrogen combines with the oxygen during combustion to form nitrogen dioxide, which can further be converted into acid rain or photochemical smog.

48. **(A)** Carbon monoxide is further oxidized in a catalytic converter to form carbon dioxide.

49. **(E)** Radon is the most important indoor pollutant listed here, although there is a chance in some homes that carbon monoxide may also be an indoor pollutant. Radon is produced underground and is trapped in tightly insulated homes. It is the second leading cause of lung cancer behind cigarette smoking.

AP Environmental Science

50. **(D)** Methane is produced by bacterial digestion of hydrocarbons. The hydrocarbons are reduced by the bacteria in the ground, in mammalian intestines, or in solid waste dumps.

51. **(C)** Nitrogen dioxide is responsible for about 30% of the acid rain that falls in the United States. The oxide mixes with water to form nitrous and nitric acid, which is a strong irritant and oxidizer.

52. **(D)** Rachel Carson's *Silent Spring* outlined the problems with DDT and the effects that the biomagnification of DDT had on non-target organisms, including humans.

53. **(D)** A staged burner first burns at a high temperature, low oxygen environment, and in the second stage it is burned at a low temperature, high oxygen environment. The net result is that less oxides of nitrogen are produced as a by-product.

54. **(A)** The biomagnification of mercury in fish causes Minimata disease, a disease that was first characterized in Minimata Bay, Japan, when methyl-mercury laden fish were ingested by several members of the population. The resulting nerve damage was crippling.

55. **(E)** The water cycle is most responsible for transferring energy from one place to another on a global scale. The evaporation of water uses energy, and the precipitation of water releases energy. Through moving water, energy that is absorbed by the oceans in tropical latitudes can be transferred—particularly through storms—to other latitudes.

56. **(A)** The National Environmental Policy Act (NEPA) established the EPA and required environmental impact statements of all government projects, or projects that take place on government land. The FFDCA provided standards for food, drug, and cosmetic safety. ESA protects endangered species. SMCRA establishes the office of surface mining and requires coal mining operations to put money in escrow to reclaim mined land. CERCLA provided the superfund for toxic waste cleanup.

57. **(A)** Temperature and available moisture are the two most important abiotic factors that dictate how a biome will develop.

58. **(D)** On a dose response curve, if the dose—or concentration—of the toxin is on the horizontal axis, and the number of organisms that have died is on the vertical axis, the concentration or dose that corresponds to 50% of the organisms dead is called the lethal dose, 50%, or LD_{50}.

Practice Exam 2 – Section I: Detailed Explanations of Answers

59. **(C)** The sigmoidal population growth curve demonstrates a population that undergoes logistic growth strategy; that would include investing more energy and resources in fewer young, and responding to environmental needs before reaching the environment's carrying capacity, rather than overshooting the carrying capacity and dying back.

60. **(A)** Assuming that the curve is a population growth curve, and that we are looking for events that could be found on such a curve, the only event that is not found is the overshoot of the carrying capacity.

61. **(B)** Of the types of toxicity listed, those compounds that cause mutagenicity are most often responsible for causing cancer. Cancer occurs when genetic changes affect the signals for cell growth. Those genetic changes are caused in individual cells by mutagens.

62. **(D)** Bacteria belong to Kingdom Monera. They are not protists, plants, fungi, or animals.

63. **(B)** Chemosynthesis using sulfur and nitrogen containing compounds is the primary way in which producers in the ocean floor capture energy—usually in warm water vents. Nonproducers obtain energy as decomposers or predators of the producers. Photosynthesis is not possible because light cannot penetrate that deep.

64. **(B)** The Department of Agriculture oversees the National Forest Service, which manages the national forest system.

65. **(C)** Survival economics, unfortunately, will probably drive the decision of a country that is having a hard time feeding its people. Survival economics focuses on short-term needs, such as obtaining food and cash.

66. **(D)** Thunderstorms involve updrafts, precipitation, low pressure zones, cumulonimbus clouds, but rarely warm fronts. They usually involve cold fronts.

67. **(A)** El Niño events involve the reversal of the usual flow of warming water across the Pacific, changing the flow so that it moves from west to east, rather than from east to west.

68. **(C)** Ozone depleting molecules are particularly dangerous because it takes so few of them; they act as catalysts, so one molecule can be used over and over again to convert ozone molecules to oxygen gas.

69. **(E)** Dioxins are not typically used in pesticides. Dioxins are toxic organic compounds that are produced from burning garbage, and as a by-product of various industrial processes.

70. **(D)** A debt for nature swap involves trading land that would otherwise be destroyed to pay off national debt, which would otherwise push a country to develop the land in order to pay the debt.

71. **(A)** The mobility of a toxin through the environment depends on the solubility of the molecule (attraction or lack thereof to water molecules) and the attraction of the toxin to soil. The other answers involve the toxicity of the molecule to living organisms, which does not ultimately determine how easily the toxin moves through the environment.

72. **(D)** There is a difference between energy and power. Energy is the ability to do work, or the amount of heat, or the energy carried by sunlight. Power is the amount of energy done per unit of time—a rate of energy used. Electrical power, while often associated with the word "energy," is actually a rate at which energy is used.

73. **(D)** Watts = joules/sec, or a joule = 1.0 watt × 1.0 sec. Therefore, 100 watts × 100 seconds = 10,000 joules.

74. **(D)** It is feared that GMOs may pollinate weeds nearby the crops and actually impart pesticide-resistant genes, or other genes that would confer an advantage to the weeds.

75. **(E)** Large industrial monoculture operations are not considered sustainable because of the heavy dependence on chemical fertilizers, pesticides, and energy. The other methods listed could be incorporated by industrial monoculture, but it would no longer be monoculture, and the profit margin would be significantly reduced—so it would no longer be as attractive to corporate investors.

76. **(A)** It takes 16 pounds of grain to produce one pound of beef. Therefore, by eating beef, a person is actually consuming 16 times the energy that they would otherwise consume if they ate the grain directly.

77. **(C)** Cyanobacteria (photosynthetic blue-green bacteria) gets the credit for creating the oxygen in the early atmosphere. Although now that the flora has developed, forests—particularly rainforests—also play a significant role.

78. **(A)** Nitrogen gas composes nearly 80% of Earth's atmosphere. Oxygen gas takes a distant second at just under 20%.

79. **(B)** The stream is weakly acidic because a pH of 7 is neutral, and pH 6 is slightly less than 7. A pH lower than 7 is acidic; higher than 7 is basic.

Practice Exam 2 – Section I: Detailed Explanations of Answers

80. **(A)** Populations that are r-strategists tend to show irruptive, or Malthusian, growth. They tend to produce many young and do not invest heavily in each one in the hope of a small percentage surviving. These species tend to be smaller than K-strategists.

81. **(D)** Food availability and disease are two density-dependent factors that can affect the size of a population. For example, the flu of 1918 was more severe in cities where it could spread easily. People succumbed more easily if they did not have proper nutrition. This flu caused a significant drop in world population.

82. **(D)** Irruptive growth is another term for Malthusian growth, which tends to demonstrate quick growth, overshoot the carrying capacity, and then grow again. Irruptive growth tends to be a trait of r-strategist populations.

83. **(A)** Solubility is the most significant factor that determines whether or not a toxin will be assimilated by organisms and biomagnify. Water soluble toxins will pass through producers and organisms low on the food chain, and therefore they will not magnify in high-trophic level organisms because they were not retained in the lower organisms. Conversely, fat-soluble toxins will be absorbed by the cells of the organisms lower on the food chain (bioaccumulation). Then, when those organisms are eaten by predators, the organisms higher on the food chain end up experiencing a higher dose of the toxin than did the lower organism (biomagnification).

84. **(E)** Different toxins are broken down in the environment in a number of ways. Oxidation (usually by bacteria), photolysis (break down by the sun), and hydrolysis (break down by water) are three mechanisms that break down toxins.

85. **(B)** Water or soil provide an essential medium in/on which bacteria may grow and come in contact with toxins.

86. **(C)** Paper is by far the most predominant component of solid waste that enters the waste stream. This is why recycling paper is an important method to reduce the waste stream.

87. **(B)** Incineration of waste is beneficial because it reduces the total volume of waste significantly. However, it turns a solid waste problem into an air pollution problem. Scrubbers can be used to retrieve the pollutants from the fly ash, but that reverts it back into a solid waste problem—with the addition of oxygen to the mass of the waste. The most significant pollutants that result from burning garbage are dioxins, which are highly carcinogenic.

88. **(C)** Carbon, oxygen, nitrogen, and water are all essential in a vibrant compost pile; only phosphorous need not be added. However, phosphorous is probably present in all of the plant material that is added to the pile. The carbon:nitrogen

ratio should be 30:1. Oxygen and water are needed to fuel the fungi and bacteria that digest the nutrients in the pile.

89. **(C)** Of those organisms and diseases listed, only the malaria mosquito is a vector. A vector is defined as an organism or physical object that carries a disease from one victim to another. A pathogen is the organism that actually causes the disease. The malaria pathogen is actually a protist (plasmodium) that lives in the digestive gland of the mosquito. Plasmodium has coevolved with the mosquito to migrate out of the digestive gland and into the saliva of the mosquito at dusk, which is when the mosquito is mobile looking for a blood meal.

90. **(A)** The shaded portion of the graph refers to the biotic potential, which is what the growth rate would be if environmental resistance did not suppress the growth of the population.

91. **(C)** Channelization improves navigation but it reduces the wetlands along the banks. Channelization creates a deeper, straighter, faster flowing waterway. The wetlands normally retain water in shallow areas; this works against navigation because the water must be deep enough to accommodate large vessels. The path of the waterway must be straighter so that the vessels can negotiate turns without running aground.

92. **(E)** Pesticides are an example of organic pollutants. Organophosphates and chlorinated hydrocarbons are two major categories of pesticides, both of which are organic molecules. If they are fat-soluble, like DDT, they will accumulate in living cells and run the risk of undergoing the process of biomagnification. Remember that the term organic refers to whether or not the molecule is a covalently bound carbon-based molecule, not whether or not it is good for the environment.

93. **(A)** The grasshopper effect refers to the situation in which toxins evaporate from soil and water in warmer climates, then precipitate and remain in cooler climates.

94. **(E)** Runoff water that fills lakes often carries with it dissolved nutrients and salts. If that water is then removed from the lake and placed over a crop, some of the water evaporates and leaves the salt on the plants and soil. When this happens over time, the salt accumulates and prohibits further growth. This is called salinization. Likewise, when the water is taken from the lake, many salts remain in the lake. If the lake is drained, the shore can be a combination of salt and dust that, if taken up by the wind, can be transported and cause respiratory problems many miles away. The California Water Project drained Mono Lake as a result of water diversion much the same way that the then Soviet Union drained most of the Aral Sea. Airborne salt residue has caused respiratory problems in the surrounding populations in both cases.

Practice Exam 2 – Section I: Detailed Explanations of Answers

95. **(B)** Peanut butter contains natural oils that putrefy and contaminate the compost pile.

96. **(D)** Garrett Hardin's essay "The Tragedy of the Common" cites the real tragedy as the loss of personal freedom of choice—freedom to choose to use resources, reproduce, or pollute.

97. **(A)** It takes 20 times more energy to produce an aluminum can from virgin materials as it does to produce a can from recycled materials. Therefore, only 5% of the energy is required to produce the can from recycled materials. Since the refining of aluminum is so energy intensive, energy savings represent the largest benefit of recycling aluminum.

98. **(B)** Gold is refined by the process called heap-leach extraction, which involves putting the gold ore on a heap, or pile, and spraying a caustic cyanide solution over it. The solution is then electrolyzed and the dissolved gold is reduced at the cathode into the solid metal. The solution can be used again, but it is often left at the mine site after the mining operation is over, resulting in a very significant water and land pollution problem.

99. **(A)** Less than one percent of the material that is mined makes it through the milling process and on to the fuel production process. The other 99% remains on the surface of the Earth and becomes radioactive waste.

100. **(D)** In both environmental economics and ethics, a sense of social fairness and justice would dictate that consumers of resources pay for indirect costs, rather than having those costs shouldered unjustly by marginalized people.

AP Environmental Science

Detailed Explanations of Answers

Practice Exam 2

Section II

1. (10 points)

    a. (2 points; +1 for correct units, +1 for correct answer)

    $$4,500 \text{ tons} \times \frac{2,000 \text{ lb}}{\text{ton}} \times \frac{5,000 \text{ BTU}}{\text{lb}} \times \frac{1 \text{ kW-hr}}{34,000 \text{ BTU}} = 1.3 \times 10^7 \text{ kW-hr}$$

    b. (1 point for each step, up to 2 points)
    - Coal
    - Combustion heats water
    - Water pushes turbine
    - Turbine turns generator
    - Turbine produces electricity

    c. (2 points; +1 for correct calculation, +1 for correct calculation and a different number of steps)

    $0.8^4 = 0.24$, or 24%

    d. (2 points each)

    i. First stage burns at high temperature with little oxygen; second stage burns at low temperature with oxygen. This minimizes the combination between oxygen and nitrogen at high temperatures.

    ii.

    $$\frac{0.75 \text{ lb NO}_x}{10^6 \text{ BTU}} \times 1.3 \times 10^7 \text{ kW-hr} \times \frac{3,400 \text{ BTU}}{1 \text{ kW-hr}} = 33,000 \text{ lb NO}_x$$

Practice Exam 2 – Section II: Detailed Explanations of Answers

2. (10 points total)

    a. (3 points maximum)

    +1 for Point A is the healthiest point.

    +1/2 for each of the following:

- DO is higher
- pH is physiologic range
- Low nitrate level suggests free of nutrient pollution
- Higher biodiversity
- Higher abundance

    b. (2 points possible)

    +1 each for

- Abundance
- Diversity

    c. (2 points possible)

    +1 each for

- Dissolved solids are measured with a conductivity meter.
- Undissolved solids are measured by measuring turbidity with a secchi disc.

    d. (3 points total)

The high nitrate level suggests that nutrient pollution caused an algal bloom, and then the decomposer bloom depleted the oxygen. This is supported by a low DO, and a drop in diversity and abundance. The higher temperature suggests that there is also thermal pollution, which amplifies the effect of nutrient pollution by decreasing the solubility of oxygen in the water.

AP Environmental Science

3. (10 points)

   a. (2 points total, +1/2 for each of the following)
      - Hunting
      - Rare species is valuable remedy.
      - Habitat is in demand for human development
      - Extreme weather
      - Predation
      - Pollution

   b. (4 points; 1 for each)
      - Long-lived
      - Large body size
      - Large roaming range
      - Habitat is threatened.
      - Genetic isolation, such as having an island population be the only representation of the species
      - Sensitive to pollution
      - Being displaced by exotic species or species that outcompete the endangered species
      - Environmental change
      - Specialist species
      - Market exists for species (hunting, food, fur, etc.).

   c. (2 points)
      - Endangered Species Act protects habitat.
      - Convention on the International Trade of Endangered Species

   d. (2 points)

      Captive breeding programs do not tend to be as effective as habitat protection programs. Rare species do not tend to reproduce in captivity as well as they do in their natural habitat. Habitat protection must be large enough so that populations do not become segmented. When small preserves are the only land that is available, genetic diversity can be maintained if breeding corridors are established so that populations in different areas are connected.

Practice Exam 2 – Section II: Detailed Explanations of Answers

4. (10 points total)

   a. (2 points)
      - Continue selling power to aluminum plants
      - Increase in costs to area farmers
      - Fishing in reservoirs
      - Exposing bank of river would increase air pollution

   b. (2 points; +1 each for any two of the following)
      - Re-establishing natural flow of river will allow the natural flora and fauna to return.
      - Wetlands near the river can be restored; this will re-create the habitat of migratory fish and birds.
      - Restored wetlands will also provide flood control.
      - Restored wetlands will also provide treatment of water and protect it from nutrient pollution.

   c. (2 points)

   GNP increases when purchases increase. That means that hiring the construction crews used to breach the dam, remove the silt, and restore the river will require payment, which increases GNP. Individuals who earn money will then purchase goods and services. GNP increases when money changes hands, regardless of the environmental benefit, or whether or not the infrastructure is improved in such a way as to build wealth in the future.

   d. (3 points; +1 each for up to three reasons)
      - Indigenous people displaced by dam construction
      - Migratory species decrease in population.
      - Loss of wetlands, decreased habitat, water cleansing, and natural flood control
      - Sand and gravel pits created nearby in order to make concrete
      - Health effects to workers for breathing dust, sunburn
      - Loss of scenic beauty

Practice Exam 3

Practice Exam 3

AP Environmental Science

Practice Exam 3

Section I

TIME: 90 minutes
100 multiple-choice questions

(Answer sheets appear in the back of this book.)

> **DIRECTIONS:** Each of the questions or incomplete statements below is followed by five suggested answers or completions. Select the best answer for each question and then fill in the corresponding oval on the answer sheet.

Questions 1–3 refer to the following answers. Select the one lettered choice that best fits each statement.

    (A) Solid

    (B) Liquid

    (C) Gas

    (D) Plasma

    (E) At least two of the above

1. What state of matter does most of the Earth's water occupy?

2. Condensation creates this state of matter.

3. This state of matter in the atmosphere contains the greatest kinetic energy.

Questions 4–7 refer to the following answers. Select the one lettered choice that best fits each statement.

(A) Cells
(B) Communities
(C) Populations
(D) Species
(E) Ecosystems

4. Contains nonliving matter

5. Contains a group of organisms from the same species

6. The smallest unit of life

7. Members of this group breed to create fertile offspring.

Questions 8–11 refer to the following map. Select the one lettered choice that best fits each statement.

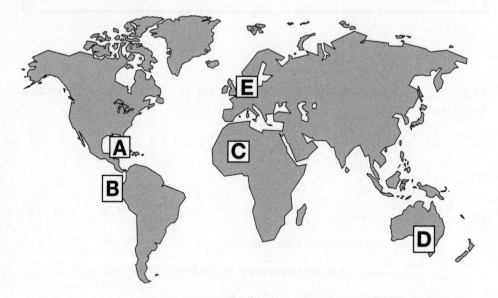

8. Area of the world that contains warm moisture during El Niño

9. High pressure area creates a desert.

10. The heaviest levels of ozone depletion have caused skin cancer in this area.

11. Fertilizer moving down a massive river has created a hypoxic zone where the river empties into the ocean.

Questions 12–15 refer to the following answers. Select the one lettered choice that best fits each statement.

(A) Carbon dioxide
(B) Soluble carbonate ions
(C) Lignite coal
(D) Limestone
(E) Organic molecules

12. Form of carbon responsible for global warming

13. A sedimentary rock formed when bodies of calcareous organisms are put under extreme pressure

14. In equilibrium with carbon dioxide in the sea

15. The form of carbon most often found in living tissue

Questions 16–19 refer to the following answers. Select the one lettered choice that best fits each statement.

(A) Nitrogen cycle
(B) Carbon cycle
(C) Sulfur cycle
(D) Phosphorous cycle
(E) Water cycle

16. Does not involve a chemical reaction

17. Used in sewage treatment process

18. Involves respiration

19. Includes both the production of coal and acid rain

Questions 20–23 refer to the following answers. Select the one lettered choice that best fits each statement.

(A) Heat of vaporization
(B) Specific heat capacity
(C) Varied densities with temperature and salinity gradients
(D) Polar nature of the water molecule
(E) At least two of the above

20. The property of water that is responsible for carrying large amounts of water vapor thousands of miles across an ocean

21. The property of water that is responsible for driving subsurface ocean currents

22. The property of water that allows it to dissolve organic molecules and ions

23. The property of water that keeps coastal climates in a moderate range

Questions 24–27 refer to the following answers. Select the one lettered choice that best fits each statement.

(A) Ecotone
(B) Demographic transition
(C) Succession
(D) Divergent evolution
(E) Convergent evolution

24. Niche portioning due to competition would tend to result in this type of natural selection.

25. One combination of flora and fauna yields to another combination during this process.

26. In this process, a similar environment will cause one species to develop traits that are similar to those of another species in the same environment.

27. The boundary between two distinct ecosystems

Questions 28–31 refer to the following answers. Select the one lettered choice that best fits each statement.

(A) Symbiosis
(B) Parasitism
(C) Commensalism
(D) Mutualism
(E) Competition

28. Demonstrated by a pathogen

29. Demonstrated by photosynthetic algae and a marine invertebrate in forming coral reefs

30. Demonstrated by remora swimming near sharks

31. The general term for organisms that co-evolve together

Questions 32–35 refer to the following answers. Select the one lettered choice that best fits each statement.

    (A) Complexity

    (B) Constancy

    (C) Inertia

    (D) Renewal

    (E) Succession

32. Refers to the number of species at each trophic level and the number of trophic levels in the community

33. Refers to a biological community's tendency toward a lack of fluctuation that creates changes in the ecosystem

34. Refers to a biological community's ability to repair ecological damage

35. Refers to a change in the composition of a biological community, with one dominant population yielding to another until a climax community becomes established

Questions 36–38 refer to the following answers. Select the one lettered choice that best fits each statement.

    (A) Genetic biodiversity

    (B) Species biodiversity

    (C) Ecosystem biodiversity

    (D) More than one of the above

    (E) None of the above

36. Demonstrated during the Green Revolution, when common strains of rice made traditional crops less resistant to attack from pests

37. Demonstrated when wolves were exterminated in the Yellowstone area during the 1930s

38. Demonstrated by the steady elimination of the Everglades as a result of nearby development

Questions 39–42 refer to the following chart. Select the one lettered choice that best fits each statement.

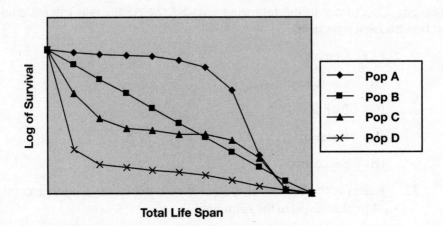

39. Which population most likely represents a developed country?

40. Which population demonstrates a species in which each age group experiences potentially lethal environmental hazards?

41. Which population most likely represents an undeveloped country?

42. Which population represents a culture with a well-developed health care system?

Questions 43–46 refer to the following answers. Select the one lettered choice that best fits each statement.

(A) Confined aquifer

(B) Unconfined aquifer

(C) Salinization

(D) Cone of depression

(E) Saltwater intrusion

43. Another name for an artesian well

44. A reservoir with a water table

45. Drop in water table near a heavily used inland well
46. Causes deposits on plants and soil during irrigation

Questions 47–50 refer to the following chemical reactions. Select the one lettered choice that best fits each statement.

(A) $SO_2 + CaO \rightarrow CaSO_3$
(B) $2O_3 \rightarrow 3O_2$
(C) $C + O_2 \rightarrow CO_2$
(D) $NH_3 + 2O_2 \rightarrow NO_3^{-1} + H_3O^{+1}$
(E) $SO_2 + H_2O \rightarrow H_2SO_3$

47. This reaction produces acid rain from coal-fired power plants.
48. This reaction increases human exposure to ultraviolet radiation.
49. This reaction occurs when wastewater undergoes secondary treatment.
50. This reaction produces heat that boils water and produces electricity, but also contributes to the greenhouse effect.

51. This environmental crisis involved buried hazardous and toxic wastes.

   (A) Love Canal
   (B) Bhopal Crisis
   (C) *Exxon Valdez*
   (D) DDT
   (E) Chernobyl

52. In this air pollution control device, platinum and palladium are used to further oxidize nitrogen and carbon.

   (A) Electrostatic precipitator
   (B) Fluidized bed combustion
   (C) Catalytic converter
   (D) Staged burner
   (E) Bag filters

53. Which section of a lake contains the portion that does not freeze in the winter and receives nutrients that drop from above during the summer?

    (A) Littoral zone
    (B) Epilimnion
    (C) Hypolimnion
    (D) Hyperlimnion
    (E) Thermocline

54. Which of the following represents the kingdom or phyla of a likely food chain in a marine biome?

    (A) moneran—insecta—pachiderm
    (B) protista—mollusk—echinoderm
    (C) protista—mammalia
    (D) plantae—insecta—aves
    (E) Two of the above

55. This type of species is a critical link in an ecosystem, upon which the survival of many other species depends.

    (A) Indicator species
    (B) Keystone species
    (C) Critical factor
    (D) Symbiot
    (E) Batesian mimic

56. The Coriolis effect is made possible by

    (A) the turning of the Earth and atmospheric convection currents
    (B) the movement of tectonic plates
    (C) the evaporation of water
    (D) ocean currents
    (E) the condensation of water

57. Which of the following nutrient cycles has the greatest effect in regulating climate?

    (A) Carbon cycle
    (B) Sulfur cycle
    (C) Nitrogen cycle
    (D) Phosphorous cycle
    (E) Arsenic cycle

58. Which of the following government agencies is NOT managed by the Department of the Interior?

    (A) National Park Service
    (B) National Forest Service
    (C) Bureau of Land Management
    (D) Office of Surface Mining
    (E) Fish and Wildlife Service

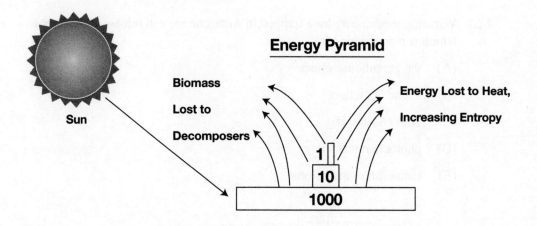

59. One trophic level that is missing on the biomass pyramid above is

    (A) a primary carnivore
    (B) a secondary carnivore
    (C) fungi that decompose plants
    (D) herbivore
    (E) producer

AP Environmental Science

60. Which of the following costs would be included in a true cost-benefit analysis, but not a traditional cost-benefit analysis?

    (A) The salaries of the people completing the project

    (B) The health care costs of people made ill by the project

    (C) The materials needed to complete the project

    (D) The health care insurance premiums of people completing the project

    (E) Administrative costs of the project

61. Which of the following is increased by stream channelization?

    (A) Stream bank erosion

    (B) Wetland biodiversity

    (C) Stream flow

    (D) Two of the above

    (E) All of the above

62. Warming methane hydrate trapped in Antarctic ice can release methane gas, which is responsible for

    (A) the greenhouse effect

    (B) ozone depletion

    (C) both (A) and (B)

    (D) photochemical smog

    (E) temperature inversions

63. When toxic substances are ingested by humans, they will be broken down in the _____ and excreted through the _____ .

    (A) gall bladder; kidneys

    (B) liver; intestines

    (C) stomach; intestines

    (D) liver; kidneys

    (E) heart; lungs

64. Which of the following types of toxicity have the capability of altering base sequences in human DNA?

    (A) Carcinogenicity

    (B) Mutagenicity

    (C) Teratogenicity

    (D) All of the above

    (E) None of the above

65. Radiation can cause severe damage to biological tissues. Which of the following units takes into account the level of damage done to biological tissue from the energy of electromagnetic radiation?

    (A) Rems

    (B) Rads

    (C) Joules

    (D) Watts

    (E) Ergs

66. Chlorinated hydrocarbon pesticides inhibit the uptake of cholinesterase. This category of pesticide takes action on cells in which of the following body systems?

    (A) Nervous system

    (B) Respiratory system

    (C) Skeletal-muscular system

    (D) Excretory system

    (E) Lymphatic system

67. Biomagnification occurs most easily with toxins that are soluble in _____, and has the effect of _____.

    (A) hydrophilic molecules; reducing toxicity in higher trophic levels

    (B) hydrophobic molecules; reducing toxicity in higher trophic levels

    (C) hydrophilic molecules; compounding toxicity in higher trophic levels

    (D) hydrophobic molecules; compounding toxicity in higher trophic levels

    (E) water; increasing the light sensitivity of an organism

68. Which of the following government agencies would be working to maintain the health and safety of people in an area that is recently ravaged by a hurricane?

    (A) Red Cross

    (B) FDA

    (C) USGS

    (D) EPA

    (E) FEMA

69. A farmer owns land among rolling hills and is concerned about reducing soil erosion. Which of the following techniques would reduce soil erosion?

    (A) Drip irrigation

    (B) Terracing of the land

    (C) Crop rotation

    (D) Decreasing dependence on chemical pesticides and herbicides

    (E) At least two of the above

70. A person who has focused primarily on making a lot of money suddenly donates a huge sum to purchase land to protect the habitat of an endangered species. In this case, the human participant is acting as the _____, the endangered species is the _____, and the growth in awareness experienced by the human could be considered an example of _____.

   (A) moral subject; moral agent; cornucopian fallacy

   (B) moral subject; moral agent; universalism

   (C) moral subject; moral agent; moral extensionism

   (D) moral agent; moral subject; moral extensionism

   (E) moral agent; moral subject; universalism

71. To eat beef rather than chicken, a consumer is using _____ as much total energy to gain the calories they need.

   (A) four times

   (B) twenty times

   (C) equally

   (D) one-half

   (E) one-fourth

72. The production of shrimp in coastal pools is

   (A) part of the Green Revolution

   (B) an environmentally responsible method for growing protein

   (C) part of the Blue Revolution

   (D) an example of a GMO

   (E) a method that produces healthier shrimp that contain fewer toxins

AP Environmental Science

73. Which of the following is NOT a trait of organisms that would make them more susceptible to extinction?

   (A) Longer life span

   (B) Tends to invest a large amount of resources in the next generation

   (C) Tends to demonstrate a Malthusian growth strategy

   (D) Larger body dimensions

   (E) Specialized diet

74. Which of the following is NOT a benefit of wetlands protection?

   (A) Provides a nursery for the young of many species

   (B) Cleans pollution out of water

   (C) Decreased flooding

   (D) Recreational opportunities

   (E) Less property damage

75. The Coriolis effect is driven by the following two events:

   (A) currents in the ocean and the turning of the Earth

   (B) convection cycles in the atmosphere and the turning of the Earth

   (C) differential heating of the Earth and thermohaline currents

   (D) volcanic activity and movement of crustal plates

   (E) convection cycles and thermohaline currents

76. Which of the following is an example of a jet stream?

   (A) The circumpolar vortex

   (B) The Gulf Stream

   (C) The Greenhouse Effect

   (D) El Niño

   (E) Northeastern trade winds

77. Energy that is needed to drive all global processes and movement of matter comes from

    (A) the sun
    (B) the sun and the moon
    (C) the sun and the center of the Earth
    (D) the sun, the moon, and the center of the Earth
    (E) cosmic radiation

78. This ecosystem is a transitional ecosystem between two others.

    (A) Salt marsh
    (B) Tundra
    (C) Grasslands
    (D) Desert
    (E) Abyssal

79. When a continental plate collides with another continental plate,

    (A) a subduction zone will be created
    (B) the two plates will push each other up into a mountain range
    (C) a constructive boundary will be created
    (D) a rift valley will be formed
    (E) a deep submarine trench will be formed

80. Which of the following Richter Scale values represents an earthquake that has 100 times the ground displacement of an earthquake of magnitude 5.2?

    (A) 3.2
    (B) 5.4
    (C) 7.2
    (D) 520
    (E) 0.52

AP Environmental Science

81. Which of the following represents the correct order of the layers of the atmosphere, starting at the surface of the Earth and moving to higher altitudes?

    (A) Troposphere, stratosphere, mesosphere, thermosphere

    (B) Stratosphere, troposphere, mesosphere, thermosphere

    (C) Mesosphere, troposphere, stratosphere, thermosphere

    (D) Mesosphere, stratosphere, troposphere, thermosphere

    (E) Troposphere, mesosphere, stratosphere, thermosphere

82. Which of the following can be caused by El Niño?

    (A) Landslides in the Rocky Mountains

    (B) More rain in Arizona

    (C) Hurricanes in the Gulf of Mexico

    (D) Decreased income from commercial fisheries in South America

    (E) All of the above

83. The evidence that Alfred Wegener cited when he originally proposed the theory of continental drift was

    (A) the changes in magnetic polarity within iron ore that occurred equidistant from the mid-Atlantic rift

    (B) the earthquakes that occurred along the edges of tectonic plates

    (C) the subduction zones, where one plate dives underneath another

    (D) the way coastlines fit together like a puzzle and similar fossils occur on both sides of an ocean

    (E) the location of mountain ranges, which occur where continental plates collide

84. Which of the following will help prevent erosion?

    (A) Tilling the ground less and leaving some material from the previous crop between the rows of the newly planted crop

    (B) Planting strips of trees between strips of crop land

    (C) Tilling and planting the ground in a manner that follows the contour of a hillside

    (D) Eliminating irrigation that uses brackish water

    (E) All of the above

85. When a proton is removed from a water molecule, the remaining ion is

    (A) basic

    (B) acidic

    (C) neutral

    (D) One cannot tell from this information.

    (E) another proton

86. What is the annual percent rate of growth of a population that doubles in size every 12 years?

    (A) 1%

    (B) 3%

    (C) 6%

    (D) 12%

    (E) 24%

87. Which of the following is NOT an example of indoor air pollution?

    (A) Radon

    (B) Dry-cleaned clothing

    (C) Aerosols

    (D) Carbon monoxide

    (E) Ozone

AP Environmental Science

88. Which of the following water compartments is most responsible for providing the greatest amount of freshwater to human cultures?

(A) Ice and snow

(B) Rivers and streams

(C) Lakes and reservoirs

(D) Groundwater

(E) Estuaries

89. What forces are most responsible for moving water through the hydrologic cycle?

(A) Solar heating and gravity

(B) Solar and geothermal heating

(C) Gravity and hydrostatic pressure

(D) Hydrostatic pressure and electromagnetic forces

(E) Nuclear forces and electromagnetic forces

90. As air moves down the downwind side of a mountain after having passed over the summit, the air

(A) warms because it experiences greater pressure

(B) cools because water is evaporating from the air

(C) warms because it is in a rain shadow

(D) cools because water is released on the downwind side

(E) becomes more cloudy, but the temperature doesn't change

91. Which of the following is NOT caused by a water-borne pathogen?

(A) Schistosomiasis

(B) River blindness

(C) Cholera

(D) Malaria

(E) Giardia

92. Put these zones in the order that represents the oxygen sag that occurs in a river after a discharge of an oxygen-demanding waste.

    (A) Decomposition zone, recovery zone, septic zone

    (B) Decomposition zone, septic zone, recovery zone

    (C) Septic zone, recovery zone, decomposition zone

    (D) Septic zone, decomposition zone, recovery zone

    (E) Recovery zone, decomposition zone, septic zone

93. Which of the following is NOT an indicator of a high stream DO?

    (A) Stonefly larvae

    (B) Mosquito larvae

    (C) Caddisfly larvae

    (D) Mayfly larvae

    (E) Trout

94. Which of the following represents the type of domestic water use that requires the LEAST amount of water?

    (A) Flushing toilets

    (B) Watering lawns

    (C) Drinking and cooking

    (D) Washing dishes

    (E) Washing clothes

95. Which of the following sectors in society represents the largest consumer of fresh water?

    (A) Cooling electric power plants

    (B) Agriculture

    (C) Domestic water use

    (D) Industrial water use

    (E) Transportation

AP Environmental Science

96. Which of the following is NOT a method of reducing our society's use of fresh water?

    (A) Recycling paper

    (B) Taking shorter showers

    (C) Preventing runoff from parking lots and construction sites

    (D) Watershed management

    (E) All of the above are methods of reducing our society's use of fresh water.

97. Secondary sewage treatment oxidizes waste ammonia into nitrates, which is still a form of water pollution because

    (A) nitrates are mutagenic.

    (B) nitrates are carcinogenic.

    (C) nitrates are neurotoxins.

    (D) nitrates ultimately decrease the dissolved oxygen in streams.

    (E) nitrates remain in the ecosystem for a very long time.

98. Oxygen solubility in water _____ as the temperature of the water increases.

    (A) increases

    (B) decreases

    (C) remains the same

    (D) Oxygen gas does not dissolve in liquid water.

    (E) increases when decomposers are present

99. Which of the following is NOT an example of an inorganic water pollutant?

    (A) Mercury
    (B) Sulfide ions
    (C) Calcium
    (D) Dioxins
    (E) Acid deposition

100. Thermal pollution is most often caused by _____ and changes a river by _____ .

    (A) power plants; decreasing dissolved oxygen
    (B) recreational boats; decreasing dissolved oxygen
    (C) power plants; increasing dissolved oxygen
    (D) recreational boats; increasing dissolved oxygen
    (E) sewage treatment; increasing dissolved phosphates

AP Environmental Science

AP Environmental Science

Practice Exam 3

Section II

Time: 90 minutes

DIRECTIONS: You have 90 minutes to answer all four of the following questions. It is suggested that you spend approximately half your time on the first question and divide the remaining time equally among the next three. In answering these questions, you should emphasize the line of reasoning that generated your results; it is not enough to list the results of your analysis. Include correctly labeled diagrams, if useful or required, in explaining your answers. A correctly labeled diagram must have all axes and curves labeled and must show directional changes.

---

1. A major coal-fired electrical power plant produces 13,000 MW-hr of electrical energy per day.

   a. Assuming that 1.0 MW-hr corresponds to 3,400,000 BTUs, how many BTUs are produced by the plant each day?

   b. Assuming that one pound of coal can produce 5,000 BTUs, how many pounds of coal are used by the plant each day?

   c. The power plant is 36% efficient. What amount of energy, in BTUs, is lost as waste heat each day?

   d. Assume that 80% of the waste heat is absorbed by the nearby river, which is diverted to cool the power plant. Local regulations limit the utility to a 5°F increase in the temperature of the river.

      i. Describe the environmental consequences of using the river water as a coolant for the power plant.

      ii. What weight of water, in pounds, is needed to absorb one day's waste heat from the plant?

2. The following questions refer to the following two age-structure diagrams.

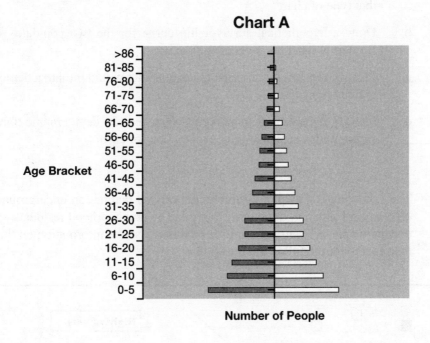

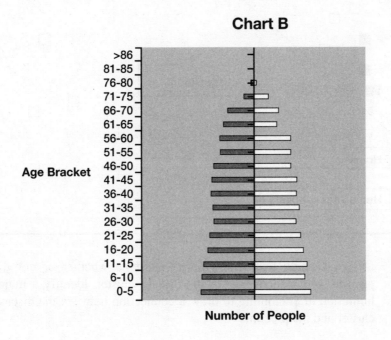

AP Environmental Science

a. Which of the above populations most likely demonstrates a developing country, and why would a developing country show that type of chart?

b. Draw a hypothetical survivorship curve for the two populations. Be sure to label the units of each axis.

c. Identify and briefly describe the events that would initiate a demographic transition.

d. Identify three environmental consequences of a demographic transition, with a sentence-long explanation for each.

3. The following is a map of a rural neighborhood that has an underground oil transport pipeline running through it. The neighborhood residents all depend on private wells as a source of water. Use the information on this map to answer the following questions.

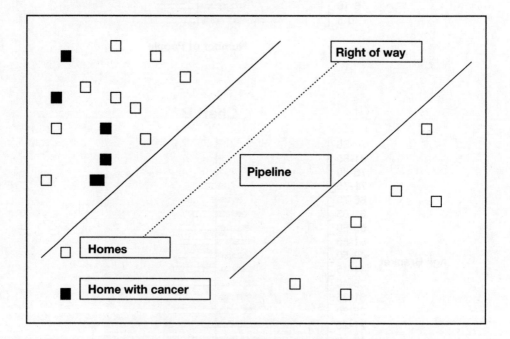

a. Design a study that would help determine whether or not the pipeline was responsible for the disease cluster. Identify a major limitation of attempting to draw a connection between the disease cluster and the pipeline.

b. Assume for a moment that there is a leak in the pipeline.

i. What does the disease cluster suggest about the geology between the pipeline and the neighborhood?

ii. Suggest a method of remediating the underground oil spill.

c. What federal laws would be related to the cleanup of this oil spill? Describe how the law would require and/or support action on the part of the owner of the pipeline.

4. The following is an article produced by an organization that is building public awareness about the re-introduction of wolves into Yellowstone National Park.

Help Us Save the Wolf

Wolves are natural inhabitants in the greater Yellowstone ecosystem that were annihilated by hunting early in this century. They belong in Yellowstone Park, and you can help them return.

The current plan to re-introduce wolves to Yellowstone is to set up a protected breeding group in a secluded corner of the Park. After the first year, the gates of the 4-acre pen will be opened, and the wolves will be allowed to roam naturally.

There are many misconceptions about wolves. Some of those misconceptions include the following:

1) wolves are dangerous to people,

2) wolves will compete with grizzlies—a prized carnivore in the Park,

3) wolves will kill cattle, and

4) wolves aren't really needed in the ecosystem.

Wolves aren't really any more dangerous to people than elk, moose, or buffalo—and they are certainly less dangerous than grizzlies. In fact, very few unprovoked wolf attacks on people have been documented.

Wolves can co-exist with grizzly bears and did so successfully for many thousands of years. Wolves are strict carnivores; they feed off of slow grazing animals, mice, and rabbits. Grizzly bears eat berries, roots, termites, and fish. There is no competition between the two species.

While some are worried about wolves killing cattle, your donation will allow ranchers to be reimbursed for any cattle lost to wolves that have strayed from the Park. Your donation contributes to restoring this vital cog in the Yellowstone ecosystem.

a. Use two or three sentences to assess the reliability of the claim of no interspecific competition between wolves and grizzlies.

b. There are three types of biodiversity: genetic, species, and ecosystem. What type of biodiversity is represented by the wolf's re-introduction? Use a sentence to explain your answer.

c. Write a short paragraph to analyze the ecological effects of the wolf's presence on the Yellowstone ecosystem.

d. In a short paragraph, analyze the strengths and weaknesses of the re-introduction plan.

AP Environmental Science

Practice Exam 3
Answer Key

Section I

1.	(B)	26.	(E)	51.	(A)	76.	(A)
2.	(B)	27.	(A)	52.	(C)	77.	(D)
3.	(C)	28.	(B)	53.	(C)	78.	(A)
4.	(E)	29.	(D)	54.	(E)	79.	(B)
5.	(C)	30.	(C)	55.	(B)	80.	(C)
6.	(A)	31.	(A)	56.	(A)	81.	(A)
7.	(D)	32.	(A)	57.	(B)	82.	(E)
8.	(B)	33.	(B)	58.	(B)	83.	(D)
9.	(C)	34.	(D)	59.	(B)	84.	(E)
10.	(D)	35.	(E)	60.	(B)	85.	(A)
11.	(A)	36.	(A)	61.	(D)	86.	(C)
12.	(A)	37.	(B)	62.	(C)	87.	(E)
13.	(D)	38.	(C)	63.	(D)	88.	(D)
14.	(B)	39.	(A)	64.	(D)	89.	(A)
15.	(E)	40.	(B)	65.	(A)	90.	(A)
16.	(E)	41.	(D)	66.	(A)	91.	(D)
17.	(A)	42.	(A)	67.	(D)	92.	(B)
18.	(B)	43.	(A)	68.	(E)	93.	(B)
19.	(C)	44.	(B)	69.	(E)	94.	(C)
20.	(A)	45.	(D)	70.	(D)	95.	(B)
21.	(C)	46.	(C)	71.	(A)	96.	(E)
22.	(D)	47.	(E)	72.	(C)	97.	(D)
23.	(B)	48.	(B)	73.	(C)	98.	(B)
24.	(D)	49.	(D)	74.	(E)	99.	(D)
25.	(C)	50.	(C)	75.	(B)	100.	(A)

Detailed Explanations of Answers

Practice Exam 3

Section I

1. **(B)** Most of the Earth's water exists in oceans.

2. **(B)** The process of condensation pulls water out of the vapor phase and into the liquid phase.

3. **(C)** Kinetic energy is marked by thermal motion and also temperature. Of the states of matter listed here, the gas phase contains the highest thermal motion and temperature, and therefore the highest amount of kinetic energy.

4. **(E)** Of the levels of organization listed, only ecosystems contain non-living matter. The other levels of organization all involve collections of living material. Cells are the smallest unit of life. Cells come together to form individuals of a species. Members of a species are collected in a population. Populations of different species come together to form a community.

5. **(C)** Members of a species are collected in a population.

6. **(A)** Cells are the smallest unit of life. Cells operate with organic and inorganic molecules, which come together to form a living cell.

7. **(D)** Members of a species breed to create fertile offspring. It is possible to have members of two different species breed and have offspring. For example, horses and donkeys can mate and create a mule, but mules are sterile—as are other progeny that result from trans-species unions.

8. **(B)** El Niño/Southern Oscillation events bring moisture across the Pacific Ocean to the western coast of South America. From here, the moisture often moves north with the Coriolis effect and confronts cool arctic air over North America. The warm, moist air from El Niño may end up dumping snow on the Rocky Mountains, or it may end up flooding the Mississippi—depending on where the tropical air collides with the arctic air.

Practice Exam 3 – Section I: Detailed Explanations of Answers

9. **(C)** The areas that are about 30 degrees north and south latitude receive the hot, dry descending air from the Hadley cell. This phenomenon tends to create hot, dry desert biomes at that latitude.

10. **(D)** The first ozone hole over the Antarctic region forms an oblong area of depletion that rotates about the polar region. One lobe of the ozone hole would pass over New Zealand, causing periodic severe exposure to UV light. This has been documented to correspond with rising skin cancers in the predominantly light-skinned population.

11. **(A)** The agricultural products and fertilizer that run off into the Mississippi are carried by that mighty river until it empties into the Gulf of Mexico. At that point, there is a large bolus of ocean that has become hypoxic, or oxygen free. This has destroyed the fisheries in this area and created a very bad smell.

12. **(A)** Carbon dioxide is a greenhouse gas that is responsible for the bulk of global warming that is taking place.

13. **(D)** Calcareous creatures include snails and diatoms, the bodies of which form limestone—which is mostly calcium carbonate.

14. **(B)** The carbon dioxide from the air dissolves in the aquatic bodies below (by Henry's Law). In the water, the combination of water and carbon dioxide creates an equilibrium in the formation of the bicarbonate ion and carbonic acid. This collection of weak acids and conjugate bases forms a buffer system that keeps ocean pH within physiological ranges. The same equilibrium, by the way, is also one of the two buffer systems that moderates pH in humans.

15. **(E)** Organic molecules include DNA, carbohydrates, proteins, and fats—all of which are carbon-based. These molecules make up our bodies and represent the form of carbon most often found in living organisms.

16. **(E)** The water cycle only involves changes in state, such as melting, freezing, vaporization, condensation, and sublimation. The water cycle does not involve a chemical reaction, or the rearranging of atoms.

17. **(A)** The sewage treatment process uses the nitrogen cycle to oxidize ammonia into nitrites and nitrates.

18. **(B)** Respiration is the process used by animals to convert carbon dioxide into carbohydrates, and is part of the carbon cycle.

19. **(C)** The sulfur cycle includes the building of proteins in plants and animals. The sulfur creates the di-sulfide linkages that hold proteins in their three-dimensional configuration. When the plants die and form coal, the sulfur remains. When the coal is burned, the sulfur in the coal combines with oxygen to form sulfur dioxide. When the sulfur dioxide combines with water vapor in the atmosphere, it forms acid rain. Sulfur dioxide from coal combustion is responsible for about 70% of acid rain.

20. **(A)** The latent heat of vaporization determines how much heat is absorbed when water is vaporized. The vaporized water stores that heat, which is returned when the water condenses. The vaporization of water in the tropical Atlantic off the coast of Africa fuels the formation of hurricanes in the summer. After it is formed, a hurricane carries the vaporized water across the Atlantic to the eastern coast of North or South America.

21. **(C)** Subsurface ocean currents are driven by thermohaline currents, which are currents that move water from high-density areas to low-density areas. The density of the water increases as the salinity increases or the temperature decreases.

22. **(D)** The polar nature of the water molecule helps it to be a good solvent for ionic compounds and polar covalent compounds. The slight negative charge on the oxygen side of the molecule is attracted to the positive charge on ionic compounds. The slightly positive hydrogen end is attracted to the negative side of ionic compounds.

23. **(B)** The specific heat capacity is the property of water that helps to moderate climate and weather. Water has an exceptionally high specific heat, which means that it holds a lot of heat for every degree that it is warmed. This means that a bay or a lake will provide a heat sink for a region. Rather than heat energy going into raising the temperature, it is absorbed by the water and does not increase the temperature as much.

24. **(D)** As a result of competition for resources, two populations will respond by partitioning the resources. Over time, each of the two populations will more closely adapt to using the separate resources, and divergent evolution will occur.

25. **(C)** Succession is the process by which one combination of flora and fauna in a given habitat yields to another combination of flora and fauna. A demographic transition is the process by which human populations undergo the transformation from high birth and death rates to low birth and death rates. The other two processes mentioned, convergent and divergent evolution, are types of natural selection, which is not the same as succession.

Practice Exam 3 – Section I: Detailed Explanations of Answers

26. **(E)** Convergent evolution is the type of natural selection that forms similar structures on different species that are solving the same environmental challenge by evolving similar traits. The traits of the two populations are converging, or becoming more similar. Divergent evolution results in two populations becoming less similar.

27. **(A)** The boundary between two biomes is called an ecotone. The other answers are biological processes, not a location in the environment.

28. **(B)** A pathogen obtains energy from its host at the expense of the host, which qualifies it as a parasite.

29. **(D)** Photosynthetic algae and the coral invertebrate participate in a mutualistic relationship in which both species benefit. The coral obtains the organic compounds produced by the algae, and the algae receives the benefit of having a protected habitat that grows at a depth that allows it to capture the sun's rays.

30. **(C)** Remora are small fish that eat the scraps left by sharks. They receive benefits, but the sharks do not receive any benefit or harm—making this a commensalistic relationship.

31. **(A)** When organisms co-evolve, it establishes a relationship between them where at least one of the species benefits. This is called a symbiotic relationship.

32. **(A)** Complexity refers to the number of ecological interactions that are available in a biological community. The greater number of trophic levels, and the greater number of species at each level, both contribute to the ecological complexity.

33. **(B)** Constancy refers to a biological community's ability to resist changes in the niches that are available for living things. If those niches do not change—as with a fire, glacier, severe storm, or succession—then the ecosystem has a high level of constancy.

34. **(D)** A biological community's renewal ability refers to its ability to repair ecological damage. For example, when the Alaska pipeline ruptures over the Arctic tundra, the growing season is so short that it will take decades for the damaged area to repair. However, succession is so fast in a temperate rainforest, clearing a small patch will be healed very quickly.

35. **(E)** Succession is the process by which one biological community yields to another until a climax community is reached.

AP Environmental Science

36. **(A)** The Green Revolution decreased the genetic diversity of crops by planting a small number of varieties where there had been many planted before. The drop in diversity of the gene pool made the crops more susceptible to pests.

37. **(B)** The decline of the wolves was an example of a drop in species diversity; one species was eliminated from that ecosystem.

38. **(C)** As the Everglades deteriorate, region after region loses habitat that could otherwise have been used to support wildlife. This is an example of a decline in ecosystem biodiversity.

39. **(A)** Population A has undergone a demographic transition because it experiences the highest survivability of children, young adults, and mature adults. It would correspond to an age-structure diagram that is more "block-like," and less pyramidal. That type of age-structure diagram and this survivability chart indicate a population that tends to live longer lives, with lower birth and death rates.

40. **(B)** Population B demonstrates a constant rate of death throughout the life span of the members in the population. There are no plateaus in the chart; a plateau would suggest continued survival through that age bracket.

41. **(D)** Population D shows the highest infant mortality because it shows the steepest decline in survivability early in life.

42. **(A)** Population A—the same population that has undergone a demographic transition—demonstrates the lowest infant mortality. Again, the plateau in the chart indicates that the population experiences continued survival at that age bracket.

43. **(A)** A confined aquifer is held between two impermeable layers of rock and is charged at some point from runoff, a lake, or a stream. Hydrostatic pressure pushes on the aquifer so that when it is tapped into it may flow out more forcefully than an unconfined aquifer.

44. **(B)** In contrast to an artesian well, an unconfined aquifer has impermeable rock underneath, but is not confined by impermeable rock from above. Water reaches the aquifer by percolating through the soil, through the zone of aeration, to build up on top of the impermeable rock. The top boundary between the water-saturation porous rock and the zone of aeration is the water table.

45. **(D)** When a well is used, it draws water from the nearby porous rock. If the well is used too heavily, the nearby rock is not recharged fast enough from the neighboring rock, and a cone of depression develops. A cone of depression is really just a drop in the water table near a well.

Practice Exam 3 – Section I: Detailed Explanations of Answers

46. **(C)** Salinization is the process by which brackish water has been used to irrigate crops, and as the water evaporates it leaves a residue of salt on the crops and soil. This residual salt upsets the osmolality of the soil and plants, both of which will lose water to the saltier environment.

47. **(E)** The hydration of sulfur dioxide, created from the combustion of coal, forms sulfuric acid, which contributes significantly to acid rain.

48. **(B)** The conversion of ozone into atmospheric oxygen in the stratosphere removes the protective blanket of ozone that protects humans from UV rays.

49. **(D)** Secondary treatment of wastewater oxidizes ammonia from human waste to form nitrate ions.

50. **(C)** The combustion of carbon (in the form of coal) will form carbon dioxide, which is a greenhouse gas and contributes to global warming.

51. **(A)** Love Canal was a chemical dump site that was later sold to a local school board and developed for housing. In the aftermath of illness and litigation, the area was closed and cleaned up. Love Canal ignited national outrage and catalyzed the passing of CERCLA, which provided funds for toxic waste remediation.

52. **(C)** Catalytic converters contain metal catalysts that further oxidize nitrogen and carbon, which reduces auto emissions significantly.

53. **(C)** The hypolimnion is the bottom layer of water in a lake that does not freeze in the winter. Likewise, it receives nutrients that drop from shallower depths during the summer. The littoral zone is the edge of the lake. The epilimnion is the shallow portion of the middle of the lake. The thermocline is between the hypolimnion and the epilimnion. There is no such thing as a hyperlimnion.

54. **(E)** Marine algae are protists, which can provide food for mollusks, such as snails, which can then be eaten by echinoderms, or star fish. Large mammals, such as whales, will eat phytoplankton, or microscopic marine algae; therefore, two answers are correct.

55. **(B)** A keystone species is a critical link in an ecological chain, without which many other links would not survive.

56. **(A)** The Coriolis effect is made possible by the turning of the Earth underneath the circular convection cells. The convection cells are, in turn, created by the differential heating of the Earth at different latitudes by the sun.

57. **(B)** Dimethylsulfide (DMS) is produced by marine phytoplankton when the sun is out and the water is warm. As more DMS is produced, it acts as a condensation nucleus and promotes cloud formation, which blocks the sun and cools the ocean, which decreases DMS production. The cycled production of DMS—which is a component of the sulfur cycle—therefore acts as a feedback loop to control global climate.

58. **(B)** Of those agencies listed, only the National Forest Service is not managed by the Department of the Interior. The National Forest Service is managed by the Department of Agriculture.

59. **(B)** A secondary carnivore is missing. The large rectangle at the base represents producers. On top of that are herbivores and decomposers. On top of the herbivores are primary carnivores.

60. **(B)** While the other answers would be included in the overall costs of a project in a traditional cost-benefit analysis, the health costs of people made ill by a project would be included in a true cost analysis.

61. **(D)** Channelization is the making of a straighter, deeper channel in an otherwise winding stream or river, and it will increase both stream flow and bank erosion. Channelization will also decrease wetlands and the accompanying biodiversity associated with wetlands.

62. **(C)** The release of methane from heating methane hydrate will increase both the greenhouse effect and the depletion of ozone. Methane has the notable distinction of affecting both global problems, which means that the warming of ice-encased methane creates a positive feedback loop that results in releasing more methane, which warms the planet further and depletes more ozone, and so on.

63. **(D)** The liver serves as the body's chemical laboratory and breaks down toxins through chemical reactions. The kidneys are part of the excretory system and work to excrete toxins that remain in the urine.

64. **(D)** Carcinogenicity, mutagenicity, and teratogenicity all affect the DNA of cells. Carcinogenicity only affects the DNA that codes for the cellular functions that deal with cell growth. Teratogenicity affects the DNA that directs normal fetal growth in utero, and results in birth defects. Mutagenicity is a more general term that affects any DNA.

65. **(A)** The unit rem stands for "radiation equivalent man" and takes into account the amount of cell damage that electromagnetic radiation can inflict. Rads only refer to an amount of energy from radiation. Joules and ergs are also units of energy. Watts is a measure of power, not energy.

Practice Exam 3 – Section I: Detailed Explanations of Answers

66. **(A)** Cholinesterase is the chemical that recycles acetylcholine, an important neurotransmitter in the nervous system. Chlorinated hydrocarbons inhibit cholinesterase, which results in a slow down in the production of the essential neurotransmitter. All of this takes place in nerve cells, or neurons, in the nervous system.

67. **(D)** Toxins that are fat-soluble or hydrophobic (fear water) are more easily absorbed by cells and accumulated (bioaccumulation). The accumulation of the toxin at a low trophic level means that those predators at higher trophic levels will ingest the toxin with each meal, and in turn accumulate the toxin in their cells. This has the effect of compounding the amount of toxin in the higher trophic levels and increasing the effect of the toxin on the predator.

68. **(E)** The Federal Emergency Management Agency (FEMA) would coordinate emergency services in the event of severe storms or other emergencies. The Red Cross is not a government agency. The FDA (Food and Drug Administration) determines standards of safety for food, drugs, medical devices, and cosmetics. The USGS (U.S. Coast and Geological Survey) makes maps and helps track weather. The EPA (Environmental Protection Agency) monitors environmental issues and enforces environmental standards.

69. **(E)** Drip irrigation decreases erosion because less water flows over the soil before it is absorbed, preventing the formation of small rivulets that eventually increase in size to form gully erosion on hills. Terracing the land also decreases erosion because the crop no longer grows on a slope, but on a flat area that steps up a slope.

70. **(D)** A moral agent—in this case, the human—is one who takes responsibility for a moral subject—in this case, the endangered species. The act of expanding one's sense of moral responsibility is called moral extensionism.

71. **(A)** Beef requires 16 pounds of grain to create one pound of meat. Chicken requires 4 pounds of grain to create one pound of meat. Therefore, the chicken is four times more efficient at converting grain into meat than is the steer.

72. **(C)** The production of shrimp in coastal ponds—which implies artificial aquaculture—is part of the "Blue Revolution" that attempted to look at ways to feed the world's hungry from the sea.

73. **(C)** The traits that make an organism susceptible to extinction also tend to be traits of K-strategists. Malthusian growth is an r-strategist trait and is not typical of rare and endangered species. Malthusian, or irruptive, growth would help a species bounce back quickly when conditions were favorable, and it would probably not become rare in the first place.

74. **(E)** All of these are benefits that result in protecting wetlands. Probably the least recognizable answer would be (D) recreational opportunities, but many people enjoy the bird watching or hunting that is available in wetlands.

75. **(B)** The Coriolis effect is driven by convection cycles in the atmosphere creating vertical cycles of air, which are made more horizontal as the Earth turns underneath them.

76. **(A)** The circumpolar vortex is the northern jet stream that circles the North Pole in an elliptical path.

77. **(D)** The sun provides energy for photosynthesis and moves the air and water due to the Coriolis effect. The moon provides gravitational potential energy that drives the tides. The center of the Earth provides heat energy from nuclear reactions that moves the tectonic plates.

78. **(A)** The salt marsh is a transitional ecosystem between the marine environment and terrestrial biomes, and so it contains characteristics of both. It is affected by the tides and must be able to adapt to high osmolarity levels like marine biomes, and yet it maintains a soil and terrestrial plants.

79. **(B)** When two continental plates collide, they push each other up into a mountain range. Both plates resist subduction because they are less dense than oceanic plates. However, one plate tends to slide under the other to some degree, which creates a high plateau on one side of the mountain range. The Tibetan plateau next to the Himalayas, caused by the collision of the Indian plate with the Asian plate, is a good example.

80. **(C)** The Richter Scale is a logarithmic scale, where one point represents a tenfold increase in ground displacement. (This is similar to the pH scale, where one point represents a tenfold increase in acid concentration.) Therefore, an earthquake with a ground displacement that is 100 times that of a 5.2 quake would have a value of 7.2.

81. **(A)** The troposphere is next to the ground, then the stratosphere, then the mesosphere, then the thermosphere.

82. **(E)** All of these events could ultimately be caused by an ENSO event. ENSO brings warm moist air toward the South American coasts, which is swept up toward North America, where it comes in contact with arctic air. The resulting precipitation can create landslides from over-soaked land and rain in Arizona, and contribute moisture to hurricanes in the Gulf. ENSO was originally discovered when anchovy fisheries off the coast of Peru experienced a drop in anchovy production. It turns out that the normal, non-ENSO situation creates seasonal upwelling that provides nutrients for the anchovies. When ENSO reversed the surface currents, the upwelling ceased and the anchovy harvest was diminished.

Practice Exam 3 – Section I: Detailed Explanations of Answers

83. **(D)** Wegener's original evidence involved biogeographical data and the analysis of the mapped coastlines. He advanced his hypothesis long before scientists were able to determine that the magnetic domains in iron ore on the sea floor not only supported his theory, but provided a method to determine the speed of plate movement.

84. **(E)** All of the responses will help prevent erosion. Contour tilling and planting helps to keep water from running off a hillside. Reducing the tillage and leaving material on the field helps to hold in moisture and prevents wind erosion. Eliminating irrigation that uses brackish water will prevent salinization, which is a type of chemical erosion of the soil.

85. **(A)** The water molecule is made of a proton and a hydroxide ion. The proton by itself is acidic. The hydroxide ion by itself is basic. Together they form a neutral molecule. However, if the proton is removed for some reason, the remaining hydroxide ion is basic.

86. **(C)** The rule of 72 helps to solve this problem quickly. 72 = annual growth rate × doubling time. Seventy-two divided by the doubling time (12) would equal the annual growth rate as a percentage (6%).

87. **(E)** Ozone is not an example of indoor air pollution. It tends to be created by photochemical oxidation of nitrogen dioxide outdoors.

88. **(D)** While the greatest amount of fresh water exists in ice and snow, most of that is bound up at the poles and is not usable. Humans rely most heavily on groundwater as a source of freshwater, even though it only contains about 0.3% of the total water on the planet.

89. **(A)** Solar heating is used to evaporate water from lakes and oceans. Once it is in the vapor phase, solar-driven winds move it from place to place. When it precipitates, gravity moves it downhill as a glacier, runoff, stream, or river, or even within the ground.

90. **(A)** The ideal gas law, $PV = nRT$, reminds us of what happens to air when it goes from a low pressure to a high pressure. As the pressure increases on one side of the equation, the temperature increases on the other. Therefore, as the air is compressed, it gets warmer.

91. **(D)** Malaria is the only disease listed that is not caused by a waterborne pathogen. Malaria is caused by a protist called plasmodium, which lives in the digestive gland of the vector mosquito.

92. **(B)** When oxygen-demanding wastes are dumped into a stream, they catalyze the growth of algae. Once the algae population peaks, decomposing organisms begin to feed off it (decomposition zone). The decomposing organisms undergo respiration, which uses up the oxygen dissolved in the water, and no other organisms can survive (septic zone). Eventually, the decomposers also die. As the water flows downhill, it picks up more dissolved oxygen and recovers (recovery zone).

93. **(B)** The mosquito has a much higher tolerance for low oxygen environments than do the other organisms. The stonefly, caddisfly, and mayfly are all good indicator species of a healthy, well-oxygenated stream.

94. **(C)** Drinking and cooking require the least amount of water of all the uses listed. Toilets and perhaps lawns (depending on the amount of lawn and location) require the most water. Low flow toilets go a long way to conserve water.

95. **(B)** Agriculture is by far the largest user of fresh water. Among industrial users, power plant cooling and paper production are the thirstiest. Recycling paper not only saves trees, but it saves fresh water as well.

96. **(E)** All of the answer choices contribute in some way to reducing our society's use of fresh water.

97. **(D)** Nitrates are a form of oxygen-demanding wastes. When oxygen-demanding wastes are dumped into a stream, they catalyze the growth of algae. Once the algae population peaks, decomposing organisms begin to feed off it. The decomposing organisms undergo respiration, which uses up the oxygen dissolved in the water, and no other organisms can survive. Eventually, the decomposers also die. As the water flows downhill, it picks up more dissolved oxygen and recovers.

98. **(B)** Oxygen solubility in water decreases as the temperature of the water increases. You can test this for yourself by comparing cold water out of the tap with hot water out of the tap. The hot water is clouded by bubbles of gas that are coming out of solution. (You can tell the cloudiness is caused by gas because it settles to the top. If it settles to the bottom, you have another problem.) Therefore, the hot water dissolves less gas in it than the cooler water.

99. **(D)** Dioxins are organic, carbon-based molecules that are produced from a number of sources—burning trash being one of them. The other choices—including acids—are inorganic ions.

100. **(A)** Thermal pollution caused by power plants decreases a stream's ability to dissolve oxygen. River water is needed to cool steam coming out of the generator so that it can pick up energy again in the boiler. In some locations, river water can be increased by as much as 5°F, which results in a significant change in the abiotic conditions of the stream.

Detailed Explanations of Answers

Practice Exam 3

Section II

1. (10 points)

   a. (2 points; +1 for set up and units, +1 for correct answer)

   $$13,000 \text{ MW-hr} \times \frac{3,400,000 \text{ BTU}}{1.0 \text{ MW-hr}} = 4.4 \times 10^{11} \text{ BTU}$$

   b. (2 points; +1 for set up and units, +1 for correct answer)

   $$4.4 \times 10^{11} \text{ BTU} \times \frac{1 \text{ lb coal}}{5,000 \text{ BTU}} = 8.8 \text{ million lb coal}$$

   c. (2 points; +1 for set up and units, +1 for correct answer)

   $$\frac{0.64}{0.36} \times 4.4 \times 10^{11} \text{ BTU} = 7.8 \times 10^{11} \text{ BTU}$$

   d. i. (2 points)

   Warming the river water results in thermal pollution, which decreases the ability of the river to dissolve oxygen. Thermal pollution competes with other utilities, such as municipal sewage treatment plants, for the oxygen-dissolving capacity of the river.

   ii. (2 points; +1 for writing the specific heat capacity equation, +1 for correct answer)

   $$\frac{7.8 \times 10^{11} \text{ BTU} \times 0.8 \times °F \times \text{lb}}{5°F \times \text{BTU}} = 1.2 \times 10^{11} \text{ lb water per day}$$

2. (10 points total)

   a. (2 points total; one for correct identification, one for a reasonable description.)

   +1 point: Chart A most likely represents a population in a developing country.

   +1 point for at least one of these explanations:

   - There are many more members of the population at the younger ages than the older ages. This means that the birth rates are high and the death rates are high.

   - More children are born with the hope that some will survive to reach productive years, so that the parents may be supported in old age.

   - Low levels of technology for food production and medical care mean that not all people survive until adulthood. Life span is lower than population B.

   b. (2 points total; +1 for a correct representation that resembles each of the two populations on the curve. Partial credit of +1 for correct axes, if nothing else is correct.)

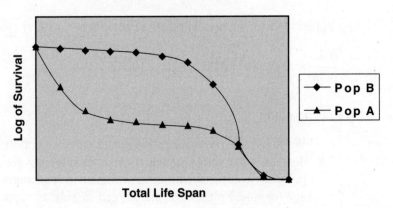

Survivorship Curves

Practice Exam 3 – Section II: Detailed Explanations of Answers

c. (3 points total; +1 for each of the following)

- Improved source of wealth decreases the need to rely on children for support in old age
- Improved medical care increases survivability of children
- Improved technology in food production or food distribution increases survivability of children
- Increased opportunities for women
- Increased educational opportunities for population

d. (3 points total; +1 for each of the following)

- Improved wealth can improve infrastructure that helps the environment (e.g., sewage treatment)
- Increased industrialization may have environmental consequences (e.g., increased pollution)
- Steady state population has less chance of exceeding the carrying capacity in the future
- Better chance for government officials to plan for the future, especially with respect to food production, medical care, energy usage, and pollution control

3. (10 points total)

   a. (3 points total)

   The pipeline could be considered responsible only if two conditions are both met. First, a statistically significant correlation between the location of the cancer incidents and the existence of oil in the groundwater would connect the illnesses to the pollution. However, it is difficult to blame the illnesses on the pollution unless a retrospective study of the scientific literature produced case studies in which oil exposure specifically caused the observed illness. Also, dose-response curves for oil can be established using laboratory animals.

   b. (4 points total; +2 each for i. and ii.)

   i. The existence of the cluster in that location suggests that the groundwater flows downhill in the direction of the disease cluster.

ii. Possible methods of remediation could include the following:

- Air stripping: pumping warm air into the ground to vaporize the oil and allow it to come to the surface

- French drains cut deep enough to interrupt the plume, have the oil flow into the drain, then pump out the oil

c. (2 points) CERCLA is the law that requires these spills to be remediated (+1), and the law provides financial support (the Superfund) to help the remediation process (+1).

4. (10 points total)

a. (2 points)

The niche overlaps to the degree that the habitat requires the same ranging area, but the ecological role (food sources) does not overlap.

b. (2 points)

The removal of the wolf is an example of a threatened species biodiversity, where individuals of a species are being removed. (Contrasted with genetic biodiversity, which refers to the number of different copies of genes existing in the population's gene pool; or ecosystem biodiversity, which refers to the availability of habitat.)

c. (3 points)

The role of a top predator such as the wolf has a strengthening effect on the prey populations. Because wolves will first attack weaker members of the population, those prey with weaker genotypes have less of a chance to survive. This has a stabilizing effect on the natural selection of the prey population—strengthening those genotypes that adapt well.

d. (3 points)

The re-introduction program is strong because it combines the benefits of a captive breeding program—without the limitations—with the benefits of establishing a preservation. One of the normal limitations of captive breeding programs is that the species do not reproduce successfully under artificial conditions. The wolf re-introduction program puts the wolves in a natural habitat and allows them to expand their range when each new pack is ready.

Practice
Exam 4

Practice
Exam 4

AP Environmental Science

Practice Exam 4

Section I

TIME: 90 minutes
100 multiple-choice questions

(Answer sheets appear in the back of this book.)

> **DIRECTIONS:** Each of the questions or incomplete statements below is followed by five suggested answers or completions. Select the best answer for each question and then fill in the corresponding oval on the answer sheet.

Questions 1–4 refer to the following answers. Select the one lettered choice that best fits each statement.

    (A)   Nitrogen gas

    (B)   Oxygen gas

    (C)   Ozone

    (D)   Chlorinated fluorocarbons

    (E)   Carbon dioxide

1. Exists in the stratosphere; protects living organisms from UV radiation

2. Oxide of this gas contributes to acid rain problem

3. Product of respiration that contributes to global warming

4. Acts as a catalyst to deplete the ozone layer

AP Environmental Science

Questions 5–8 refer to the following answers. Select the one lettered choice that best fits each statement.

    (A) Nitrogen dioxide

    (B) Sulfur dioxide

    (C) Ozone

    (D) Radon

    (E) Volatile organic compounds

5. Major indoor pollutant with nonanthropogenic source

6. NOT a criteria pollutant

7. Major source is from coal-fired power plants.

8. Irritant formed as a result of photochemical oxidation

Questions 9–12 refer to the following answers. Select the one lettered choice that best fits each statement.

    (A) Bacterial pathogen

    (B) Anthropogenic toxin

    (C) Animal vector

    (D) Animal toxin

    (E) Protist pathogen

9. Causes giardia, amebic dysentery, and sleeping sickness

10. Gypsy moth, malaria mosquito, and rats

11. Mercury, lead, dioxins, and CFCs

12. Causes tetanus, pneumonia, cholera, botulism, and gonorrhea

Questions 13–16 refer to the following answers. Select the one lettered choice that best fits each statement.

    (A) Neurotoxin

    (B) Allergen

    (C) Mutagen

    (D) Teratogen

    (E) Corrosive

13. Ozone is an example.

14. Carcinogens are a subset of this category.

15. Causes birth defects

16. Minimata disease is caused by this type of toxicity.

Questions 17–20 refer to the following answers. Select the one lettered choice that best fits each statement.

    (A) Tiaga
    (B) Tundra
    (C) Temperate deciduous forests
    (D) Savannah
    (E) Cloud forests

17. Tend to occur on the windward side of equatorial mountains

18. The driest of these biomes

19. Forest of the far north

20. Its trees lose their leaves in the fall

Questions 21–24 refer to the following answers. Select the one lettered choice that best fits each statement.

    (A) Euphotic zone
    (B) Limnetic zone
    (C) Pelagic zone
    (D) Benthic zone
    (E) Littoral zone

21. The section of a freshwater lake that is near the surface, where algae might grow

22. The bottom of either a lake or an ocean

23. The open water section of a lake where the sun does not easily penetrate

24. The shallow portion of the lake that can allow rooted vegetation to grow

AP Environmental Science

Questions 25–28 refer to the following answers. Select the one lettered choice that best fits each statement.

    (A) Carbon cycle

    (B) Nitrogen cycle

    (C) Phosphorous cycle

    (D) Sulfur cycle

    (E) Water cycle

25. Includes fermentation and respiration

26. Nutrient is always in the form of the same polyatomic ion

27. Helps to drive the formation of hurricanes

28. Phytoplankton production in this cycle can cool global climate

Questions 29–32 refer to the following answers. Select the one lettered choice that best fits each statement.

    (A) Gifford Pinchot

    (B) Rachel Carson

    (C) Garrett Hardin

    (D) George Perkins Marsh

    (E) John Muir

29. Wrote "Tragedy of the Commons"

30. Helped to found the National Forest Service, which embodied his philosophy of deriving the greatest good for the greatest number of people

31. Wrote *Silent Spring*

32. Biocentric conservationist who helped found the National Park Service

Questions 33–36 refer to the following answers. Select the one lettered choice that best fits each statement.

(A) Department of the Interior

(B) Department of Commerce

(C) Department of Agriculture

(D) Department of Health and Human Services

(E) Environmental Protection Agency

33. Manages the BLM and enforces the SMCRA

34. Established by NEPA in 1970

35. Manages the National Forest Service

36. Operates marine research vessels that monitor ocean and atmospheric conditions on a global scale

Questions 37–40 refer to the following answers. Select the one lettered choice that best fits each statement.

(A) Specific heat

(B) Latent heat of vaporization

(C) Heat of entropy

(D) Heat of combustion

(E) Thermohaline gradient

37. Responsible for keeping coastal cities within a moderate temperature range

38. Responsible for providing the energy to help construct a hurricane

39. Waste heat that is not useful

40. Responsible for causing subsurface ocean currents

AP Environmental Science

Questions 41–44 refer to the following answers. Select the one lettered choice that best fits each statement.

(A) Troposphere
(B) Lithosphere
(C) Stratosphere
(D) Thermosphere
(E) Mesosphere

41. Contains molten magma

42. The jet stream exists just below this layer.

43. Most of the Earth's weather occurs here.

44. Contains disperse, ionized gas

Questions 45–48 refer to the following answers. Select the one lettered choice that best fits each statement.

(A) Active solar heating
(B) Passive solar heating
(C) Photovoltaic cells
(D) Geothermal energy from heat pumps
(E) Wind generators

45. The form of heat that does not originate from the sun

46. Solar heating that requires a pump

47. Produces electricity using no moving parts

48. Heats and cools by virtue of house design and orientation

Questions 49–52 refer to the following answers. Select the one lettered choice that best fits each statement.

(A) Volcanic eruption
(B) Fecundity
(C) Competition
(D) Predation
(E) Disease

49. Ability to reproduce and have offspring

50. A type of symbiosis

51. A density independent factor that affects population size

52. Leads to regular spacing between plants

Questions 53–56 refer to the following answers. Select the one lettered choice that best fits each statement.

    (A) Heap-leach extraction

    (B) Electrolysis

    (C) Electrostatic precipitation

    (D) Reverse osmosis

    (E) Bioremediation

53. Used to obtain fresh water from sea water

54. Used to refine aluminum metal from bauxite ore

55. Process that uses cyanide solutions to extract gold from ore

56. Uses bacteria to digest toxic wastes

Questions 57–60 refer to the following answers. Select the one lettered choice that best fits each statement.

    (A) Desertification

    (B) Eutrophication

    (C) Reclamation

    (D) Succession

    (E) Remediation

57. A change in a lake that will eventually fill in the lake

58. Accelerated by overgrazing dry grasslands

59. One biological community replaces another

60. Clean up of a toxic chemical dump

AP Environmental Science

61. Settling of a ground surface due to over-drawing water from nearby wells is called

    (A) upwelling
    (B) subsidence
    (C) vertical stratification
    (D) plume
    (E) downburst

62. Stone fly nymphs live in beds of streams that contain a high level of dissolved oxygen (DO). When the dissolved oxygen decreases in the stream, stone fly nymphs can no longer live there. Stone fly nymphs are an example of

    (A) a keystone species
    (B) an indicator species
    (C) an endangered species
    (D) a marine species
    (E) a threatened species

63. Which of the following represents the smallest use of fresh water?

    (A) Residential
    (B) Agricultural
    (C) Paper and pulp mills
    (D) All industrial uses
    (E) Cooling power plants

64. Which of the following occurs as a result of acid rain?

    (A) Global warming
    (B) Ozone depletion
    (C) Aluminum poisoning of plants
    (D) An increase in stream pH
    (E) At least two of the above

65. Which of the following is an adaptation that plants use in hot, dry desert regions?

    (A) Broad leaves

    (B) Waxy leaves and thorny stems

    (C) Thin leaves on plants that are low to the ground

    (D) Broad, shallow web of roots

    (E) High leaf to woody stem ratio

66. Which technique of air pollution remediation turns an air pollution problem into a solid waste problem?

    (A) Fluidized bed combustion

    (B) Catalytic converter

    (C) Staged burner

    (D) Electrostatic precipitators

    (E) At least two of the above

67. Which of the following is an initial step in accomplishing a demographic transition?

    (A) Decrease in death rates

    (B) Decrease in birth rates

    (C) Increase in birth rates

    (D) Increase in death rates

    (E) Lack of change in either birth or death rates

68. The epicenter of an earthquake exists

    (A) on the surface of the Earth

    (B) at the actual spot underground where two tectonic plates adjust position relative to one another

    (C) under the ocean

    (D) where the earthquake does the most damage

    (E) Two of the above

69. A seismically active arc of islands, such as Alaska's Aleutian Island chain, is an indicator of

   (A) the convergence of oceanic and continental plates

   (B) a mid-plate hotspot

   (C) the convergence of two oceanic plates

   (D) a divergent boundary

   (E) a mid-oceanic rift

70. How long does it typically take to produce one millimeter of top soil?

   (A) About one week

   (B) About one year

   (C) About a decade

   (D) About a hundred years

   (E) About a month

71. Most of the initial oxygen in the atmosphere was created by

   (A) volcanoes

   (B) bacteria

   (C) algae

   (D) plants

   (E) animals

72. Which of the following types of energy does not originate in a stellar nuclear reaction?

   (A) Nuclear power

   (B) Tidal power

   (C) Fossil fuels

   (D) Wind power

   (E) Hydroelectric power

Practice Exam 4 – Section I

73. Which of the following is the most prevalent gas in the atmosphere?

    (A) Nitrogen

    (B) Oxygen

    (C) Ozone

    (D) Carbon dioxide

    (E) Water vapor

74. Ice ages that occur about every 100,000 years are due to

    (A) Milankovitch cycles

    (B) volcanoes

    (C) El Niño

    (D) the Coriolis effect

    (E) sun spots

75. Which of the following does NOT contribute to global warming?

    (A) Burning fossil fuels

    (B) Heating peat

    (C) Melting polar ice

    (D) Volcanic eruptions

    (E) Methane release from melting permafrost

76. At which portion of the nuclear fuel cycle is the highest volume of radioactive waste produced?

    (A) Mining

    (B) Milling

    (C) Fuel cell fabrication

    (D) Fuel reprocessing

    (E) Fuel disposal

AP Environmental Science

77. Which of the following nuclear reactions does NOT take place inside a nuclear reactor?

    (A) Fission
    (B) Fusion
    (C) Beta decay
    (D) Alpha decay
    (E) Neutron capture

78. A transfer of energy involves three steps, each of which is 80% efficient. What is the approximate total efficiency of all three steps?

    (A) 24%
    (B) 51%
    (C) 66%
    (D) 80%
    (E) 98%

79. Which of the following supplies the greatest amount of energy in developed countries?

    (A) Oil
    (B) Coal
    (C) Natural gas
    (D) Hydroelectric
    (E) Solar

80. A population experienced a birth rate of 23 people per thousand and a death rate of 11 deaths per thousand. What is the approximate growth rate of the population?

    (A) 0.12%
    (B) 1.2%
    (C) 2.4%
    (D) 3.8%
    (E) None of the above

Practice Exam 4 – Section I

81. A stream with a pH of 4.0 is

    (A) strongly acidic
    (B) weakly acidic
    (C) neutral
    (D) weakly basic
    (E) strongly basic

82. The period within the Paleozoic Era during which most of the world's coal reserves began forming is the

    (A) Cambrian period
    (B) Cretaceous period
    (C) Carboniferous period
    (D) Jurassic period
    (E) Quarternary period

83. Denitrifying bacteria

    (A) convert nitrates into atmospheric nitrogen
    (B) convert atmospheric nitrogen to ammonia
    (C) convert ammonia into nitrites
    (D) convert nitrites into nitrates
    (E) convert nitrites into sulfates

84. The underground section between the surface of the Earth and a water table is called the

    (A) recharge zone
    (B) confined aquifer
    (C) transform boundary
    (D) zone of aeration
    (E) divergent boundary

387

85. The process by which plants release water is called

    (A) condensation

    (B) evaporation

    (C) transpiration

    (D) vaporization

    (E) perspiration

86. A population takes about seven years to double in size. This population's annual percentage growth is approximately

    (A) 2%

    (B) 7%

    (C) 10%

    (D) 15%

    (E) 22%

87. Which of the following is NOT a trait demonstrated by K-strategist organisms?

    (A) Population tends to overshoot the carrying capacity

    (B) Adults invest considerable energy in the next generation.

    (C) Higher on the food chain

    (D) Tend to be long-lived

    (E) Tend to mature slowly

88. Which of the following is the most prevalent form of water pollution?

    (A) Oxygen-demanding wastes

    (B) Pathogens

    (C) Methylated mercury

    (D) Sediment

    (E) Acid deposition

89. Which federal law provided for a fund to clean up massive toxic spills?

    (A) NEPA
    (B) FFDCA
    (C) ESA
    (D) SMCRA
    (E) CERCLA

90. Milpa agriculture involves clearing a portion of rainforest and harvesting diverse crops from the area as the forest undergoes succession. Milpa agriculture is different from modern "slash-and-burn" farming in that

    (A) Milpa agriculture involves small patches of forest at a sustainable rate
    (B) Milpa agriculture involves larger patches of land than "slash-and-burn" farming
    (C) "slash-and-burn" farming involves planting seeds and using irrigation
    (D) Milpa allows for the growth of fewer crops than "slash-and-burn" agriculture
    (E) There is no difference between Milpa agriculture and "slash-and-burn" deforestation; they are the same.

91. This environmental crisis involved the meltdown of a nuclear facility.

    (A) Love Canal
    (B) Bhopal Crisis
    (C) *Exxon Valdez*
    (D) DDT
    (E) Chernobyl

92. A Habitat Conservation Plan guarantees that

    (A) no endangered species are killed

    (B) natural resources can be used as long as the species experiences some overall benefit

    (C) an Environmental Impact Statement must be produced

    (D) all endangered species receive in situ management

    (E) the habitat can only be used for the conservation of endangered species

93. Integrated Pest Management involves the use of

    (A) pesticides

    (B) natural predators

    (C) sterilized males to breed unsuccessfully with females

    (D) sex attractants to lure insects away from crops

    (E) All of the above

94. Environmental consequences of raising beef in feedlots include

    (A) nutrient pollution of groundwater and nearby waterways

    (B) increased bacterial resistance to antibiotics

    (C) increased levels of hormones in our food

    (D) higher cholesterol in those who eat beef that is higher in fat content

    (E) All of the above

95. Which of the following is NOT a renewable source of energy?

    (A) Organic material compressed to form coal

    (B) Corn crops fermented to form ethanol

    (C) Solar energy

    (D) Hydroelectric energy

    (E) Tidal energy

96. How many 100 W light bulbs could be powered for two hours by 1.0 kW-hr of energy?

    (A) 2
    (B) 5
    (C) 10
    (D) 50
    (E) 100

97. Normally, rain is slightly acidic due to

    (A) sulfurous acid created from combustion of coal
    (B) metallic acid deposition from mining wastes
    (C) carbonic acid created by carbon dioxide in the air
    (D) nitrous acid created from oxides of nitrogen
    (E) a combination of anthropogenic sources

98. Which of the following is NOT a way to disinfect waterborne pathogens in the water treatment process?

    (A) UV light
    (B) Ozone
    (C) Chlorine
    (D) Soap
    (E) Combination of two or more of the above

99. Ozone in the stratosphere is important to life on Earth because

    (A) it blocks the sun's intensity
    (B) it absorbs microwaves
    (C) it is broken down by CFCs
    (D) it absorbs ultraviolet rays
    (E) it absorbs greenhouse gases

100. Which of the following does NOT represent pronatalist pressure to a Third World family who lives in a country that has not yet undergone a demographic transition?

   (A) Status of being wealthy and fertile enough to have children

   (B) Educational and occupational opportunities for parents

   (C) One must have many children so that some survive into adulthood.

   (D) Children are a source of cheap labor.

   (E) Children are needed to support parents in old age.

AP Environmental Science

Practice Exam 4

Section II

Time: 90 minutes

DIRECTIONS: You have 90 minutes to answer all four of the following questions. It is suggested that you spend approximately half your time on the first question and divide the remaining time equally among the next three. In answering these questions, you should emphasize the line of reasoning that generated your results; it is not enough to list the results of your analysis. Include correctly labeled diagrams, if useful or required, in explaining your answers. A correctly labeled diagram must have all axes and curves labeled and must show directional changes.

---

1. Answer the questions that relate to the following article.

    Developers Eager to Create Habitat

    In recent weeks, Johnson and Smith, Inc., a major Bay Area developer of residential neighborhoods, has been jumping at any chance to buy real estate. That is nothing new, but it seems odd that they are not going to use the land to build houses, but to preserve a pristine habitat for endangered frogs, foxes, and fish.

    The beneficiaries are the California red-legged frog, the Delta smelt, and the San Joaquin fox. The U.S. Fish and Wildlife Service has stated that developers will be allowed to use the habitat of these species for housing projects if they offset the damage by preserving a place for these species somewhere else.

Critics of this arrangement state that this will create an active real estate market for land that needs to be left alone. A representative from Johnson and Smith responded by saying that the land will be put into a trust once it is purchased. This will keep it off of the secondary real estate market.

This type of scheme has prompted other developers to begin buying ecologically significant land that previously had no commercial value, such as wetlands. Tad Pole, owner of 320 acres of wetland near the delta, states "This is a real windfall for me. I could not have afforded to preserve these lands. Now someone is coming to me with an offer that is many times what I could have gotten on the market at this time last year. Now I'm thinking about retiring."

While Pole and several species of wildlife are probably feeling pretty good about this, some developers complain that this arrangement increases costs that they have to then pass on to the buyers of the homes they are building. "Well, at least everyone is getting a place to live," says Pole.

a. Identify a major law that is being enforced by the U.S. Fish and Wildlife Service. What cabinet-level department oversees this agency, and to which branch of the government—legislative, executive, or judicial—does this department belong?

b. Use your understanding of environmentally healthy cost-benefit analyses to respond to the developer's complaint about having to pass on the cost of the land preserves to consumers.

c. Of the species mentioned, pick one and identify three traits of the species that predispose it to be rare or endangered.

d. Even though land has been set aside for wildlife, it doesn't necessarily mean that they are safe from environmental hazards. Pick three remaining risks to survival for these species; briefly explain why you chose these risks.

Practice Exam 4 – Section II

2. A new plan has been made to create an aqueduct system to transport water from the Columbia River in Washington State to the parched lands of eastern Oregon. The plan involves the construction of a channel so that water can flow to inexpensive land that is marginally productive. This would expand the local economy and add to the already abundant areas that grow wheat in Washington and Oregon.

   a. Identify two environmental concerns that you have about this project.

   b. Pick one of the concerns you cited in the previous question and identify another major water project in the world where this concern has proven to be valid.

   c. The growers in the region would like to alternate their wheat crop—which would be used for flour—with another crop that would be used to feed cattle. Comment on the ecological consequences of this plan.

   d. Describe three agricultural techniques that would conserve fresh water.

3. Several car companies have produced versions of electric cars and "hybrid" automobiles. The electric car plugs into a regular wall socket at home and operates off of electricity that is stored in batteries within the car. The hybrid car uses gasoline for long-distance driving, but a new type of braking system directs the resistance in the braking process to a generator that produces electricity. The electricity is then stored in a battery and used when the hybrid drives at low speeds.

   a. Does the electric car really save energy or reduce pollution? Why or why not?

   b. Which of these two cars gives off fugitive emissions? Explain your answer briefly.

   c. Compare the environmental consequences of using the electric car in Seattle—which uses predominantly hydroelectric and nuclear power to produce electricity—versus Atlanta, which uses coal to produce electricity.

   d. Propose a policy or law that would promote the use of electric or hybrid vehicles.

AP Environmental Science

4. You have inherited a parcel of land in an undeveloped portion of a moderately sunny climate. You would like to build a home there, but the land is somewhat isolated and it is far from power, phone, water, and sewer systems.

   a. You can bring power to the house, but the utility says that you have to pay for the cable that comes to your home. For the price that you would have to pay the power company, you can put in your own alternative energy system. Identify four ideas that you would implement to provide energy, electric or otherwise.

   b. Every household needs to be able to eliminate septic waste from sinks, toilets, showers, and washing. Since you are too far from the town's sewer line, you have to put in a septic system. Describe how the nitrogen cycle is involved with septic treatment systems.

   c. You want to move to the country to escape the air pollution of urban life, but indoor air pollutants in a rural home can be just as unhealthy. Identify three indoor air pollutants and one method for reducing each pollutant.

AP Environmental Science

Practice Exam 4
Answer Key

Section I

1.	(C)	26.	(C)	51.	(A)	76.	(B)
2.	(A)	27.	(E)	52.	(C)	77.	(B)
3.	(E)	28.	(D)	53.	(D)	78.	(B)
4.	(D)	29.	(C)	54.	(B)	79.	(A)
5.	(D)	30.	(A)	55.	(A)	80.	(B)
6.	(D)	31.	(B)	56.	(E)	81.	(A)
7.	(B)	32.	(E)	57.	(B)	82.	(C)
8.	(C)	33.	(A)	58.	(A)	83.	(A)
9.	(E)	34.	(E)	59.	(D)	84.	(D)
10.	(C)	35.	(C)	60.	(E)	85.	(C)
11.	(B)	36.	(B)	61.	(B)	86.	(C)
12.	(A)	37.	(A)	62.	(B)	87.	(A)
13.	(E)	38.	(B)	63.	(A)	88.	(D)
14.	(C)	39.	(C)	64.	(C)	89.	(E)
15.	(D)	40.	(E)	65.	(B)	90.	(A)
16.	(A)	41.	(B)	66.	(E)	91.	(E)
17.	(E)	42.	(C)	67.	(A)	92.	(B)
18.	(B)	43.	(A)	68.	(A)	93.	(E)
19.	(A)	44.	(D)	69.	(C)	94.	(E)
20.	(C)	45.	(D)	70.	(B)	95.	(A)
21.	(A)	46.	(A)	71.	(B)	96.	(B)
22.	(D)	47.	(C)	72.	(B)	97.	(C)
23.	(B)	48.	(B)	73.	(A)	98.	(D)
24.	(E)	49.	(B)	74.	(A)	99.	(D)
25.	(A)	50.	(D)	75.	(D)	100.	(B)

AP Environmental Science

Detailed Explanations of Answers

Practice Exam 4

Section I

1. **(C)** Ozone is the gas that exists in the stratosphere and protects living organisms from harmful UV radiation. Most likely, life evolved on Earth with the ozone layer intact, so all organisms exist with a tolerance range that would be overcome if the ozone layer were to be depleted. One might argue that humans would figure out a way to survive without this UV protection, but we are very dependent on other organisms for our food; they would need to survive as well.

2. **(A)** The oxide of nitrogen gas is nitrogen dioxide, which combines with water vapor in the atmosphere to form acid rain. Nitrogen sources are responsible for about 30% of the acid rain, with the remainder coming from oxides of sulfur.

3. **(E)** Carbon dioxide contributes to global warming.

4. **(D)** Chlorinated hydrocarbons are the most prevalent ozone-depleting gas. The chlorine strips an oxygen atom off the ozone molecule, turning the ozone molecule into oxygen gas. The remaining oxygen atom combines with another to form more oxygen gas.

5. **(D)** Radon is a major indoor pollutant with a nonanthropogenic source. Radon is responsible for about half the lung cancer cases in the United States, second to smoking. Radon is produced in the Earth and in earthen materials (like concrete) during the decay of radioactive heavy elements. Radon is the only decay product that is a gas, so it will bubble out of the ground when it is formed, then get lodged in a well-insulated home and reside there unless ventilation is adequate to remove it.

6. **(D)** Radon is the only gas on the list that is not a criteria pollutant. The criteria pollutants were designated by the Clean Air Act as the most important gases to regulate. They are the oxides of carbon, nitrogen, and sulfur; photochemical oxidants such as ozone; heavy metals and halogens—such as lead, mercury, and CFCs; particulate matter; and volatile organic compounds (VOCs).

Practice Exam 4 – Section I: Detailed Explanations of Answers

7. **(B)** Sulfur dioxide mostly comes from coal-fired power plants. The reason why coal contains sulfur is related to how coal was formed. Most sulfur was created from anaerobic decomposition of plants. Sulfur creates the di-sulfide linkages that hold proteins in their three-dimensional configuration. When the plants died and formed coal, the sulfur remained. Now when the coal is burned, the sulfur in the coal combines with oxygen to form sulfur dioxide. When the sulfur dioxide combines with water vapor in the atmosphere, it forms acid rain. Sulfur dioxide from coal combustion is responsible for about 70% of acid rain.

8. **(C)** Ozone is the irritant formed from photochemical smog. You wouldn't be the only one who thinks, "Why don't they just pipe the unwanted ozone near the Earth's surface up to the stratosphere to repair the ozone hole?" It might work if they could find a long enough pipe.

9. **(E)** All of these diseases are caused by one protist pathogen or another.

10. **(C)** All of these organisms represent vectors from the animal kingdom. Vectors are disease carriers, not the actual organism that causes the disease. Sometimes, eradication of a disease begins with eradicating the vectors.

11. **(B)** All of these items listed are toxins of anthropogenic origin. Although some mercury is also emitted from volcanoes, most of it is emitted from coal-fired power plants.

12. **(A)** All of these organisms are caused by one bacterial pathogen or another.

13. **(E)** Ozone is an example of a corrosive molecule. If you are an athlete in an urban area and had the taste of blood in your mouth after a particularly strenuous race or game, then you have probably experienced the corrosive effects of ozone.

14. **(C)** Mutagens alter the genetics of a cell. We now know that all cancer is based upon some alteration of the genetics of a cell, so all carcinogens are also mutagens. Mutagens that cause cancer come in one of two types: they either turn off the gene (tumor suppressor gene) that keeps the cell from growing out of control, or they turn on a gene (oncogene) that causes the cell to undergo uncontrolled growth. Usually a single cell needs four or five hits by a mutagen in order for the cell to become cancerous.

15. **(D)** Teratogens are chemicals that cause birth defects. A notorious example is thalidomide, a drug that was prescribed in the 1950s. While it worked well as a pain-reliever, if an expecting mother took the drug, it would cause a severe shortening—or even elimination—of the arms or legs of her baby.

16. **(A)** Minimata disease was caused by mercury poisoning (a neurotoxin).

17. **(E)** The windward side of equatorial mountains receive almost constant rainfall or mist as the moist air releases water to decrease in pressure enough to pass over the mountains.

18. **(B)** Of the biomes listed, a tundra is the most dry. While it may have some water, most of it is bound into ice or permafrost. The next driest biome listed is a savannah, but savannahs tend to have a rainy season at some time during the year.

19. **(A)** Tiaga forests are the northernmost forest. Tiaga borders on both arctic and alpine tundras.

20. **(C)** Deciduous trees lose their leaves in the fall.

21. **(A)** The euphotic zone is the section of the lake near the surface where algae typically grow.

22. **(D)** The bottom of a lake or ocean is called the benthos, and benthic organisms live there.

23. **(B)** Underneath the euphotic zone in the middle of a lake is the limnetic zone, where light does not easily penetrate.

24. **(E)** The littoral zone is the shallow portion of the lake where rooted vegetation grow. This is an important nursery zone where small fish and invertebrate larvae grow to adulthood. It may also be the area where a moose walks through to eat the tender roots of littoral grasses.

25. **(A)** The carbon cycle includes fermentation and respiration. Respiration involves the complete combustion of carbon containing carbohydrates to carbon dioxide, which then enters the atmosphere. Fermentation is the incomplete combustion of carbon containing carbohydrates into alcohols and ketones. Primitive microorganisms called yeasts undergo fermentation.

26. **(C)** In the phosphorous cycle, the phosphorous atom remains in the form of a phosphate ion (PO_4^{3-}) throughout the cycle. In the carbon, nitrogen, and sulfur cycles, the element is oxidized, reduced, and recombined with different atoms, and also exists by itself. Water is not really a nutrient cycle and does not exist as an ion.

27. **(E)** The water cycle helps drive the formation of hurricanes because it is the evaporation of water in the equatorial Atlantic that builds the energy in a hurricane.

Practice Exam 4 – Section I: Detailed Explanations of Answers

28. **(D)** Dimethylsulfide—which is part of the sulfur cycle—is produced by marine plankton. When the ocean is warm, marine phytoplankton release dimethylsulfide, which is oxidized to sulfur dioxide and then to sulfate ions in the atmosphere. The sulfate ions act as condensation nuclei that help produce clouds and cool the Earth.

29. **(C)** Garrett Hardin wrote "Tragedy of the Commons."

30. **(A)** Gifford Pinchot was a utilitarian conservationist who helped found the National Forest Service, which oversees the national forests. The national forests are sometimes called "the land with many uses" because they are used for commercial operations (such as logging), recreation, and preservation.

31. **(B)** Rachel Carson wrote *Silent Spring*, which many credit as catalyzing the modern environmental movement. *Silent Spring* outlined the drawbacks of using DDT to fight the vector of Dutch Elm disease. Carson increased the public's awareness of bioaccumulation and biomagnification.

32. **(E)** John Muir was a biocentric conservationist who was a writer, activist, and the first president of the Sierra Club. His influence was instrumental in establishing the National Park Service and, in particular, Yosemite National Park.

33. **(A)** The Department of the Interior manages the Bureau of Land Management and enforces the Surface Mining Reclamation and Control Act. It also manages other public land-related agencies, such as the Fish and Wildlife Service, the National Park Service, and the United States Geological Survey (USGS).

34. **(E)** The Environmental Protection Agency was established by the National Environmental Policy Act in 1970. The EPA is charged with protecting the environment through research and enforcement of many environmental laws.

35. **(C)** The Department of Agriculture manages the National Forest Service, and also the Farm Service Agency and the National Resource Conservation Agency.

36. **(B)** The Department of Commerce operates the National Oceanic and Atmospheric Administration (NOAA). The National Weather Service is a part of NOAA.

37. **(A)** The specific heat capacity is the property of water that helps to moderate climate and weather in coastal cities. Water has an exceptionally high specific heat, which means that it holds a lot of heat for every degree that it is warmed. This means that a bay or a lake will provide a heat sink for a region. Rather than heat energy going into raising the temperature, it is absorbed by the water and does not increase the temperature as much.

38. **(B)** The latent heat of vaporization determines how much heat is absorbed when water is vaporized. The vaporized water stores that heat, which is returned when the water condenses. The vaporization of water in the tropical Atlantic off the coast of Africa fuels the formation of hurricanes in the summer. After it is formed, a hurricane carries the vaporized water across the Atlantic to the eastern coast of North or South America.

39. **(C)** Vaguely put, entropy is a measure of disorder. The Second Law of Thermodynamics states that all spontaneous processes lead to an increase in the entropy of the universe. More specifically, every transfer of heat allows the escape of some low-quality heat that can't be used for anything. This low-quality heat is entropy.

40. **(E)** Subsurface ocean currents are driven by thermohaline currents, which are currents that move water from high-density areas to low-density areas. The density of the water increases as the salinity increases or the temperature decreases.

41. **(B)** The lithosphere is the only choice that is not in the atmosphere, and is composed primarily of molten magma. The lithosphere exists just below the crust.

42. **(C)** The jet stream exists just below the stratosphere in an area called the tropopause.

43. **(A)** Most of the Earth's weather, and most of the molecules in the atmosphere, occur in the troposphere—which is the layer of atmosphere closest to the Earth.

44. **(D)** The thermosphere contains ionized gas of very low density. This is the area where the aurora borealis occurs near the poles, as magnetic fluctuations accelerate the free electrons surrounding the plasma and emit photons.

45. **(D)** One doesn't have to go very deep into the ground to find a constant temperature. This geothermal energy is ultimately fueled by the nuclear reactions in the center of the Earth, and can be tapped from any place in the Earth by using a geothermal heat pump. A geothermal heat pump is much like a refrigerator, which pumps heat from one place to another. When it is cold, a geothermal heat pump pumps heat from the Earth to the home; when it is hot, it works the other way around.

46. **(A)** Active solar heating refers to the type of solar heat that requires energy and a pump (moving parts) at some point. For example, one active scheme would pump water from the hot water heater to be heated on the roof. Whatever amount of heating the water experiences from the sun on the roof is an amount of heat that doesn't need to come from gas or electricity.

Practice Exam 4 – Section I: Detailed Explanations of Answers

47. **(C)** Photovoltaic solar cells capture photons, which set mobile electrons into motion. Conductors within the cell capture the moving electrons, and they pass along a wire as electricity.

48. **(B)** Passive solar heating is based on design, rather than moving parts and an input of electrical energy. For example, planting a deciduous tree on the south side of the house is a passive design. The tree shades the house in the summer and allows sun to shine through in the winter. Having a trombe wall behind a window is a passive technique. The wall absorbs heat during the day, then releases heat in the evening.

49. **(B)** Fecundity is the ability to have offspring. Fertility is the number of births—as measured usually by the crude birth rate, which is equal to the number of births per 1,000 people per year.

50. **(D)** Predation is a type of symbiosis, just as is parasitism. One species benefits (predator), and the prey does not.

51. **(A)** Volcanic eruption is an abiotic, density-independent factor that could affect population size.

52. **(C)** Regular spacing between plants suggests that they are in competition with one another for resources. Random distribution suggests that individuals are not experiencing intraspecific competition.

53. **(D)** Reverse osmosis (RO) is the technique that is used to remove dissolved and suspended particles from seawater. It works by pushing the water at great pressure through a fine, heavily folded filter.

54. **(B)** Bauxite ore is crushed, melted, and then put into a vat where a strong electrical current is passed through it. The aluminum cations are attracted to the negatively charged anode, given more electrons, and converted into the metal. This process is called electrolysis.

55. **(A)** Gold is refined using heap-leach extraction. Heap-leach extraction refers to putting gold ore in a big heap, or pile, and letting an acidified cyanide solution drip through the pile. The solution dissolves the gold, which can then be retrieved from the solution through electrolysis.

56. **(E)** Bioremediation is the process of using bacteria to digest one kind of waste or another. Generally, there are bacteria in nature that are able to digest just about anything. For example, the oil spill in Alaska that resulted when the *Exxon Valdez* ran aground was ultimately cleaned most effectively when they allowed the natural bacteria to digest the oil.

57. **(B)** Eutrophication is a change in a lake that begins with an excess of nutrients. Algal blooms begin, which die and form a mat that littoral grasses can use as a source of nutrients. The littoral zone becomes steadily more littered with organic material, and it steadily evolves into a boggy area, and then a more solid shoreline. This process gradually decreases the size of the lake until it is filled in.

58. **(A)** Overgrazing grasslands, such as is the case in the Sahel, leads to desertification. Overgrazing, particularly with sheep—which eat the roots of plants—will prevent the plants from replenishing themselves. The soil does not retain as much moisture, which leads to further reduction of plant growth. A strong wind can then blow away the top soil, leaving hard pan and broken parent material underneath.

59. **(D)** Succession is the process by which one biological community yields to another until a climax community is established. Primary succession occurs when land begins from bare rock. Secondary succession occurs when the land is stripped and it returns to a formerly held climax community.

60. **(E)** Remediation is the process of cleaning up chemical spills. Sometimes remediation is confused with the other "r" words, like restoration (restoring an area to its former state), reclamation (putting a former mine site back into a condition that is usable and pleasing), and rehabilitation (repairing the land for some new use, but not its original condition).

61. **(B)** Subsidence is the settling of the ground when the subsurface ground has compressed or chemically eroded.

62. **(B)** A stone fly nymph is an indicator species because it provides an indication of oxygen levels in the stream. The nymphs, like other indicator species, provide qualitative evidence of an important environmental parameter, without having to make a quantitative measurement.

63. **(A)** Residential consumers represent the smallest user of fresh water of all the users mentioned. Agriculture is the largest user, and paper and pulp mills are the largest industrial user.

64. **(C)** When acid rain falls on rocks and the ground, aluminum ions are leached out of the earth, which damage plants and people.

65. **(B)** Waxy leaves and thorny stems are two adaptations that help plants retain water. The waxy leaves help hold water in, and the thorny stems are a competitive scheme to repel animals who attempt to steal the water from the plant. Broad leaves are an adaptation under rainforest canopies among plants that compete for sun. Thin leaves low to the ground are tundra adaptations. Broad shallow roots are another rainforest adaptation, where the soil is thin and the few nutrients are close to the surface.

Practice Exam 4 – Section I: Detailed Explanations of Answers

66. **(E)** Electrostatic precipitators and fluidized bed combustion both remove pollutants from the air by turning them into solids, but then the solid waste must be disposed of. The fly ash from electrostatic precipitators in coal-fired plants contains heavy metals. The sludge from fluidized bed combustion is many times heavier than the pollutant was as a gas.

67. **(A)** The first step in a demographic transition is a decrease in death rates. This may actually result in an increase in population growth for a time, until the birth rates decline.

68. **(A)** The epicenter of an earthquake is actually on the surface of the Earth or ocean, just above the focus—or origin—of the quake.

69. **(C)** The convergence of two ocean plates creates a subduction zone—with islands that are formed above the edge of the overlapping plate. Indonesia, Japan, and the Aleutians are all examples of this.

70. **(B)** Under most typical conditions, it takes about a year to produce one millimeter of top soil. With little biological abundance and lack of moisture, it would take much longer.

71. **(B)** Cyanobacteria, or blue-green algae, get the credit for giving us our first oxygen-rich atmosphere. However, trees now help out significantly—particularly rainforests.

72. **(B)** Tidal power is a result of the gravitational potential energy of the moon acting on the Earth. Nuclear power actually comes to us because a star in some other solar system blew up long ago and produced the heavy metals at the instant of nova. Those heavy metals roamed the galaxy until they precipitated into our Earth.

73. **(A)** Nitrogen composes 80% of the atmosphere, with oxygen a distant second at 20%.

74. **(A)** Milankovitch cycles are currently getting the credit for the climate changes associated with periodic ice ages. Milankovitch cycles are created from the wobbling of the plane of axis over time.

75. **(D)** Volcanic eruption actually does the opposite; everything else listed contributes to global warming. The particulate matter from the eruption blocks the sun's rays, and less solar radiation reaches the surface of the planet. Melting polar ice, permafrost, and peat all release methane, which joins the carbon dioxide from burning fossil fuels as a greenhouse gas and contributes to global warming.

76. **(B)** Less than one percent of the total amount of material mined makes it through the milling process and progresses on to be processed as fuel.

77. **(B)** Fusion takes place on the surface of the sun or in a thermonuclear explosion (hydrogen bomb), but fusion is not controllable and does not take place in a nuclear reactor. Neutron capture catalyzes the process of fission, which is the process of splitting larger nuclei into smaller nuclei. Various nuclei are all undergoing different types of decay, which would include alpha and beta decay.

78. **(B)** If each step is 80% efficient, then it allows 80% (or 0.8) of the energy to be used at the next step. The amount of energy that makes it through all three steps is: $0.8 \times 0.8 \times 0.8 = 0.51$, or 51% of the original amount of energy.

79. **(A)** Oil gets the award as the largest supplier of energy among all developed countries. However, our coal reserves are more extensive. We think that we have enough coal for about 400 years. However, coal is the dirtiest of fuels. Natural gas gives the cleanest burn, but wouldn't it be nice if we could depend more on renewable resources, like solar energy?

80. **(B)** $(23/1{,}000 - 11/1{,}000) \times 100 = 1.2\%$ annual growth rate

81. **(A)** pH 4 is a strongly acidic solution. pH is a measure of acidity. Since pH is a logarithmic scale, each pH level corresponds to a tenfold increase in acid concentration. A pH of 4 corresponds to a hydrogen ion (acid) concentration of 10^{-4} moles acid per liter of solution. Any pH less than 7.0 is acidic.

82. **(C)** The Carboniferous period was characterized by rapid plant growth, which is the most likely origin of the coal.

83. **(A)** Nitrification is the process of oxidizing—or adding oxygen to—nitrogen; denitrification is the process of reducing—or removing oxygen from—the nitrogen. Oxygen must be removed from the nitrate ion in order to become nitrogen gas.

84. **(D)** The zone of aeration is the portion of the ground through which water percolates as it leaches through the ground on its way to an unconfined aquifer.

85. **(C)** Transpiration is the process that plants undergo when they release water vapor.

86. **(C)** The rule of 72 works for this one. 72 = annual growth rate × doubling time. Seventy-two divided by the doubling time (7) would equal the annual growth rate as a percentage (about 10%).

Practice Exam 4 – Section I: Detailed Explanations of Answers

87. **(A)** K-strategists do not generally tend to overshoot carrying capacity. They are more sensitive to environmental resistance and adjust population growth before the carrying capacity is reached. This results in an S-curve on a growth curve chart, rather than a J-curve—which tends to belong to an r-strategist population.

88. **(D)** Sediment is by far the most prevalent form of water pollution, or spoiling.

89. **(E)** CERCLA was passed in the wake of the Love Canal debacle and provided tax money for clean up of toxic spills, which was later dubbed "the superfund."

90. **(A)** Milpa agriculture works because it involves small sections of forest at a rate that is sustainable. Modern slash and burn farming is not sustainable because it involves large areas, and the land erodes away before natural succession can reseed and replant the area.

91. **(E)** The Chernobyl disaster in 1986 involved the meltdown of a nuclear reactor in Chernobyl, a town outside the Ukranian city of Kiev.

92. **(B)** Habitat Conservation Plans are a form of compromise with the Endangered Species Act that allows some habitat to be destroyed as long as the species experiences some overall benefit. One example of this kind of solution would be a deal made by a developer and the U.S. Fish and Wildlife Service that would create a preserve for an endangered species in exchange for being able to destroy habitat in another place.

93. **(E)** Integrated Pest Management (IPM) involves an integrated approach of many pest-fighting techniques, which could include chemicals, natural predators, sterilizing males, and luring pests away with pheremones (sex attractants). IPM usually leads to less use of chemical pesticides.

94. **(E)** Unfortunately, all of the items mentioned on this list are consequences of raising beef cattle on feedlots. While it takes less real estate to raise beef this way, it is a chemical-intensive industry that has many adverse effects on the environment.

95. **(A)** Don't be fooled by the word "organic." In this sense, it means that it involves carbon in the molecule. It takes many centuries for organic material—such as proteins and carbohydrates—to be pressed into first peat, then coal. Therefore, it is not considered a renewable process within a time frame that does our species any good.

AP Environmental Science

96. **(B)** Starting with the given information (1.0 kW-hr), convert to W, then divide by the number of hours and the W per bulb.

$$1.0 \text{ kW-hr} \times \frac{1,000 \text{ W}}{1.0 \text{ kW}} \times \frac{1}{2 \text{ hr}} \times \frac{1 \text{ bulb}}{100 \text{ W}} = 5 \text{ bulbs}$$

97. **(C)** The carbon dioxide given off by respiration reacts with atmospheric water vapor to form a carbonic acid equilibrium, which is very weakly acidic.

98. **(D)** Soap is not a way to kill waterborne pathogens, but UV light, ozone, and chlorine (all of which are used in many water treatment plants) will kill pathogens.

99. **(D)** Ozone in the stratosphere absorbs UV rays. That is important to life on Earth because UV rays are mutagenic, and can create burns—both of which would make it difficult to have life on Earth.

100. **(B)** Pronatalist pressure refers to the forces acting on a family that encourage them to have children. Only (B) represents a birth reduction pressure; all the other answers are examples of pronatalist pressures.

Detailed Explanations of Answers

Practice Exam 4

Section II

1. (10 points total)

    a. (2 points total)

    +1 for mentioning one of the following:

    - Endangered Species Act
    - Fish and Wildlife Conservation Act
    - Habitat Conservation Plans

    OR:

    +1/2 for mentioning the following instead:

    - Fish and Wildlife Act
    - Fish and Wildlife Improvement Act
    - Species Preservation Act

    PLUS:

    +1/2 each for mentioning either or both of the following:

    - Federal Fish and Wildlife Service is part of the Department of the Interior.
    - The Department of the Interior is a part of the Executive Branch of the federal government OR the Department of the Interior is led by the President.

    b. (2 points total)

    Full credit should be given for a detailed description of full-cost pricing, or internalizing external costs. This essay is an example.

AP Environmental Science

While some might be concerned that increasing the price of each home results in excluding low-income families (social injustice), the most important point is that the new housing developments are approaching their true cost. Normally, an external cost of a house that homeowners rarely bear is the cost of a loss of biodiversity and habitat. However, with this scheme, the external cost of lost habitat becomes an internal cost that is paid by the person who is gaining the privilege of displacing rare species.

c. (3 points)

+1 for any one of these answers, to a maximum of 3 points:

- Species uses a logistic growth strategy
- Long life span
- Large body
- Top carnivore
- Low reproductive rate
- Requires a large amount of land for roaming
- Migratory animal has one target biome endangered
- Narrowly defined niche
- Experienced genetic assimilation by exotic species
- Outcompeted by a more dominant species for resources
- Particularly sensitive to pollution

d. (3 points)

+1 each for any of these answers, to a maximum of 3 points (give +1/2 if reason is given without explanation):

- Water pollution, because the protected real estate still obtains water from outside the protected area, such as from runoff or groundwater
- Air pollution, because the protected real estate still obtains air from outside the protected area (clearly)
- Segmentation or genetic isolation, because the organism may not have a large enough choice of breeding partners in the preserve in order to sustain a minimum viable population

2. (10 points)

   a. (2 points)

   +1 for each of these answers, to a maximum of 2 points:
   - Evaporation will waste precious fresh water
   - Salinization of the land may result
   - Decreased flow of the river may affect fish habitat
   - Channelization may lead to increased sediment or erosion
   - Downstream users may be pressed into environmentally poor decisions because of decreased flow

   b. (2 points)
   - Aral Sea (Russia) or Mono Lake (California) water diversion for crops led to depletion of the source and salinization

   c. (3 points) While crop rotation may be desirable if the alternative crop fixed nitrogen into the soil (such as with a legume), it ultimately is less ecologically sound to use resources—particularly precious water—to feed cattle, who require 16 pounds of grain for every pound of meat.

   d. (3 points)

   +1 for each of these techniques, up to a maximum of 3 points:
   - Low or no till farming, to allow crop residue to retain moisture, recycle nutrients, and prevent wind erosion
   - Drip irrigation, rather than sprinkling, to decrease evaporation (although this is less practical with grains—better to just plant "dry wheat," which does not need irrigation)
   - Contour plowing
   - Terracing

3. (10 points)

   a. (2 points)

   +1 for YES, +1 for an explanation, such as

   Electric power plants are more efficient than obtaining the oil, refining it, and then combusting it in an automobile.

AP Environmental Science

b. (2 points)

+1 for answering HYBRID, +1 for responding with an explanation, such as

The electric vehicle results in emissions at the power plant, which can be tightly controlled using electrostatic precipitators or other point-source controls. The hybrid vehicle still burns gasoline, so the use of it still creates many sources of emissions (a fugitive source).

c. (4 points)

This question is really asking you to compare the environmental effects of hydroelectric, nuclear, and coal-fired power plants.

Seattle (hydroelectric and nuclear) +1/2 each, up to 2 points

Hydro power results in

- decreased biodiversity through loss of habitat
- loss of land due to water impoundment
- pollution due to concrete used to make the dam

Nuclear power results in

- contaminated soil from mining
- waste generated through mining, milling, processing, use, and final disposal
- decreased health of those working at all stages of the fuel "cycle"
- risk of malfunction
- thermal pollution because plant cooling increases eutrophication
- steam production from cooling towers causes change in local ecosystem

Atlanta (coal) +1 for each, to a maximum of 2 points

- Thermal pollution due to plant cooling, increases eutrophication
- Acid rain due to airborne sulfates
- Particulate matter that either goes into the air, or is captured and becomes a solid waste problem
- Mining of coal degrades the land.

Practice Exam 4 – Section II: Detailed Explanations of Answers

- Transport of coal requires burning of fossil fuel.
- Transport of coal down the Ohio River requires channelization of that river, with associated environmental drawbacks.

4. (10 points)

   a. +1 for each of the following, to a maximum of 4 points

   - Passive design; any one example counts for points, such as extended eaves, deciduous trees, trombe wall, floor heating, passive convection.
   - Active solar heating, such as direct heating of water by solar collectors
   - Photovoltaic cell system
   - Geothermal system
   - Wind generator
   - Energy savings; any number of different examples may be used, such as energy-saving appliances, fluorescent lights, etc.

   b. (3 points)

   Ammonia from human waste is digested by nitrifying bacteria and converted into nitrites, then nitrates, which are dispersed through the soil and used as nutrients by plant roots.

   c. +1 for each of the following, to a maximum of 3 points (+1/2 if no reason given)

   - VOCs from carpets, solvents, paints, adhesives, new material; solution: use natural fibers, wood floors, eco-friendly paints
   - Smoking; solution: don't smoke or allow people to smoke in your house
   - Radon; solution: ventilate basement or crawlspace
   - Carbon monoxide; solution: do not use combustion appliances or have a CO alarm in your house
   - CFCs from dry cleaned clothes; solution: use clothing that does not require dry cleaning
   - Fiberglass fibers from insulation; solution: use insulation from recycled materials or cellulose

Answer Sheets

AP ENVIRONMENTAL SCIENCE — Exam 1

AP ENVIRONMENTAL SCIENCE
EXAM 1 – PART II

AP ENVIRONMENTAL SCIENCE
EXAM 1 – PART II

AP ENVIRONMENTAL SCIENCE — Exam 2

AP ENVIRONMENTAL SCIENCE
EXAM 2 – PART II

AP ENVIRONMENTAL SCIENCE
EXAM 2 – PART II

AP ENVIRONMENTAL SCIENCE — Exam 3

AP ENVIRONMENTAL SCIENCE
EXAM 3 – PART II

AP ENVIRONMENTAL SCIENCE
EXAM 3 – PART II

AP ENVIRONMENTAL SCIENCE — Exam 4

1. Ⓐ Ⓑ Ⓒ Ⓓ Ⓔ
2. Ⓐ Ⓑ Ⓒ Ⓓ Ⓔ
3. Ⓐ Ⓑ Ⓒ Ⓓ Ⓔ
4. Ⓐ Ⓑ Ⓒ Ⓓ Ⓔ
5. Ⓐ Ⓑ Ⓒ Ⓓ Ⓔ
6. Ⓐ Ⓑ Ⓒ Ⓓ Ⓔ
7. Ⓐ Ⓑ Ⓒ Ⓓ Ⓔ
8. Ⓐ Ⓑ Ⓒ Ⓓ Ⓔ
9. Ⓐ Ⓑ Ⓒ Ⓓ Ⓔ
10. Ⓐ Ⓑ Ⓒ Ⓓ Ⓔ
11. Ⓐ Ⓑ Ⓒ Ⓓ Ⓔ
12. Ⓐ Ⓑ Ⓒ Ⓓ Ⓔ
13. Ⓐ Ⓑ Ⓒ Ⓓ Ⓔ
14. Ⓐ Ⓑ Ⓒ Ⓓ Ⓔ
15. Ⓐ Ⓑ Ⓒ Ⓓ Ⓔ
16. Ⓐ Ⓑ Ⓒ Ⓓ Ⓔ
17. Ⓐ Ⓑ Ⓒ Ⓓ Ⓔ
18. Ⓐ Ⓑ Ⓒ Ⓓ Ⓔ
19. Ⓐ Ⓑ Ⓒ Ⓓ Ⓔ
20. Ⓐ Ⓑ Ⓒ Ⓓ Ⓔ
21. Ⓐ Ⓑ Ⓒ Ⓓ Ⓔ
22. Ⓐ Ⓑ Ⓒ Ⓓ Ⓔ
23. Ⓐ Ⓑ Ⓒ Ⓓ Ⓔ
24. Ⓐ Ⓑ Ⓒ Ⓓ Ⓔ
25. Ⓐ Ⓑ Ⓒ Ⓓ Ⓔ
26. Ⓐ Ⓑ Ⓒ Ⓓ Ⓔ
27. Ⓐ Ⓑ Ⓒ Ⓓ Ⓔ
28. Ⓐ Ⓑ Ⓒ Ⓓ Ⓔ
29. Ⓐ Ⓑ Ⓒ Ⓓ Ⓔ
30. Ⓐ Ⓑ Ⓒ Ⓓ Ⓔ
31. Ⓐ Ⓑ Ⓒ Ⓓ Ⓔ
32. Ⓐ Ⓑ Ⓒ Ⓓ Ⓔ
33. Ⓐ Ⓑ Ⓒ Ⓓ Ⓔ
34. Ⓐ Ⓑ Ⓒ Ⓓ Ⓔ
35. Ⓐ Ⓑ Ⓒ Ⓓ Ⓔ
36. Ⓐ Ⓑ Ⓒ Ⓓ Ⓔ
37. Ⓐ Ⓑ Ⓒ Ⓓ Ⓔ
38. Ⓐ Ⓑ Ⓒ Ⓓ Ⓔ
39. Ⓐ Ⓑ Ⓒ Ⓓ Ⓔ
40. Ⓐ Ⓑ Ⓒ Ⓓ Ⓔ
41. Ⓐ Ⓑ Ⓒ Ⓓ Ⓔ
42. Ⓐ Ⓑ Ⓒ Ⓓ Ⓔ
43. Ⓐ Ⓑ Ⓒ Ⓓ Ⓔ
44. Ⓐ Ⓑ Ⓒ Ⓓ Ⓔ
45. Ⓐ Ⓑ Ⓒ Ⓓ Ⓔ
46. Ⓐ Ⓑ Ⓒ Ⓓ Ⓔ
47. Ⓐ Ⓑ Ⓒ Ⓓ Ⓔ
48. Ⓐ Ⓑ Ⓒ Ⓓ Ⓔ
49. Ⓐ Ⓑ Ⓒ Ⓓ Ⓔ
50. Ⓐ Ⓑ Ⓒ Ⓓ Ⓔ
51. Ⓐ Ⓑ Ⓒ Ⓓ Ⓔ
52. Ⓐ Ⓑ Ⓒ Ⓓ Ⓔ
53. Ⓐ Ⓑ Ⓒ Ⓓ Ⓔ
54. Ⓐ Ⓑ Ⓒ Ⓓ Ⓔ
55. Ⓐ Ⓑ Ⓒ Ⓓ Ⓔ
56. Ⓐ Ⓑ Ⓒ Ⓓ Ⓔ
57. Ⓐ Ⓑ Ⓒ Ⓓ Ⓔ
58. Ⓐ Ⓑ Ⓒ Ⓓ Ⓔ
59. Ⓐ Ⓑ Ⓒ Ⓓ Ⓔ
60. Ⓐ Ⓑ Ⓒ Ⓓ Ⓔ
61. Ⓐ Ⓑ Ⓒ Ⓓ Ⓔ
62. Ⓐ Ⓑ Ⓒ Ⓓ Ⓔ
63. Ⓐ Ⓑ Ⓒ Ⓓ Ⓔ
64. Ⓐ Ⓑ Ⓒ Ⓓ Ⓔ
65. Ⓐ Ⓑ Ⓒ Ⓓ Ⓔ
66. Ⓐ Ⓑ Ⓒ Ⓓ Ⓔ
67. Ⓐ Ⓑ Ⓒ Ⓓ Ⓔ
68. Ⓐ Ⓑ Ⓒ Ⓓ Ⓔ
69. Ⓐ Ⓑ Ⓒ Ⓓ Ⓔ
70. Ⓐ Ⓑ Ⓒ Ⓓ Ⓔ
71. Ⓐ Ⓑ Ⓒ Ⓓ Ⓔ
72. Ⓐ Ⓑ Ⓒ Ⓓ Ⓔ
73. Ⓐ Ⓑ Ⓒ Ⓓ Ⓔ
74. Ⓐ Ⓑ Ⓒ Ⓓ Ⓔ
75. Ⓐ Ⓑ Ⓒ Ⓓ Ⓔ
76. Ⓐ Ⓑ Ⓒ Ⓓ Ⓔ
77. Ⓐ Ⓑ Ⓒ Ⓓ Ⓔ
78. Ⓐ Ⓑ Ⓒ Ⓓ Ⓔ
79. Ⓐ Ⓑ Ⓒ Ⓓ Ⓔ
80. Ⓐ Ⓑ Ⓒ Ⓓ Ⓔ
81. Ⓐ Ⓑ Ⓒ Ⓓ Ⓔ
82. Ⓐ Ⓑ Ⓒ Ⓓ Ⓔ
83. Ⓐ Ⓑ Ⓒ Ⓓ Ⓔ
84. Ⓐ Ⓑ Ⓒ Ⓓ Ⓔ
85. Ⓐ Ⓑ Ⓒ Ⓓ Ⓔ
86. Ⓐ Ⓑ Ⓒ Ⓓ Ⓔ
87. Ⓐ Ⓑ Ⓒ Ⓓ Ⓔ
88. Ⓐ Ⓑ Ⓒ Ⓓ Ⓔ
89. Ⓐ Ⓑ Ⓒ Ⓓ Ⓔ
90. Ⓐ Ⓑ Ⓒ Ⓓ Ⓔ
91. Ⓐ Ⓑ Ⓒ Ⓓ Ⓔ
92. Ⓐ Ⓑ Ⓒ Ⓓ Ⓔ
93. Ⓐ Ⓑ Ⓒ Ⓓ Ⓔ
94. Ⓐ Ⓑ Ⓒ Ⓓ Ⓔ
95. Ⓐ Ⓑ Ⓒ Ⓓ Ⓔ
96. Ⓐ Ⓑ Ⓒ Ⓓ Ⓔ
97. Ⓐ Ⓑ Ⓒ Ⓓ Ⓔ
98. Ⓐ Ⓑ Ⓒ Ⓓ Ⓔ
99. Ⓐ Ⓑ Ⓒ Ⓓ Ⓔ
100. Ⓐ Ⓑ Ⓒ Ⓓ Ⓔ

AP ENVIRONMENTAL SCIENCE
EXAM 4 – PART II

AP ENVIRONMENTAL SCIENCE
EXAM 4 – PART II

Glossary

Glossary

Glossary

Abiotic factors: conditions in the physical, nonliving environment to which organisms are subjected, such as temperature, salinity, light intensity, etc.

Abiotic: nonliving; usually applied to the physical and chemical aspects of an organism's environment

Abundance (of SPECIES): a measure of the total number of organisms of each species in a particular ECOSYSTEM

Abyss: a BENTHIC biome existing at depths greater than 2,000 meters in the ocean; sometimes also applied to the zone in lakes below the depth of light penetration

Acid rain: rain or any other form of precipitation that has become acidic (PH of 4.0 or less) by absorbing air pollutants, especially nitrogen oxides and sulfur dioxides

Acid: a substance that releases hydrogen ions (H^+) in water solution, or accepts electrons in chemical reactions; a solution with a PH lower than 7

Acute toxicity: a large dose of a toxic substance that inflicts immediate harm on an organism (compare CHRONIC TOXICITY)

Adaptation: a genetically-controlled change in structure, function, or other characteristic which may make an organism better fit to survive in its environment

Aesthetic value: value placed on a commodity or capital good because it is pleasing or beautiful

Age-structure diagrams: histograms that reveal the distribution of people at different ages within the POPULATION

Alpha decay: ionizing radiation consisting of two protons and two neutrons emitted from an atom's nucleus with little penetrative power but damaging to living tissue

Amino acid: an organic molecule including one or more amino (NH_2) and acid (COOH) groups; one of the 20 nitrogen-containing molecules that are metabolically linked together to form proteins (see ESSENTIAL AMINO ACIDS)

Anthropogenic pollution: human-caused source of pollution

Antibody: a protein produced by an organism in response to the invasion of a substance not normally present and which may pose a harm to the organism. The antibody attaches to the substance, facilitating its destruction or removal from the organism

Antigen: any substance produced by an organism that stimulates the production of ANTIBODIES

Aphotic zone: deeper parts of aquatic HABITATS where light does not penetrate

Aquifer: geologic formation through which water can percolate, sometimes for long distances; porous rock or soil saturated with water

Arithmetic population growth: linear growth, one that increases by a constant amount over time (compare EXPONENTIAL GROWTH)

Asthenosphere: the less rigid, flexible layer of the Earth's interior, between 70–150 and 200–360 kilometers below the surface

Atom: the basic unit of matter; the smallest complete unit of the elements, consisting of protons, neutrons, and electrons

Bacterial resistance: the phenomena where disease producing bacteria are no longer sensitive to the antibiotics that have been used to destroy them in the past (often a consequence of NATURAL SELECTION adaptation to excessive, and often inappropriate, medical use of antibiotics)

Base: a substance that releases a hydroxyl ion (OH⁻¹) in water solution, or gives up electrons in chemical reactions. Also, a nitrogen-containing chemical compound, such as purine or pyrimidine, found in nucleic acid

Benthic: pertaining to the bottom zone, or bed, of lakes, seas, and oceans

Beta decay: ionizing radiation consisting of electrons ejected by an unstable atomic nucleus with moderate penetrative power and able to damage living tissue

Bioaccumulation: the process by which substances, especially harmful ones, concentrate in each link of the FOOD CHAIN, increasing levels in the TISSUES of each succeeding CONSUMER; sometimes also called BIOMAGNIFICATION

Bioassay: quantitative measurement, under laboratory conditions of standardization, of the effects of the DOSE of a substance on an organism or part of an organism

Biodiversity: the number of available SPECIES in an ECOSYSTEM; the amount of genetic variation present in an area

Biological control: a method of reducing the POPULATION of an undesirable species by introducing a natural enemy of that SPECIES

Biological Oxygen Demand (BOD): a measure of the amount of oxygen required to oxidize organic matter in water samples; an estimate of the load of oxygen-consuming organisms in an aquatic ECOSYSTEM

Biomagnification: increase in concentration of certain stable chemicals (for example, heavy metals or fat-soluble pesticides) in successively higher trophic levels of a FOOD CHAIN or FOOD WEB (see BIOACCUMULATION)

Biomass: the total weight of all living organisms within a particular POPULATION, COMMUNITY, ECOSYSTEM, or the Earth itself; commonly presented as weight per unit of volume or area (biomass density)

Biome: a major category of ECOSYSTEMS on Earth, part of the BIOSPHERE, which may consist of a large regional COMMUNITY of interrelated organisms and their environment, such as TUNDRA, TROPICAL RAINFOREST, or GRASSLAND

Bioremediation: the use of biological organisms to remove hazardous or toxic pollutants from the environment

Biosphere: the part of the Earth and its atmosphere in which organisms live

Biotic factors: influences on the environment and living organisms that are the result of the activities of living organisms

Biotic potential: the maximum reproductive rate of an organism, given unlimited resources and ideal environmental conditions (i.e., no ENVIRONMENTAL RESISTANCE)

Boreal: see TAIGA

Bottleneck effect: a limitation on GENETIC DIVERSITY because some cataclysmic event has created only a few survivors, so that the GENE POOL of the subsequent POPULATION is limited by the available GENES that the survivors transmit to the subsequent generations of offspring; also called GENETIC BOTTLENECK

British Thermal Unit (BTU): a unit of heat energy where 1.0 BTU equals the amount of heat needed to raise 1.0 pound of water by 1°F

Calorie: amount of energy required to raise one gram of water one degree Celsius

Carbon cycle: the chain or cycle of events by which carbon is circulated through the environment and living organisms, especially via the processes of PHOTOSYNTHESIS and RESPIRATION

Carcinogen: a substance or agent capable of causing cancer

Carrying capacity (K): the number of individuals in a POPULATION that the resources of a HABITAT can support at a given time; the asymptote, or plateau, of the SIGMOID equations for population growth

CDC: Centers for Disease Control and Prevention; an agency within the U.S. Public Health Service, Department of Health and Human Services

Cell: the basic unit of a plant of animal; an individual mass of living matter

Chaparral: a typically dry terrestrial BIOME characterized by short, wet winters and long, dry summers with thickets of low evergreen oaks or dense underbrush; term usually applied to areas in southwestern United States

Chemical synergy: the simultaneous exposure to two or more TOXINS that together have a greater effect than the sum of the effects of the two toxins separately

Chlorophyll: a green pigment found in the CHLOROPLAST of plants that is essential for PHOTOSYNTHESIS

Chloroplast: a cellular organelle (plastid) found only in plants, contains CHLOROPHYLL and is the site of PHOTOSYNTHESIS

Chromosome: long strands of nucleic acid and protein encoding hereditary information of an organism

Chronic toxicity: small doses of one or more toxic substances over a long period of time which produces harm to an organism (compare ACUTE TOXICITY)

Circumpolar vortex: a massive movement of cold air that flows from west to east around the northern portion of the globe (see JET STREAM)

Classical economics: the economics of Adam Smith, Thomas Malthus, and later followers such as John Stuart Mill. The theory concentrates on consumer and producer behaviors and the effects of resource scarcity, monetary policy, and competition on supply and demand of goods and services in the marketplace. This is the basis for the capitalist market system

Clear-cut: forestry practice where every tree, regardless of species or size, is cut down and removed

Climax community: a relatively stable state achieved when SUCCESSION of biological SPECIES has proceeded to the point where the biological COMMUNITY resists further change in composition; a self-perpetuating, community of organisms that continues as long as environmental conditions under which it developed prevail; the final stage in an ecological SUCCESSION

Cold front: moving boundary of cooler air displacing warmer air (see WEATHER FRONTS)

Commensalism: a relationship between organisms in which one individual lives close to, or on, another and benefits, and the host is unaffected (see SYMBIOSIS)

Community: an assemblage or group of interdependent POPULATIONS of different SPECIES whose NICHES overlap by geographical location and time

Competition: process in which more than one SPECIES, or individuals of the same species, attempt to make use of the same limited set of resources in the environment

Compound: a chemical substance composed of two or more ELEMENTS joined by chemical bonds and chemically united in fixed proportions

Conduction: the conveyance of energy, such as heat, sound, or electricity through a solid

Coniferous forest: a BIOME of cone-bearing evergreen trees with needle-like leaves, found in TEMPERATE ZONES and colder regions (see TAIGA)

Conservation: the planned management and wise use of natural resources for present and future generations; the use, protection and improvement of natural resources according to principles that will ensure their highest economic or social benefit (see PRESERVATION)

Constructive boundary: an elongated region where magma is pushed up from the asthenospheric MANTLE, producing new CRUST, and pushes apart TECTONIC PLATES. Also called DIVERGENT BOUNDARY (see also SEAFLOOR SPREADING)

Consumer: a heterotrophic organism that obtains nourishment through the consumption of other organisms

Continental drift: the PLATE TECTONIC process by which land masses slowly and continuously move positions on the globe, creating new land masses by colliding or splitting apart

Convection: the transfer and transport of heat by the movement of a gas or fluid; the primary mechanism of energy transfer in the atmosphere, oceans, and within the Earth

Convective mixing: the vertical mixing of water masses driven by wind stresses or density changes at the sea surface (see SPRING TURNOVER)

Conventional pollutants: any of seven major pollutants as designated by the Clean Air Act of 1970 (sulfur dioxide, carbon monoxide, particulates, hydrocarbons, nitrogen oxides, photochemical oxidants, and lead); pollutants that make up the largest volume of air quality degradation and pose the most serious threat to human health; also called criteria pollutants

Convergent evolution: the evolution of two different groups of organisms so that they come to resemble one another closely, or occupy a similar NICHE in different ECOSYSTEMS, e.g., sharks and dolphins

Convergent plate boundary: the elongated region of collision between two TECTONIC PLATES (see DESTRUCTIVE PLATE BOUNDARY)

Coral reefs: a type of MARINE ECOSYSTEM based on a mutualistic relationship between photosynthetic algae and a marine colonial coelenterate that secretes a calcareous skeleton, forming elaborate structures in shallow, submerged ocean banks or along shelves in warm, shallow, tropical seas

Coriolis Effect: the deflecting force acting on a body in motion due to the rotation of the Earth, which alters the direction of a moving body toward the right in the northern hemisphere and to the left in the southern hemisphere; the driving force for atmospheric convection cells and ocean currents

Cost-benefit analysis (CBA): economic technique used to evaluate large public projects where the desirability of a proposed course of action is estimated by listing advantages and disadvantages expressed in monetary terms, and the totals compared

Crude birth rate: the number of offspring per 1000 people in a given year

Crust: the solid outer shell of the Earth, varying in thickness from six km (under oceans) to 35–70 km in continental regions

Cultural value: the value placed on an item because it is a part of the history of a group of people

Glossary

Death rate: the number of deaths divided by the POPULATION size in a given area during a given time (see MORTALITY)

Deciduous forests: a terrestrial TEMPERATE BIOME characterized by oak, maple, birch, beech, elm, ash, and other hardwood trees, where the leaves are shed in the fall

Decomposer: an organism which obtains nutrients from feeding upon dead organisms, breaking them down into simpler substances, which become nutrients for other organisms

Demand: the amount of product desired by buyers; the want, need, or desire for a product that is backed by an ability to pay, measured over a given time period

Demographic transition: changes pertaining to maturing POPULATIONS, particularly in growth rate and age structure

Desert: a very arid terrestrial BIOME characterized by very low moisture—usually less than 25 cm of rainfall annually—and little vegetation other than cacti and desert shrubs or sagebrush, daily and seasonal temperatures fluctuate widely

Desertification: a process, usually associated with human misuse, by which fragile, semiarid ECOSYSTEMS lose PRODUCTIVITY because of loss of plant cover, soil erosion, SALINIZATION, etc.

Destructive plate boundary: elongated region of TECTONIC PLATE collision where one or more plates is destroyed (see SUBDUCTION)

Dew point: the temperature at which condensation occurs for a given concentration of water vapor in the air

Dioxin (Agent Orange): a compound that is extremely toxic to plants and animals, very persistent, and capable of causing chromosome damage; scientifically known as TCDD, and also called Agent Orange from its use as a herbicide during the Vietnam War

Divergent evolution: the situation where comparative structures on different organisms evolve to perform different functions

Divergent plate boundary: an elongated region where TECTONIC PLATES move apart (see SEAFLOOR SPREADING)

Diversity: a measure of the number of different SPECIES, genetic variation, or HABITATS in the ECOSYSTEM, also known as BIODIVERSITY

Dominant species: a SPECIES that exerts an overriding influence in determining the characteristics of a COMMUNITY

Dose: the amount of a chemical substance, microorganism, or energy that is taken into or absorbed by the body. Dose is determined by both amount and duration of exposure

Dose-response curve: a histogram of the relationship between degree of exposure to a substance or agent (DOSE) and observed biological effect or response

Ecological economics: an economic philosophy that attempts to take into account the importance of NATURAL, HUMAN, and SOCIAL CAPITAL in COST-BENEFIT ANALYSIS (see ENVIRONMENTAL ECONOMICS)

Ecology: the study of organisms in relation to one another and their environment

Ecosystem management: an integration of ecological, economic, and social goals in a unified systems approach to resource management

Ecosystem restoration: an attempt to reinstate an entire COMMUNITY of organisms to as near its natural condition as possible

Ecosystem: a holistic concept of a self-contained unit or area of biological COMMUNITIES and their nonliving physical surroundings

Ecotone: a HABITAT created by the overlapping of two distinctly different habitats; the zone of transition between habitats (see EDGE EFFECT)

Ectoparasite: a parasite living on the outside of a host, such as a tick or flea

Edge effect: the tendency towards increased variety and density of organisms in ECOTONE areas; a change in SPECIES composition, physical conditions, or other ecological factors at the boundary between two different ECOSYSTEMS

El Niño event: a large-scale change in normal weather patterns and ocean currents of the Pacific basin and adjacent regions. Nutrient-rich upwelling currents along the coast of South America are stalled, limiting PRODUCTIVITY and causing fisheries to fail. An El Niño event normally is accompanied by droughts in Australia and Southeast Asia together with heavy rain and snow in western North America

Element: substance that is composed of one kind of atom and cannot be broken down further by chemical change; each element is composed of just one type of atom

Emergent diseases: diseases that are either new to a POPULATION or have been absent for some time

Emigration: the movement of members of a POPULATION to another locale

Endangered species: a SPECIES that is in danger of becoming extinct because of small population sizes, poor reproduction, reduced available HABITAT, changes in the natural environment, human intervention, or a combination of these factors

Endoparasite: a parasite living within the body of a host, such as a tapeworm

Energy budget: an accounting of the way in which energy coming into an ecosystem from the sun is lost or processed by organisms of the ecosystem

Energy efficiency: a measure of energy produced compared to energy consumed in an ECOSYSTEM

Energy pyramid: a representation of the loss of useful energy at each step in a FOOD CHAIN

Entropy: the internal energy of a system that cannot be converted to mechanical work; a measure of the disorder in a system; the loss of available energy at each TROPHIC LEVEL of an ECOSYSTEM

Environment: the sum total of all the external conditions within which an organism lives

Environmental economics: the application of neoclassical economics to environmental issues with particular emphasis on the concept of EXTERNAL COSTS, in which some effects of an activity are not taken into account when it is priced

Environmental ethics: a philosophical search for moral values and ethical principles in human relations with the natural world

Environmental Impact Statement (EIS): a report based on detailed studies that assesses the environmental consequences of a course of action, used as an aid to decision-making; an analysis, required by provisions in the National Environmental Policy Act of 1970, of the effects of any major program a U.S. federal agency plans to undertake

Environmental justice: philosophy combining civil rights with environmental protection to demand a safe, healthy, life-giving environment for everyone

Environmental management: the process of directing or controlling production or the use of resources

Environmental racism: racial discrimination in environmental policymaking, law, and regulation enforcement, and the deliberate targeting of people of color for toxic waste facilities

Environmental resistance: all the limiting factors (BIOTIC and ABIOTIC forces) that tend to reduce POPULATION growth rates and set the maximum allowable population size or CARRYING CAPACITY of an ecosystem

Epilimnion: The upper layer of warm, oxygen-rich water in a thermally stratified lake or reservoir, extending from the surface to a depth of 5–10 meters, lying above the THERMOCLINE

Essential amino acids: the eight AMINO ACIDS that cannot be synthesized by the human body and must be consumed; tryptophane, lysine, methionine, phenylalanine, threonine, valine, leucine, and isoleucine

Estuary: the interface between a river ECOSYSTEM and a coastal MARINE ECOSYSTEM; often a river mouth, or semienclosed coastal environment, which has a high influx of freshwater and great variation in salinity

Euphtotic zone: surface layer of water where PHYTOPLANKTON can live; the depth of light penetration at which PHOTOSYNTHESIS balances RESPIRATION

Eutrophic: aquatic HABITATS that are nutrient-rich and characterized by high PRODUCTIVITY (abundant plankton)

Eutrophication: the situation in which an excess of nutrients, usually fertilizers or effluent rich in phosphorus and nitrogen, is introduced into an aquatic HABITAT producing a dramatic growth of PHYTOPLANKTON. As the algae die, their DECOMPOSERS use up the oxygen in the water, which can then adversely impact other organisms in the water

Evolution: the process by which all organisms descend from common ancestors; successive generations are modified in response to changes in the environment and new species are formed via NATURAL SELECTION

Exosphere: atmospheric layer 300–6,000 miles above surface, forming the transition to interstellar space, and composed primarily of hydrogen and helium

Exponential population growth: a condition where each new generation of individuals produces more potential reproducers than the previous generation. Growth is slow at first, then rapidly accelerates beyond the CARRYING CAPACITY of the HABITAT, and stops suddenly as ENVIRONMENTAL RESISTANCE takes effect. Also called GEOMETRIC or LOGARITHMIC growth (see J-CURVE)

External cost: the negative spillover effects of production or consumption for which no compensation is paid, such as pollution, where the actions of the polluter has an adverse effect on other people and damages the environment. The addition of the private costs of production and consumption with the external costs yields the total SOCIAL COSTS. Also called externalities

Extinction: the disappearance of a SPECIES, group, or GENE, globally or locally

Famine: massive, acute incidences of UNDERNOURISHMENT that are usually catalyzed by political or economic upheaval (as in war), or environmental devastation

Fecundity: the ability to produce offspring; the rate at which an individual reproduces

Fertility: a measure of the actual number of offspring produced, often expressed statistically as the CRUDE BIRTH RATE

First Law of Thermodynamics: law of physics stating that energy is neither created nor destroyed, it can only change forms

Fixed cost: the costs paid to make a product or provide a service that does not change as production increases, such as loan repayments, security, marketing, and administration costs (compare VARIABLE COST)

Food chain: an abstract representation of the sequences of organisms from PRODUCER to CONSUMER that trace the movement of BIOMASS or energy through a series of TROPHIC LEVELS

Food web: an abstract representation of interconnected FOOD CHAINS within a COMMUNITY

Fossil fuels: fuels, such as oil, coal, peat, and natural gas, formed by geological processes of great pressure, heat, and time on ancient organic remains

Founder effect: a limit to genetic diversity that is created when a small group of organisms invades a new area and begins a new evolutionary line

Fragmentation: a situation that occurs when human development or natural catastrophe has turned a large contiguous ECOSYSTEM into a patchwork of subpopulations that are unable to interbreed, or have a limited roaming region

Fugitive emissions: air pollutant category as defined by the Clean Air Act of 1970; pollution that does not come from a point-source, such as a smoke stack, but from nonlocalized sources, such as collective automobile emissions from roadways

Fusion: the formation of heavier atomic nuclei from lighter ones with the release of great amounts of energy

Gene: a unit of genetic inheritance; the portion of DNA containing information needed to produce a specific protein or RNA molecule, including noncoding and associated regulatory sequences

Gene expression: the cellular process of converting the information contained in the sequence of BASE pairs within a GENE into a protein (not all genes are expressed)

Gene pool: the collective group of traits (expressed and nonexpressed) that exist in all the CHROMOSOMES of all the individuals in a POPULATION

Genetic assimilation: the blending of the genetic traits of a subspecies into the GENE POOL of a closely related, more hardy SPECIES, ultimately leading to the extinction of the subspecies

Genetic bottleneck: see BOTTLENECK EFFECT

Genetic isolation: an event occurring when a small number of individuals have been isolated from the rest of the population, and as a result, their GENE POOL becomes limited to the genes available in the few breeding individuals, ultimately leading to genetic incompatibility with the original species

Genetically Modified Organisms (GMOs): an organism with a genetic constitution that has been altered in some manner by human intervention for commercial purposes

Global warming: phenomenon where the average annual temperature of the Earth increases (see GREENHOUSE EFFECT)

Grasslands: a terrestrial BIOME characterized by 25–75 cm of rainfall annually, supporting extensive expanses of grass and shrub plants. Also known as SAVANNAH or PRAIRIE

Green Revolution: the development and use of high-yield, fertilizer-dependent agricultural grain varieties on a commercial scale

Greenhouse effect: the process by which solar heat radiation is trapped by gases in the atmosphere; the warming of earth's climate by the concentration of carbon dioxide and other pollutant gases in the atmosphere (see GLOBAL WARMING)

Gross Domestic Product (GDP): the market value of all goods and services produced within a nation in a given period of time (usually one year); essentially, a measure of national income

Gross National Product (GNP): the total monetary value of all goods and services produced by a nation's economy in a given period of time (usually one year); a convenient indicator of the level of economic activity. Unlike GROSS DOMESTIC PRODUCT, it includes goods and services produced abroad

Growth rate (natural): population growth due only to births and deaths (usually the crude birth rate minus the crude death rate), excluding effects of population movement

Growth rate (total): the sum of the increases to a population due to immigration and births, minus those individuals who have died or emigrated

Habitat: the place in the environment where a particular organism, or COMMUNITY of organisms, normally lives

Homeostasis: a physical state of health with a balanced operation of all the systems of the body; the maintenance of constant internal conditions despite a varying external environment

Human capital: the accumulated skill, knowledge and expertise of a group of workers; sometimes called intellectual capital

Human Development Index (HDI): an alternative to GNP, a measure of a nation's economy which assesses quality and length of life, education, and standard of living in a country. Some economists feel that the ISW and HDI combine to give a better picture of the natural, human, and SOCIAL CAPITAL, rather than simply measuring the money that changes hands as with GDP and GNP

Hydrologic cycle: movement of water through the atmosphere, bodies of water, in the earth, and through ECOSYSTEMS, including passage through living organisms

Hypolimnion: the cold, oxygen-depleted bottom layer of a lake or other body of water lying below the THERMOCLINE where the movement of water is usually stagnate

Igneous rocks: rocks formed from the solidification of MAGMA

Index of Sustainable Welfare (ISW): a measure of a nation's economy, which takes into account resource depletion, unpaid labor, and environmental damage (see HUMAN DEVELOPMENT INDEX)

Industrial Revolution: a period in human history beginning in England in the eighteenth century, when agricultural technology, the availability of coal, and technology for making steel and using steam engines coincided to produce industrialized economies

Industrialized monoculture: an agricultural practice where large tracts of land are planted with a single crop, and the same maintenance techniques are implemented over a broad region

Instrumental value: the value placed on a commodity or capital good as determined by its usefulness

Integrated Pest Management (IPM): procedures involving a combination of pest control strategies—chemical and nonchemical—that are specific to the pest, crop, and location

Internal cost: the cost of production that is directly borne by the producer or consumer of a product (raw materials, labor, distribution, etc.), and which exclude EXTERNAL COSTS (pollution, environmental damage, aesthetic degradation, etc.)

Intertidal zone: the region between low and high tides where organisms are subjected to the forces of moving water and waves, and the periodic exposure to open air and salt water immersion

Intrinsic value: the value placed on a commodity or capital good as a result of its very existence; also called inherent value

Ion: atoms or combinations of atoms that have an unbalanced electrical charge; the disassociated parts of a molecule carrying either a positive (cation) or negative (anion) charge

Ionosphere: the region of the atmosphere that extends 50–300 miles above the Earth's surface and is made of multiple layers dominated by electrically charged, or ionized, atoms

J-curve: a type of POPULATION GROWTH curve that demonstrates the maximum rate, or BIOTIC POTENTIAL, of a population, named for the shape of its histogram

Jet stream: long, narrow, high-speed wind currents at high levels in the atmosphere, generally near the TROPOPAUSE

Joule: metric unit of energy; the work done when a force of 1 newton displaces a point 1 meter

Kelp forest: a type of aquatic HABITAT characterized by dense, 5–20 meter growth of kelp (sea weed) found in shallow areas of TEMPERATE seas

Known resources: natural resources that have been identified and located, but are not currently economically realistic to recover, such as mineral deposits on the ocean floor

K-strategy: the reproductive growth patterns of POPULATIONS that tend to respond more quickly to ENVIRONMENTAL RESISTANCE and experience a more sigmoidal, or S-CURVE, growth curve

Law of Conservation of Mass: see FIRST LAW OF THERMODYNAMICS

LD_{50}: the concentration of a substance or agent that causes death in half of the exposed population within a given period of time

Life expectancy: the number of years an individual is statistically predicted to survive

Life span: the longest length of life reached by a given SPECIES

Limnetic zone: the portion of open water in a lake or other deep body of freshwater below the EUPHOTIC ZONE that cannot sustain rooted vegetation

Lithosphere: The outer solid part of the earth, including the CRUST and uppermost MANTLE, about 100 km thick

Littoral: pertaining to the shore or beach environments (in freshwater ECOSYSTEMS, the shallow area of the lake that can sustain rooted vegetation; in MARINE ECOSYSTEMS, the INTERTIDAL ZONE)

Logarithmic (population growth): see EXPONENTIAL GROWTH

Magma: molten material that originates from the CRUST and MANTLE regions of the Earth's interior; liquid rock usually composed of silica

Mangrove swamp: a semiaquatic HABITAT found in tidal ESTUARIES and muddy coasts where highly adapted, salt-secreting mangrove trees form dense thickets with roots growing in and above the water

Mantle: the layer of the Earth lying between the CRUST and the core extending between depths of about 10 kilometers (6 miles) under the oceans and 30 kilometers (19 miles) in the continental areas to approximately 2,800 km (1,790 miles)

Manufactured capital: the time and materials used in making a completed product that continues to serve a purpose after manufacture, such as tools, roads, sewer systems, completed buildings, etc.

Margin of diminishing returns: an expression of the relationship between input and output, stating that adding units of any one input (labor, capital, etc.) to fixed amounts of the others will yield successively smaller increments of output. In common usage, the "point of diminishing returns" occurs when additional effort or investment will not yield correspondingly increasing results. Also called the law of decreasing returns, the law of variable proportions, or law of diminishing marginal returns

Marginal cost: the cost of making one additional unit of a product or service, or the total costs (FIXED and VARIABLE) per item when one more unit is produced

Marine biomes: the saltwater ECOSYSTEMS in the ocean, which include the PELAGIC, BENTHIC, and coastal ecosystems

Meiosis: a series of two cell nuclear divisions destined to produce gametes where paired CHROMOSOMES of a parent cell (diploid) is duplicated and segregated into four unpaired chromosome (haploid) daughter cells

Mesolimnion: the middle-layer temperature zone of a thermally stratified lake or other deep body of water below the upper warm EPILIMNION and the lower cold HYPOLIMNION

Mesopause: atmospheric layer 50–52 miles above the Earth's surface, forming a boundary between the MESOSPHERE and the THERMOSPHERE where temperature is constant

Mesosphere: atmospheric layer 31–50 miles above the Earth's surface where temperature decreases with altitude, contains some ice-crystal clouds

Metamorphic rocks: SEDIMENTARY and IGNEOUS rocks which have been altered in texture and composition as a result of being exposed to great heat or pressure (without passing through a molten stage)

Milpa: see SWIDDEN

Minamata disease: a bay and town in Japan that gave name to a disease of the central nervous system caused by mercury poisoning from consuming fish and shellfish that were contaminated by industrial effluents

Mitigation: in the context of land use, the creation of an ECOSYSTEM of comparable health and magnitude in exchange for damage done by the development of another nearby area

Mitochondria: a cellular organelle that is the site of aerobic RESPIRATION, con-verting glucose and oxygen into carbon dioxide and water in a process that also manufactures adenosine triphosphate (ATP)

Mitosis: common process of cell nuclear division where each CHROMOSOME of a parent cell is duplicated then separated into two identical daughter cells

Morbidity: a measure of the level of illness in a POPULATION

Mortality: a measure of the actual number of individuals who die in a POPULATION

Mutagen: a chemical substance that alters the DNA in cells

Mutation: a random, sudden change in the CHROMOSOMES of the cell; most mutations affect the DNA of individual GENES, although mutations can also involve the structure or number of chromosomes

Mutualism: a type of SYMBIOSIS where both organisms in the relationship benefit

Natality: a measure of the production of new individuals

Natural capital: natural, non-manmade resources (land, minerals, water, air, energy, etc.); can be categorized as either renewable or nonrenewable, or whether their recovery is economical and technically possible

Natural population growth rate: population growth as measured only by the difference of birth rate and death rate (not including population mobility)

Natural selection: variations among individuals in a population result in differential survival chances, those with more advantageous traits are more likely to survive and pass their characteristics to offspring, where the inheritance of particular variations over generations leads to new SPECIES formation

Neolithic Revolution: a period in hominid history representing a shift from a nomadic hunting/gathering lifestyle to agriculture where some populations settled in areas near food or rich soil

Neritic zone: the MARINE ECOSYSTEM that consists of relatively shallow water extending from the LITTORAL ZONE to the edge of the continental shelves (see PELAGIC ECOSYSTEMS)

Niche: the combination of HABITAT and the ecological role an organism plays in an ECOSYSTEM; often conceived as a multidimensional space

Nitrogen cycle: the chain or cycle of events by which nitrogen is circulated through the environment and living organisms

Nonpoint sources of water pollution: pollution coming from sources that are not limited to a discreet, identifiable location, such as agricultural or stormwater runoff

Nonrenewable resource: a natural resource contained within the Earth and present in a fixed quantity which cannot be replaced during human time scales and can be used only once, such as mined minerals or FOSSIL FUELS

Nucleus: the largest organelle of a cell, it contains the CHROMOSOMES and nucleolus

Oligotrophic: aquatic HABITATS that are poor in plant nutrients and tend to have little biological PRODUCTIVITY

Oncogene: Mutated and/or overexpressed GENE that can induce a tumor by releasing the cell from normal restraints on growth

Organ systems: groups of TISSUES that together perform a specific function

Oxygen cycle: the chain or cycle of events by which oxygen is circulated through the environment and living organisms

Ozone hole: a thinning of the upper atmosphere ozone layer (which filters harmful ultraviolet radiation) due to reaction with man-made chlorinated hydrocarbons

Parasitism: an association of organisms where one organism derives all of its nutrients from a host organism, usually without killing the host immediately

Pathogen: disease-causing microscopic PARASITIC organisms or viruses

Pelagic: pertaining to the open sea; an aquatic BIOME characterized by the upper, open waters of a lake, sea, or ocean, where actively swimming creatures and suspended plankton exist and no land is available

pH: the scale of acidity (1–7) or alkalinity (7–14); the logarithm of the concentration of hydrogen ions (10^{-1} to 10^{-14})

Phosphate cycle: the chain or cycle of events by which phosphate is circulated through the environment and living organisms

Photic zone: the upper layer of a lake, sea, or ocean where there is enough light for PHOTOSYNTHESIS to take place

Photochemical oxidants: SECONDARY AIR POLLUTANTS that are synthesized with the aid of solar energy, such as smog and ground-level ozone

Photosynthesis: the process in green plants in which carbon dioxide and water is combined to produce simple sugars using the energy of light

Phytoplankton: microscopic floating plants in aquatic ecosystems; the basic organism in aquatic FOOD CHAINS

Pioneer species: the first of a successive series of SPECIES invading a newly opened HABITAT

Plate tectonics: the scientific theory stating that the Earth's CRUST is made of several large rigid plates which are moved by convective forces below the crust

Plates: rigid slabs of the Earth's CRUST and upper MANTLE that make up the LITHOSPHERE; plate boundaries can be CONSTRUCTIVE where the plates are moving apart, as in RIFT VALLEYS, or DESTRUCTIVE where one plate is forced under another along a SUBDUCTION ZONE

Point source pollutant: pollution discharged from a specific point, or location, such as a pipe from a factory

Political economics: economic philosophy based on value systems and equal relationships among social classes, originally espoused by Karl Marx

Pollution: the act of introducing into the natural environment any substance or agent that may harm that environment and which is added more quickly than the environment is able to render it safe

Population: an interbreeding group of organisms of the same SPECIES that lives in the same general area over a given period of time

Prairie: see GRASSLAND

Precipitation: moisture that reaches the Earth's surface from the atmosphere, including rain, mist, dew, sleet, snow, and hail

Preservation: the act of reserving, protecting or safeguarding a portion of the natural environment from unnatural disturbance. Preservation suggests that natural resources will be left undisturbed, while CONSERVATION usually indicates some resource management

Primary pollutants: air pollutant category as defined in the Clean Air Act of 1970; pollutants harmful to humans in the form in which they are initially released

Primary productivity: the amount of BIOMASS produced by PHOTOSYNTHETIC organisms; a measure of how much solar energy is converted into chemical energy per area per unit of time

Primary succession: initial sequence of COMMUNITIES that develop in a newly exposed HABITAT, species that colonize a new area during primary succession are called PIONEER SPECIES

Primary treatment: the removal of particulate matter and sludge from sewage by filtration and settling in tanks

Producer: organisms that manufacture nutrients from inorganic sources by processes such as PHOTOSYNTHESIS, or chemosynthesis

Productivity: the amount of BIOMASS that is produced by a COMMUNITY

Protein synthesis: the creation of protein compounds through the linkage of AMINO ACIDS

Proven resources: consumable natural resources that are thoroughly mapped and can be realistically recovered, economically and technologically (see KNOWN RESOURCES and RECOVERABLE RESOURCES)

Rad (radiation absorbed dose): the unit of power of radiation absorbed by an organism, expressed as energy per gram of absorbing material; one rad equals 10^{-2} JOULES

Radiation: energy emitted in the form of electromagnetic waves (cosmic, gamma, ultraviolet, infrared, radio, etc.); the process by which an object emits energy, such as the atmosphere and ground surface are heated by the radiant energy from the sun

Radioactivity: the spontaneous disintegration of atoms emitting radiant energy (gamma radiation) and sometimes atomic particles (ALPHA, BETA, neutron)

Radon: a naturally occurring radioactive gas that can pose a human health hazard if a building is constructed over a spot where radon seeps out of the earth. The gas can accumulate in the building and cause cell damage and eventually cancer. Radon is the second leading cause of lung cancer in the United States

Rainforest: a terrestrial BIOME characterized by heavy rainfall, no marked dry season, and with rapid, lush vegetation growth

Reclamation: in environmental management, the rehabilitation of a massively scarred, denuded, or otherwise devastated area to a condition that is environmentally useful and socially or politically acceptable

Recoverable resources: consumable natural resources that are available with current technology, but not economically feasible to extract (see also KNOWN RESOURCES and PROVEN RESOURCES)

Recycling: the use of a resource more than once, either in the same or another form

Rem: (roentgen equivalent for man) a more accurate unit than RAD when assessing biological risk of radiation dosage; one rem equals the number of rads multiplied by a constant that is related to the type of particle causing the radiation

Remediation: in environmental management, the use of chemical, biological, or physical methods to remove hazardous or toxic pollutants

Renewable resource: a resource that is continually replenished within a human lifetime, such as wood or sunlight

Respiration: the process in organisms whereby energy is produced by chemical reactions through the use of oxygen and the release of carbon dioxide

Restoration: in environmental management or land use context, returning a damaged ECOSYSTEM back to its unspoiled, natural condition

Rift valley: the place where upwelling convective forces in the Earth's MANTLE causes a separation of the CRUST into plates

Risk assessment: a step in the environmental risk management process, measuring two quantities of risk: the magnitude of the potential loss, and the probability that the loss will occur. Once risks have been identified and assessed, steps to properly deal with them are described

R-strategy: the reproductive growth pattern of a POPULATION that tends to reproduce close to BIOTIC POTENTIAL, then die back after using up too many resources

Salinization: the destruction of agricultural land by an increase in its salt content, frequently occurring from over-irrigated soil when evaporation of water at the surface draws salts from underground rocks and soils, causing salts to crystallize and interfere with root growth

Salt marsh: see ESTUARY

Savannah: a terrestrial BIOME characterized by 50–150 cm of rainfall annually, similar to TEMPERATE GRASSLANDS, but with scattered trees and existing in tropical latitudes

Scientific value: the value conferred upon a resource if it contains lessons—whether already understood or as yet to be discovered; also called educational value

S-curve: POPULATIONS that have a K-STRATEGY for reproductive growth tend to respond more quickly to ENVIRONMENTAL RESISTANCE and experience a more SIGMOIDAL, or S-CURVE, growth curve as plotted on a histogram

Seafloor spreading: The mechanism by which new seafloor CRUST is created as adjacent TECTONIC PLATES move apart at a CONSTRUCTIVE BOUNDARY and MAGMA is pushed up from the asthenospheric MANTLE

Second Law of Thermodynamics: law of physics that states no reaction involving the transformation of energy from one form to another occurs without some portion of that energy being converted to a less useful form, such as heat; i.e., no reaction is 100% efficient

Secondary pollutant: air pollutant category as defined in Clean Air Act of 1970; pollutants released in a form that is initially not harmful, but becomes toxic or hazardous after being transformed in the environment, such as acid rain

Secondary productivity: the amount of BIOMASS produced by organisms that eat PHOTOSYNTHETIC organisms, indirectly measured by the amount of waste that a consumer produces

Secondary succession: the progression of COMMUNITIES when some change creates a new HABITAT, such as when a CLIMAX COMMUNITY is severely disturbed or destroyed

Secondary treatment: a second stage of sewage treatment after PRIMARY TREATMENT where wastewater is held for a longer time in conditions that are favorable to reduce BOD by bacterial digestion of the carbon and nitrogen wastes

Sedimentary rocks: rocks made of layers of sediments formed from weathering processes that break down other rocks

Selective cut: a forestry practice that involves harvesting of a portion of the mature trees with minimal disruption to the HABITAT (compare CLEAR CUT)

Sigmoid curve: an S-SHAPED CURVE as plotted on a graph in which there is an initial acceleration phase followed by a subsequent deceleration phase leading to a plateau

Social capital: the shared values among groups of people, such as trust, morale, and community organization; the stock of society's assets which provide a service

Social costs: the total costs of an economic activity on both the individual and the spillover effects on third parties; the total of private costs and EXTERNAL COSTS

Social justice: the concept holding that governments are instituted for the benefit of the people, and those governments which fail to do so are unjust; the principle that all persons are entitled to "basic human needs," regardless of economic disparity, class, gender, race, ethnicity, citizenship, religion, age, sexual orientation, disability, or health—this includes the eradication of poverty and illiteracy, the establishment of sound environmental policy, and equality of opportunity for healthy personal and social development

Soil horizons: soil layers of different types of material, distinguished by chemical and physical properties

Soil profile: characterization of the structure of soil vertically through its layers; a cross-section of SOIL HORIZONS

Speciation: the process by which two or more new SPECIES evolve from one original species as breeding groups become separated and develop distinctive traits to the extent that the isolated populations are no longer able to breed with one another

Species: a group of organisms with similar enough genetic makeup to be able to reproduce and produce fertile offspring; a group of actually or potentially interbreeding POPULATIONS that are reproductively isolated from all other organisms

Specific heat capacity: a measurement, in CALORIES, of a material's ability to absorb heat

Spring turnover: in deep freshwater lakes and reservoirs, a process involving the mixing of surface and subsurface lake waters due to seasonal shifts of temperature. In winter, less dense ice rests on more dense, warmer water. In spring, lake surface temperature warms causing surface water to sink and mix with deeper water. The process reverses in the fall (fall turnover)

Stratopause: atmospheric layer 30–32 miles above the Earth's surface, forming a boundary between the STRATOSPHERE and the MESOSPHERE where temperature is constant

Stratosphere: atmospheric layer 13–30 miles above the Earth's surface; temperature increases with altitude because ozone absorbs shortwave solar radiation

Subduction zone: an elongated region of TECTONIC PLATE collision (CONVERGENCE) where one plate is destroyed by being forced to slide down and below the other

Subsistence farming: agricultural practice that produces only what is needed for consumption by the farmer family and is not designed to produce food for market

Succession: the nonseasonal, continual process where NICHES and the composition of SPECIES in a COMMUNITY change over time

Sulfur cycle: the chain or cycle of events by which sulfur is circulated through the environment and living organisms

Superfund: Popular name of the hazardous waste cleanup fund established by the Comprehensive Environmental Response, Compensation and Liability Act (CERCLA) as amended by the Superfund Amendment and Reauthorization Act of 1986 (SARA), which carries out hazardous waste emergency removal and long-term remedial activities. Under this law, parties found responsible for polluting a site must clean up the contamination or reimburse the government for doing so

Supply: the amount of a good or product that producers are willing and able to sell at a given price

Survival economy: the economic status of impoverished regions or nations on the brink of social, economic, or environmental disaster

Survivorship: the number of people in a given POPULATION and given age bracket who continue to remain alive each year

Sustainable development: development where consideration is given to the quality of life for future generations as well as satisfying short-term objectives

Sustainable growth: economic growth that can continue over the long-term without NONRENEWABLE RESOURCES being used up

Swidden: agricultural practice used by indigenous people in TROPICAL RAINFOREST areas where small plots of land are cleared for farming. After a few crop yields, a new area is cleared and the old area is allowed to regenerate. Also called MILPA

Symbiosis: an intimate association of two species usually involving coevolution; three major types of symbiosis are PARASITISM, COMMENSALISM, and MUTUALISM

Taiga: a terrestrial BIOME characterized by spruces and firs, with a short summer and long, snowy winter, which extends across much of north America and Eurasia; also called NORTHERN CONIFEROUS FOREST, or BOREAL FOREST

Temperate rainforest: a RAINFOREST BIOME occurring at higher latitudes where coastal mountains cause high PRECIPITATION with mosses, ferns, and large conifer trees as dominant plants, such as the Olympic Peninsula in Washington State

Temperate zone: the part of the Earth's surface characterized by temperate climate located between the Arctic Circle and the TROPIC OF CANCER, or between the Antarctic Circle and the TROPIC OF CAPRICORN

Teratogen: a substance or agent that can cause birth defects

Tertiary treatment: a third stage in a series of wastewater treatments that involves any of several methods to remove chemical and disease-causing organisms after most sewage particulate matter has been removed by settling and filtration; see also PRIMARY TREATMENT and SECONDARY TREATMENT

Thermocline: temperature gradient; the boundary between layers of water in a thermally stratified lake or other deep body of water, especially the zone of water depth within which temperature changes rapidly between upper warm surface waters (EPILIMNION) and lower cold bottom waters (HYPOLIMNION); see also MESOLIMNION

Thermohaline circulation: the CONVECTION cycle in aquatic ECOSYSTEMS created by differences in temperature and salinity

Thermohaline gradient: a difference, or gradient, that involves both temperature and salinity in aquatic ECOSYSTEMS

Thermosphere: atmospheric layer 52–300 miles above the Earth's surface where temperature increases with altitude because of high-energy solar radiation (gamma rays, UV radiation, X-rays); also includes the IONOSPHERE

Tissues: a group of cells that perform a particular function in an organism

Total population growth rate: population growth as measured by the difference of birth rate and death rate, added to the difference of immigration rate and emigration rate

Toxic colonialism: the use of Third World countries as depositories of toxic wastes (see also ENVIRONMENTAL RACISM)

Toxic: ability, or property, of a substance to produce a harmful or lethal effect on an organism and/or the environment

Toxin: a poison formed as a specific secretion product in the metabolism of an organism as distinguished from inorganic poisons—such poisons can also be manufactured by synthetic processes; any poisonous substance that can cause disease

Trade winds: the typical surface winds that are created by the CORIOLIS EFFECT. Trade winds move in a clockwise direction in the northern hemisphere and a counterclockwise direction in the southern hemisphere

Trophic level: a step along a FOOD CHAIN; a particular position an organism occupies in an ECOSYSTEM determined by the number of energy-transfer (food consumption) steps to that level

Tropic of Cancer: an imaginary circle around the Earth at the latitude 23.5 degrees north of the equator; the northern most location of the sun at the June solstice (the most northern point on the Earth where the sun can be seen directly overhead)

Tropic of Capricorn: an imaginary circle around the Earth at the latitude 23.5 degrees south of the equator; the southern most location of the sun at the December solstice (the most southern point on the Earth where the sun can be seen directly overhead)

Tropical rainforest: a RAINFOREST BIOME in the TROPICAL ZONE characterized by lush vegetative growth and substantial rainfall of about 200 cm annually

Tropical zone: the part of the Earth's surface between the TROPIC OF CANCER and the TROPIC OF CAPRICORN; characterized by a hot climate

Tropopause: atmospheric layer 7–13 miles above the Earth's surface, forming the boundary between the TROPOSPHERE and the STRATOSPHERE where JET STREAMS occur and temperature is constant

Troposphere: lowest atmospheric layer 0–7 miles above the Earth's surface, where 75% of the atmosphere exists, and where generally, temperature decreases with altitude

Tundra: a terrestrial BIOME of the arctic regions characterized by a short growing season and low temperatures

Undernourishment: not having enough food to develop or function normally, less than 2,000–2,500 calories each day for humans

Upwelling: vertical movement of water that brings nutrients from the depths to surface layers

Urbanization: the conversion of agricultural, forested, or other undeveloped areas to urban land as a result of NATURAL POPULATION GROWTH and movement of people to urban areas

Vaccine: any substance containing ANTIGENS which is injected into an animal's body to stimulate the production of ANTIBODIES that will resist infections

Variable costs: production costs that increase as the number of products produced increases, such as for raw materials to manufacture a product, i.e., if output increases then variable costs increase

Vector: an agent that transfers a PATHOGEN from one organism to another, such as tick or mosquito

Volatile Organic Compounds (VOC): a class of organic compounds mainly composed of hydrocarbons that are easily vaporized and pose health and environmental hazards. The most common volatile organic compound released into the atmosphere is methane, which is involved in the formation of PHOTOCHEMICAL smog

Waste stream: in solid waste management, the flow of matter from raw materials through manufacturing and marketing, to the consumer, and on to its final use, where if not recycled, then usually disposed in a solid waste dump

Weather fronts: massive movements of air created when an air mass of one temperature and pressure collides with an air mass of a different temperature and pressure. COLD FRONTS occur when cool air moves into a zone previously occupied by warm air. Warm fronts occur when warm air displaces cool air

Zero population growth (ZPG): a demographic situation in which a POPULATION is held in stable numbers over an extended period of time

Zooplankton: tiny floating MARINE animals; together with PHYTOPLANKTON composing the plankton life forms at the surface layers in aquatic ECOSYSTEMS

Index

A

Abundance
  of organisms, 56
  of species, threats to, 70
Abyss, 65–66
Acid-base reactions, 17
Acid rain, 21, 130–131
Acids, 14
  as water pollutants, 145–146
Active solar heating, 198
Adams, Ansel, 223
Adenosine triphosphate (ATP), 16
Aerosol effect, 131–132
Aesthetic value, 220
Agent Orange, 114–115
Age-structure diagrams, 88–90
Agricultural practices, water pollution control and, 149
Agriculture
  farming versus ranching and, 171
  forestry as, 174
  land degradation due to, 167–170
  land management and, 165–173
  "revolutions" in, 172–173
  sustainable, industrial monoculture versus, 170–171
  water use for, 139, 140
Agriculture Department, 224
A-horizon, 39
Air pollution, 120–134
  categories of, 120
  in cities, 162
  criteria pollutants and, 120–126
  ecosystem effects of, 130–131
  health effects of, 129–130
  indoor, 126–129
  legislation on, 227
  natural sources of, 120
  non-criteria pollutants and, 126
  volcanic, 133–134
Air stripping, 108
Alaska, oil transport from, 205
Aleutian islands, 36
Algae
  brown, 65
  green, 65
  photosynthetic, 39
  red, 65
Algal blooms, 143–144
Allergens, 102
Alpha decay, 194
Alpine tundra, 61
Aluminum electrolytic extraction, 181
Amino acids, essential, 165
Ammonification, 19
Andes Mountains, 36
Anemia, 166
Animals as vectors, 99
Anions, 15
Anthropocentric philosophers, 221
Antibiotics, 101
Antibodies, 101
Antifungals, 101
Antigens, 101
Antihistamines, 102
Aphotic zone, 64
Aquatic ecosystems, acid rain and, 130–131
Aquifers, 138, 151
Aral Sea, 141
Arctic National Wildlife Refuge (ANWR), 73, 206
Arctic tundra, 61
Argon in atmosphere, 42, 119
Ariboflavinosis, 166
Arithmetic population growth, 80

451

Army Corps of Engineers, 225
Artesian wells, 138
Asbestos Hazard and Emergency Response Act of 1986, 227
Asian long-horned beetle, 70
Asian Plate, 36
Asthma triggers, 127
Atmosphere, 42–48, 119–134. *See also* Air pollution
    composition of, 42–43, 119–120
    convection in, 30–33
    layers of, 43
    origin and evolution of, 42
    toxins in, treatment of, 106–107
    weather and. *See* Climate; Weather
Atoms, 13
    in molecules, 14
Audubon, John, 222
Automobiles
    energy use and, 202–203
    traffic in cities and, 162

B

Bacteria, 100
    thermophilic, 75
Bangladesh famine, 166
Barrier islands, 67
Bases, 16
Bedrock, 40
Beliefs, 221–222
Benthic marine ecosystems, 65–66
Benthic zone, 67
Beriberi, 167
Beta decay, 194
Bhopal crisis, 113–114
B-horizon, 40
Bicarbonate ions, 18
Bioassays, 103–104
Biocentric philosophers, 221
Biodiversity, 57, 68–74
    ecosystem, 71–72
    genetic, 68–69
    legislation on, 230
    managing, 72–74
    marine, 75
    species, 69–71, 75
Biological Oxygen Demand (BOD), 144
Biomagnification, 76

Biomass
    productivity and, 55
    solar energy and, 199–200
Biomes, 17, 61–68
    freshwater, 67–68
    marine, 64–67
    terrestrial, 61–64
Biotic potential, 82
Birth defects, 102
Birth rate, crude, 79
Birth reduction pressures, population size and, 86
Blue whale, 76
Body waves, earthquakes and, 36–37
Boiling water reactors (BWRs), 195
Boreal forests, 63
Bottleneck Effect, 68–69
Brain, air pollution and, 130
British Thermal Units (BTUs), 189
Brown algae, 65
Building construction, land management and, 158–161
Building materials, indoor air pollution related to, 127–128
Bureau of Land Management (BLM), 73, 225

C

Calcium ions as water pollutants, 145
California Water Project, 142
Cancer, 15
    development of, 102
Capillary action, 34
Capital, 213–215
    flow of, 214–215
Captive breeding programs, 74
Carbamates, 169
Carbon absorption of wastes, 108
Carbonate ion, 18
Carbon cycle, 18–19, 24–25
Carbon dioxide, 18–19
    in atmosphere, 42, 47–48, 119. *See also* Greenhouse effect
Carbonic acid, 18
Carbon monoxide as indoor air pollutant, 128–129
Carbon oxides as air pollutants, 123

Index

Carcinogens, 102
Carrying capacity, 83–84
    shifting, 94–95
Carson, Rachel, 223
Cations, 13
Cells, 15–16
Centers for Disease Control (CDC), 225, 232
Channelization, 178
Chaparral, 62–63
Chattahoochee River, 150
Chemical erosion, 41, 167–168
Chemical reactions, 17–18
    acid-base, 17
    equilibrium and, 18
    oxidation-reduction, 17
    precipitation, 17–18
Chemical synergy, toxicity and, 106
Chemical weathering, 37–38
Chernobyl nuclear catastrophe, 207–208
Chloride ions as water pollutants, 145
Chlorinated hydrocarbons, 169
Chlorofluorocarbons (CFCs), 47–48
Chlorophyll, 188
Chloroplasts, 16, 55
C-horizon, 40
Chromium in building construction, 158
Chromosomes, 15
Circumpolar vortex, 33
Cirrus clouds, 44
Cities. *See* Urbanization
City planning, 164–165
Classical economics, 211, 217–218
Clay, 39
Clean Air Act of 1970, 227
Clean Water Act of 1972, 148–149, 228
Clean zone, 144
Clear-cutting, 173
Climate. *See also* Weather
    global changes in, 46–48
    properties of water contributing to, 33
Climax communities, 61
Clouds, 44–45
Clustered population distribution, 57
Clustering in cities, 164
Coal, 192–193
Coastal ecosystems, 66–67
Cold fronts, 45
Colonialism, toxic, 220

Columbia River, 141
Commensalism, 59
    population size and, 84, 85
Commerce Department, 224
Commercial energy use, 203
Communities, 16
    traits of, 60–61
Competition, 57–58
    population size and, 85
Complexity of communities, 60
Composting, 24–25, 112
Compounds, 14
Comprehensive Environmental Response, Compensation and Liability Act (CERCLA) of 1980, 157, 229
Conduction, 23, 29
Cone of depression, 138
Confined aquifers, 138
Coniferous forests, northern, 63
Conservation
    definition of, 233
    of energy, 22, 204
    of land, 156
    reducing volume of waste and, 110–112
    of soil, 41
    of water, 142–143
Conservation of matter, 17–22
    chemical reactions of, 17–18
    nutrient cycles and, 18–22
Constancy of communities, 60
Constructive boundaries, 35
Continental drift, theory of, 34–35
Continental-oceanic convergence, 36
Convection, 23, 29–33
    in atmosphere, 30–33
    definition of, 35
    in Earth's mantle, 30
    in oceans, 30
Convection cells, 32
Conventional pollutants, 120–126
Convergent boundaries, 35
Convergent evolution, 59
Coral reefs, 65
    climate change and, 48
Coriolis Effect, 32, 41, 44
Corridors, establishing between fragmented populations, 73

Cost-benefit analysis (CBA), 215–216
Costs
    external, 216–217, 235
    fixed, 215
    marginal, 215–216
    true, 217
    variable, 215
Crime in cities, 163
Criteria pollutants, 120–126
Crop rotation, 169–170
Crude birth rate, 79
Crude death rate, 79
Cultural value, 220
Cumulus clouds, 44
Cyanobacteria, 39
    photosynthetic, 42

D

Dam projects, 140–141, 152
DDT, 76, 106, 112–113
Death rate, crude, 79
Debt-for-nature swaps, 73
Deciduous forests, temperate, 63
Decomposers, 39
Decomposition zone, 144
Defense Department, 225
Deforestation, 155, 174
Delaney Amendment of 1958, 229
Demand and supply. See Supply and demand
Demographic transitions, 90–92
    benefits of, 90
    characteristics of, 90
    stages of, 91–92
Desalinization of sea water, 139
Desert belts, rain in, 139
Desertification, 155, 167
Deserts, 32, 62
Destructive boundaries, 35
Developed countries, 203
Developed economies, 213
Dew point, 44
Dieback, 84
Dimethyl sulfoxide, 22
Diminishing returns, margin of, 216
Disease, climate change and, 48
Disease clusters, 103
Disinfection of wastewater, 147
Dissolved Oxygen (DO), 144

Divergent boundaries, 35
Diversity of organisms. See Biodiversity
DNA, 15, 68
    cancer and, 102
    mutation and, 58
Dose, toxicity and, 105
Dose-response curves, 104–105
Dumps, 109
Dust Bowl, 40
Dutch Elm Disease, 76

E

Earth, convection in mantle of, 30
Earthquakes, 36–37
Earthworms, 38–39
Easter Island, 94–95
Ebola virus, 114
Ecocentric viewpoints, 222
Ecofeminism, 222
Ecological economics, 218
Ecological role, 57
Ecology, 211
Economic forces, 211–218
    capital and, 213–214
    cost-benefit analysis and, 215–216
    demand and supply, 211–212
    economic systems and, 217–218
    internal and external costs and, 216–217
    market forces as, 211
Economics, 211
Ecosystem biodiversity, 71–72
Ecosystems, 17
    air pollution and, 130–131
    coastal, 66–67
    energy movement through, 23–24
    marine, 64–67
    productivity of, 55
Ectoparasites, 59
Edge effects, 72
Educational value, 220
Efficiency
    of energy transfer, 23, 188
    of lights, 25
E-horizon, 40
Ehrlich, Paul, 223
Electrical energy
    calculations involving, 192

production of, 141
Electrical power, production of,
    188–189, 200–201, 204–205, 206
Elm trees, 76
El Niño, 46–47, 51
El Niño Southern Oscillation (ENSO),
    46–47
Emergent diseases, 101, 114, 115
Emigration, population size and, 86
Endangered species, 72, 175–177. *See
    also* Wildlife
    monitoring markets for, 177
    traits of, 70–71
Endangered Species Act (ESA) of 1973,
    72, 73, 177, 230, 236
Endoparasites, 59
Energy, 187–208. *See also* Electrical
    entries; Nuclear power
    agricultural use of, 170
    for building construction,
        159, 161
    calculations for, 190–192
    cell requirements for, 16
    changes of state and, 15
    conservation of, 22, 204
    efficient use of, 25
    for farming versus ranching, 171
    geothermal, 200
    heat, transfer of, 22–23. *See also*
        Convection
    kinetic, 187–188
    legislation on, 228
    movement through ecosystems,
        23–24
    solar, 30–32, 47
    tidal, 201
    unit conversions, 190
    units of, 189
    use of, 202–204
    of vaporization, of water, climate
        and, 33
    wind, 199
Energy Department, 225
Energy sources, 187–189
    nonrenewable, 192–198
    renewable, 198–201
Entropy, 22–23
Environmental ethics, 218–219
Environmental laws, 226–230

Environmental policy and management,
    230–235
Environmental Protection Agency (EPA),
    73, 226, 232
Environmental racism, 220
Environmental resistance, 82
Epicenter of earthquakes, 36
Epilimnion, 68
Equatorial latitudes, solar energy striking,
    31–32
Equilibrium, 18
Erosion of soil, 29, 40–41, 167–168,
    183–184
Essential amino acids, 165
Estuaries, 66
Ethics, environmental, 218–219
Euphotic zone, 67
Eutrophic lakes and ponds, 67
Eutrophic rivers, 144
Everglades, 150
Evolution
    climate change and, 48
    convergent, 59
Exosphere, 43
Exponential population growth, 80–81
Ex situ management of wildlife, 178
External costs, 216–217, 235
Extinction, causes of, 176–177
Exxon Valdez, 205

F

Famines, 165–166
Fauna, soil, 39
Fecundity, 79
Federal Emergency Management Agency
    (FEMA), 225
Federal Food, Drug, and Cosmetic Act
    (FFDCA) of 1906, 229
Federal Hazardous Substances Act of
    1960, 229
Federal Insecticide, Fungicide, and
    Rodenticide Act (FIFRA) of 1947,
    229
Federal Land Policy and Management Act
    of 1976, 229
Federal Water Pollution Control Acts of
    1948, 228
Federal Water Resources Planning Act of
    1965, 228

Feedlot pollution, 171
Ferrel cells, 32
Fertility, 79
Fertilization, 15
Fertilizers, 169–170
Fiberglass in building construction, 158
Filtration of wastewater, 146
Fires, 63
    management of, 74, 174
First Law of Thermodynamics, 22
Fish and Wildlife Conservation Act of 1980, 158
Fish and Wildlife Service, 226
Fishing, 177
Fission, 194
Fixed costs, 215
Flocculation, 108, 146
Flood control, land management and, 178, 182–183
Floods, 182–183
Flue gas desulfurization, 122
Flu vaccine, 115
Focus of earthquakes, 36
Folly Island, 184
Food and Drug Administration (FDA), 225, 232
Food chains, 23, 55–56, 66
Food webs, 23, 56
Forest fires, 63
Forestry
    as agriculture, 174
    land degradation due to, 167–170
    land management and, 167–170, 173–174
Forests
    coniferous, northern (boreal), 63
    current forest harvesting practices and, 173–174
    deciduous, temperate, 63
    fire management in, 174
    multiple use of, 173–174
Fossil fuels, 193
Founder Effect, 68–69
Fragmentation of populations, 72
Freshwater biomes, 67–68
Freshwater ecosystems, 17
Frontier economies, 213
Fronts, weather, 45
Fuel cells, 198–199
Fuels, fossil, 193
Full-cost analysis, 217
Fungi, 100
Fusion, 187, 194–195

G

Gap analysis, 73
Gas, natural, 193
Gases, 15
Gene expression, 68
Gene pool, 68
Genes, 15, 68
Genetically modified organisms (GMOs), 172–173
Genetic assimilation, 71, 176
Genetic biodiversity, 68–69
Genetic bottlenecks, 71, 175
Genetic damage, radioactivity and, 196–197
Genetic isolation, 71, 176
Genetic predisposition, toxicity and, 106
Geothermal energy, 200
Glaciers, melting of, 48
Global processes, properties driving, 29
Global warming, 47–48, 50–51
Gobi Desert, 62
Goiter, 166
Gold, heap-leach extraction for, 181
Government agencies, 224–226
Graphite reactors, 195
Grasshopper effect, 132
Grasslands, 62
Gravel, 39
Great Depression, 40
Greater Yellowstone ecosystem, 63
Green algae, 65
Greenhouse effect, 47–48, 50–51, 131
Green Revolution, 172
Gross Domestic Product (GDP), 214
Gross National Product (GNP), 214
Groundwater, 138
Gully erosion, 40, 167

H

Habitat, 57
Habitat conservation plans, 177
Hadley cells, 32

Hail, 45
Halogens as air pollutants, 125
Hardin, Garrett, 155, 217, 223
Hazardous materials, 102–103
    pollution prevention and, 107
    waste disposal and, 108–112
    waste treatment and, 107–108
Health. *See* Human health
Health and Human Services Department, 225
Heap-leach extraction for gold, 181
Heat
    calculations involving, 190–192
    of vaporization, of water, climate and, 33
Heat domes, 132–133
Heat transfers, 22–23. *See also* Convection
Heavy metal ions, as water pollutants, 145
Heavy water reactors (HWRs), 195
Helium in atmosphere, 42, 119
High-temperature, gas-cooled reactors (HTGCRs), 196
Himalaya Mountains, 36
Histamine, 102
Homeland Security Department, 225
Hoover Dam, 141
Horizons in soil, 39–40
Hotspots, 30, 49–50
Housing in cities, 163, 164–165
Human capital, 214
Human Development Index (HDI), 215
Human health, 99–115
    air pollution and, 129–130
    emergent diseases and, 101, 114, 115
    environmental health hazards and, 99–103, 112–114
    identifying health risks and, 103–104
    morbidity and, 99
    mortality and, 99
    nutrition and, 165–167
    radioactivity and, 196–197
    toxic materials and, 101–102, 105–112
Human populations, 87–94
    demographic transitions and, 90–92
    growth strategies of, 92–94
    overpopulation and, 94
    structure of, 87–90

Humus, 39
Hunter/gatherers
    energy use by, 202
    population growth of, 92–93
Hunting, wildlife and, 177
Hurricane-prone areas, real estate development in, 183–184
Hurricanes, 46, 51
Hydroelectric power, 188, 200–201, 204–205, 206
Hydroelectric projects, 141
Hydrogen in atmosphere, 43, 119
Hydrogen ions, 15, 18
Hydrologic cycle, 22, 187
Hydroxide ions, 15
Hypolimnion, 68

I

Igneous rocks, 37
Immigration, population size and, 86
Immune system, 102
Incineration of wastes, 107, 109
Index of Sustainable Welfare (ISW), 214–215
Indian Plate, 36
Indoor air pollution, 126–129
Industrial monoculture, 170–171
Industrial Revolution
    energy use and, 202
    population growth and, 93–94
Industrial waste, treatment at point sources, 149
Industrial water use, 139, 140
Inertia of communities, 60
Infrared light, 31
Inherent value, 219
Inorganic wastes as water pollutants, 144
In situ management of wildlife, 177–178
Instrumental value, 219
Integrated pest management (IPM), 169
Interior Department, 225–226
Internal costs, 216
Interspecific competition, 57–58
Intertidal zone, 66
Intraspecific competition, 57
Intrinsic value, 219–220
Inversions, 132–133
Ionosphere, 43

Ions, 13–14
  bicarbonate, 18
  calcium, as water pollutants, 145
  carbonate, 18
  chloride, as water pollutants, 145
  dissolved, as water pollutants, 144
  heavy metal, as water pollutants, 145
  hydrogen, 15, 18
  hydroxide, 15
  iron, as water pollutants, 145
  sulfide, as water pollutants, 145
Iron ions, as water pollutants, 145

J

James Bay Power Project, 206
J-curve, 82, 83
Jet streams, 33
Joules, 189, 196

K

Kelp, 65
Khian Sea (ship), 109–110
Kilauea, 133–134
Kinetic energy, 187–188
Known resources, 213
Krakatau, eruption of, 50
k-strategy, 82, 83
Kudzu, 70
Kwashiorkor, 166
Kyoto Protocol, 227

L

Labor Department, 226
Lacey Act of 1990, 230
Lakes, 67–68
  upwelling in, 34
Land, 155–184
  abuse of, 155–156
  conservation of, 156
  legislation on, 228–229
  management of. *See* Land management
  mitigation of problems affecting, 158
  preservation of, 156–157
  reclamation of, 157
  remediation of, 157
  restoration of, 157

Landfills, 109
Land management, 158–182
  agriculture and forestry and, 165–174
  building construction and, 158–161
  flood control and, 178, 182–183
  mineral resources and mining and, 179–182
  solid waste disposal and, 182
  urbanization and, 161–165
  wildlife and, 175–178
Land reclamation, 74
Landslides, El Niño and, 51
Land use mitigation, 74
La Niña, 47
Law of Conservation of Mass, 17
Lead, 125, 130, 158
Legislation, wildlife and, 177
Leopold, Aldo, 223
Lethal dose for 50% (LD50), 104
Life, properties of water contributing to, 34
Life expectancy, 80
Life span, 80
Light
  infrared, 31
  ultraviolet, 31
Lights, energy efficiency of, 25
Limnetic zone, 67
Liquids, 15
Littoral zone, 67
Loam, 39
Lock projects, 140–141
Logarithmic population growth, 80–81
Love Canal, 113
Low-flow utilities, 142
Lung cancer, 130
Lungs, irritation by pollutants, 130

M

Magma, 30, 37
Malaria, 101
Malnourishment, 166–167
Malthus, Thomas, 216
Mangrove swamps, 65
Mantle, convection in, 30
Manufactured capital, 214
Marasmus, 166
Marginal costs, 215–216

Margin of diminishing returns, 216
Marianas Trench, 36
Marine biodiversity, 75
Marine biomes, 64–67
Marine ecosystems, 17
    benthic, 65–66
    pelagic, 64
Market forces, 211
Marsh, George Perkins, 222
Marshes, 67
Marx, Karl, 218
Mass, Law of Conservation of, 17
Matter
    conservation of, 17–22
    living, 15–17
    nonliving, 13–15
    states of, 15
Mead, Lake, 141
Mechanical weathering, 37
Meiosis, 15
Mental illness in cities, 163
Mercury, 125, 130, 158
Mesolimnion, 68
Mesopause, 43
Mesosphere, 43
Metals
    as air pollutants, 124–125
    heavy, as water pollutants, 145
Metamorphic rocks, 38
Methane
    in atmosphere, 42, 119
    global warming and, 47–48
Methylene chloride as indoor air pollutant, 129
Methyl isocyanate, 114
Mid-Atlantic rift, 35
Milankovitch cycles, 46
Mill, John Stuart, 218
Mineral resources, 179
    land management and, 179–182
Minimum viable populations, 69
Mining, 179–182
    environmental consequences of, 182
    land management and, 179–182
Mississippi River, 142, 151–152, 182–183
Mitigation, 235
    of problems affecting land, 158
    of problems affecting wildlife, 177–178

Mitochondria, 16
Mitosis, 15
Molecules, polar, 14
Monoatomic ions, 13–14
Mono Lake, 142
Montreal Protocol, 227
Moral agents, 219
Moral extensionism, 219
Morals, 218–219
Moral subjects, 219
Morbidity, 79
Mortality, 79
Muir, John, 222
Multiuse design for cities, 164
Mutagens, 102
Mutation(s), 58, 68, 102
Mutualism, 59–60
    population size and, 85

N

Natality, 79
National Environmental Policy Act (NEPA) of 1969, 227, 232
National forests, 72
National Forest Service (NFS), 224
National Institutes of Health (NIH), 225
National Oceanic and Atmospheric Association (NOAA), 224
National parks, 72
National Park Service, 226, 229
National Park Service Act of 1916, 229
National Resource Conservation Agency, 224
Natural capital, 213
Natural growth rate, 80
Natural selection, 58–59
Nature Conservancy, 157
Nazca Plate, 36
Neolithic Revolution, 93, 202
Neo-Malthusians, 222
Neon in atmosphere, 42, 119
Neritic zone, 64
Nervous system, 16
Neurotoxins, 101–102
Neutron capture, 194
Niche, 57
Nihilists, 221
Nimbus clouds, 44

Nitrate ions, 19, 20, 24
Nitrification, 19, 20
Nitrite ions, 19, 20, 24
Nitrogen cycle, 19–20, 24–25
Nitrogen fixation, 19, 20
Nitrogen in atmosphere, 42, 119
Nitrogen oxides
    as pollutants, 121–122
    reducing emission of, 149
Nitrous oxide
    in atmosphere, 43, 119
    greenhouse effect and, 48
Non-criteria pollutants, 126
Nonpoint source pollutants, 143
Nonrenewable resources, 213
Nontoxic materials in building
    construction, 159–161
North American Plate, 36
Northern coniferous forests, 63
Northern jet stream, 33
Northern spotted owl, 236
Nuclear fuel cycle, 207
Nuclear power, 193–198, 206–208
    pollution risks of, 197–198
    radioactivity and, 193, 196–197
Nuclear reactions, 194–195
Nuclear reactors, 195–196
Nuclear Regulatory Commission (NRC), 226
Nucleus, 15
Nutrient cycles
    carbon, 18–19, 24–25
    nitrogen, 19–20, 24–25
    phosphate, 20
    sulfur, 21–22
    water (hydrologic), 22
Nutrition, 165–167

O

Occupational Safety and Health Act (OSHA) of 1970, 227
Occupational Safety and Health Administration (OSHA), 226, 232
Ocean, 41–42
Ocean currents, 41
    properties of water contributing to, 34
Ocean Dumping Ban Act of 1988, 228
Oceanic-continental convergence, 35–36
Oceanic-oceanic convergence, 36
Oceans
    climate change and, 48
    convection in, 30
    global warming and, 50–51
    sea floor spreading and, 35
    tides in, 41–42
Office of Environmental Management, 225
Office of Surface Mining, Reclamation, and Enforcement, 226
Ogallala Aquifer, 151
Ohio River, locks on, 140–141, 204
O-horizon, 39
Oil, 193, 205
Oligotrophic lakes and ponds, 67
Oligotrophic rivers, 144
Open space in cities, 165
Ordered population distribution, 57
Organic wastes, toxic, as water pollutants, 146
Organophosphates, 169
Organ systems, 16
Overcrowding, population size and, 85
Overnutrition, 166
Overpopulation, 94
Overshoot, 84
Oxidants, photochemical, as air pollutants, 126
Oxidation-reduction reactions, 17
Oxygen
    in atmosphere, 42, 119
    formation of, 42
Oxygen-demanding wastes, 121
    as water pollutants, 143–144
Oxygen sag, 144
Ozone
    in atmosphere, 43, 119
    depletion of, 132

P

Pacific Plate, 35, 36
Para-dichlorobenzene as indoor air pollutant, 129
Paralytic shellfish poisoning, 146
Parasitism, 59
Park Service Act of 1916, 156
Particles, suspended, as water pollutants, 144

Particulate matter as air pollutant, 124
Passive solar heating, 198
Pathogens, 99
    as water pollutants, 143
Pelagic marine ecosystems, 64
Pellagra, 167
Permafrost, melting of, 48
Persian Gulf wars, 205
Persistence, chemical, 106
Pesticides, 101, 168–169
    as water pollutants, 146
pH, 15
Philippine Plate, 36
Philosophies, 221–222
Phosphate cycle, 20
Photic zone, 64
Photochemical oxidants as air pollutants, 126
Photons, 187–188
Photosynthesis, 16, 18, 19, 55, 187
Photosynthetic algae, 39
Photosynthetic cyanobacteria, 42
Photovoltaic cells, 198–199
Physical factors, erosion due to, 168
Phytoplankton, 64, 76
Pinchot, Gifford, 222, 223
Pioneer species, 60
Plants
    acid rain and, 131
    as pathogens, 100
Plasmodium, 59
Plate tectonics, 34–37
    continental drift theory and, 34–35
    earthquakes and, 36–37
    plate boundaries and, 35–36
    rate of movement and, 35
Plato, 222
Point source pollutants, 143
Poivre, Pierre, 222
Polar cells, 32
Polar molecules, 14
Political economics, 218
Pollution. *See also* Air pollution; Water pollution
    at Love Canal, 113
    for nuclear power production, 197–198
    preventing, 107
    Superfund and, 113, 235–236
Polyatomic ions, 13–14

Ponds, 67–68
Poor countries, sending waste to, 109–110
Population density, population size and, 84–86
Population growth, 79–82
    biotic potential and, 82
    calculating rates of, 81–82
    carrying capacity and, 83–84
    rates of, 80–82
Populations, 16
    abundance of, 56
    diversity in, 57
    fragmentation of, 72
    human. *See* Human populations
    identifying health risks in, 103–104
    minimum viable, 69
    natural selection and, 58–59
    niches and competition and, 57
    size of, factors affecting, 84–86
    symbiosis and, 59
    tolerance limits and, 58
Prairies, 62
Precipitation, 44
Precipitation reactions, 17–18
Preservation, 233–234
Preservation of land, 156–157
Pressure, atmospheric, 44
Pressurized water reactors (PWRs), 195
Primary productivity, 23, 55
Primary succession, 60
Productivity, 23, 55
Pronatalist pressures, population size and, 86
Protein synthesis, 15
Protists, 100
Protons, 15, 17
Proven resources, 213
Public lands, overgrazing, 171–172
P waves, earthquakes and, 37

R

Racism, environmental, 220
Radiation, 29
Radioactivity, 193
    health risks due to, 196–197
Radon
    as air pollutant, 128
    in building construction, 158

Rads, 196
Rain, 138–139
    acid, 21, 130–131
    in deserts, 139
    rain shadows and, 138
Rainforests
    deforestation of, 174
    temperature, 64
    tropical, 64
Rain shadows, 138
Random population distribution, 57
    population size and, 85
Range management, 73
Recharge zones, 138
Reclamation, 234
    of land, 157
    mining and, 181–182
Recoverable resources, 213
Recovery zone, 144
Recycled materials in building construction, 159–161
Recycling, 24–25, 111–112
Red algae, 65
Red tide, 146
Re-filtration of wastewater, 146
Relativists, 221
Remediation, 234
    of land, 157
Rems, 196
Renewable energy sources, 198–201
Renewable resources, 213
Renewal of communities, 60
Reservoirs, 141
Residential energy use, 203
Residential water use, 139, 140
Resource Conservation and Recovery Act (RCRA) of 1976, 229
Respiration, 16
Restoration, 234
    of land, 157
Reuse, reducing volume of waste through, 111
Reverse osmosis, 139
Richter Scale, 37
Rickets, 166
Rift valleys, 35
Rill erosion, 40, 167
Risk assessment, 230–233
    to determine policy or action, 232–233

risk perception and, 231–232
River bank management, 142
River channelization, reducing, 149–150
Rivers, 67
RNA, 15
Rock cycle, 37–38
Rocks
    igneous, 37
    metamorphic, 38
    sedimentary, 37–38
Roosevelt, Theodore, 157, 222, 223
r-strategy, 82, 83

S

Sacramento River, 142
Safe drinking Water Act of 1974, 228
St. Helens, Mount, eruption of, 49
Salinization
    of soil, 41
    of water, 138
Salt marshes, 66
Saltwater intrusion, 138
San Andreas fault, 36
Sand, 39
Savannahs, 62
Schumacher, E. F., 218
Scientific value, 220
S-curve, 82, 83
Scurvy, 167
Sea floor spreading, 35
Secondary productivity, 23, 55
Secondary succession, 61
Second Law of Thermodynamics, 22, 188
Sedimentary rocks, 37–38
Selective cutting, 173
Septic zone, 144
Sheet erosion, 40, 167
Sierra Club, 222
Silent Spring (Carson), 112–113
Silt, 39
Siltation, decreasing, 143
Silt runoff, decreasing, 149
Sinclair, Upton, 223
Smith, Adam, 217
Snake River, 141
Social capital, 214
Social justice value, 220
Society, 211–236

economic forces in, 211–218
environmental ethics and, 218–219
environmental laws and, 226–230
government agencies and, 224–226
government organization and, 223–226
history of environmental movement and, 222–223
Soil, 38–41
    acid rain and, 131
    composition of, 39
    conservation of, 41
    erosion, 29, 40–41, 167–168, 183–184
    fauna of, 39
    formation of, 38–39
    profiles of, 39–40
    toxins in, treatment of, 107
Soil Conservation Act of 1935, 40, 229
Soil Erosion Service, 40
Soil profiles, 39–40
Solar energy, 30–32
    greenhouse effect and, 47–48
Solar heating, 198–199
Solids, 15
Solid waste disposal
    environmental consequences of, 182
    land management and, 156, 182
Solubility, chemical, 106
Somatic cells, radioactivity and, 196–197
South American Plate, 36
Soviet nuclear devices, 208
Speciation, 58
Species, 16
    endangered, 72
    pioneer, 60
    threatened, 72
Species biodiversity, 69–71, 75
    benefits of, 69
    endangered species' traits and, 70–71
    minimum viable populations and, 69
    threats to species abundance and, 70
Species Conservation Act of 1966, 230
Specific heat of water, climate and, 33
Spring turnover, 67

Stability of communities, 60
States of matter, 15
Storm runoff, separation from septic treatment, 149
Stratopause, 43
Stratosphere, 43
Stratus clouds, 44
Streams, 67
Subduction zones, 35
Subsidence of aquifers, 138
Subsidized housing in cities, 164–165
Subsistence farming, 170
Substance abuse in cities, 163
Subsurface mining, 180
Subtropical jet stream, 33
Succession of communities, 60–61
Sulfate ions, 21, 22
Sulfide ions, 22
    as water pollutants, 145
Sulfite ions, 21
Sulfur cycle, 21–22
Sulfur dioxide, 21, 133–134
Sulfuric acid, 21
Sulfurous acid, 21
Sulfur oxides
    as pollutants, 122
    reducing emission of, 149
Superfund, 113, 235–236
Supply and demand, 211–212
    creating demand using policy, 212
    technology's effect on, 212–213
Supratidal zone, 66–67
Surface mining, 180
Surface Mining Control and Reclamation Act (SMCRA) of 1977, 157, 181, 228
Survival economy, 218
Survivorship, 80, 87
Suspended particles as water pollutants, 144
Sustainable farming, 170–171
Swamps, 67
S waves, earthquakes and, 37
Swideen agriculture, 173–174
Symbiosis, 59
    population size and, 84
Synergy, chemical, toxicity and, 106

T

Taiga, 63
Tambora, eruption of, 50
Taylor Grazing Act of 1934, 229
Technological optimists, 222
Technology, supply and demand and, 212–213
Temperate deciduous forests, 63
Temperate rainforests, 64
Temperature. *See also* Heat
 of atmosphere, 44
Tennessee Valley Authority (TVA), 226
Teratogens, 102
Terrestrial biomes, 61–64
Tetrachlorethylene as indoor air pollutant, 129
Thermal pollution of water, 146
Thermocline, 68
Thermodynamics
 heat transfers and, 22–23. *See also* Convection
 laws of, 22
 Second Law of, 188
Thermohaline circulation, 30
Thermohaline currents, 34, 41
Thermohaline gradient, 30
Thermophilic bacteria, 75
Thermosphere, 43
Thoreau, Henry David, 222
Threatened species, 72
Three Gorges Project, 152
Three Mile Island, 206–207
Thunderstorms, 45
Tibetan plateau, 36
Tidal bulges, 42
Tidal energy, 201
Tides, 41–42
Tissues, 16
Tolerance limits, 58
Tornadoes, 45
Total growth rate, 80
Toxic colonialism, 220
Toxicity, 105–107
 environmental factors affecting, 106–107
 human factors affecting, 105
Toxic materials, 101–102, 105–112
 in building construction, 158
 legislation on, 229
 pesticides, 168–169
 types of toxicity and, 101–102
Toxic wastes
 organic, as water pollutants, 146
 removal from water, 148
Toxins
 in atmosphere, treatment of, 106–107
 in soil, treatment of, 107
 in water, treatment of, 107
Trade winds, 32, 41
Traffic in cities, 162
Tragedy of the commons, 155
Trans-Alaska pipeline, 61
Transform boundaries, 36
Transportation in cities, 162, 164
Trenches, 36
Trophic levels, 23, 56, 76
Tropical rainforests, 64
Tropic of Cancer, 139
Tropic of Capricorn, 139
Tropopause, 43
Troposphere, 43
True cost, 217
True-cost pricing, 217
Tundra, 61

U

Ultraviolet (UV) light, 31, 42
Unconfined aquifers, 138
Undernourishment, 165
Undeveloped countries, energy use by, 203
Union Carbide, 113–114
United States Geological Survey (USGS), 226
United States Public Utility Regulatory Policy Act (PURPA) of 1978, 228
Universalists, 221
Upwelling in lakes, 34
Uranium processing, 181
Urbanization, 161–165
 causes for, 161–162
 city planning and, 164–165
 current problems related to, 162–163
 land management and, 156, 161–165
 trends toward, 161
 urban renewal and, 163
Utilitarians, 221

V

Vaccinations, 115
Vaccines, 101
Value, 218–219
    aesthetic, 220
    cultural, 220
    educational, 220
    instrumental, 219
    intrinsic (inherent), 219–220
    scientific, 220
    social justice, 220
Vaporization, energy of, of water, climate and, 33
Variable costs, 215
Vectors, 99
Viruses, 99–100
Vitamin A, 167
Volatile organic compounds (VOCs)
    as air pollutants, 123–124
    in building construction, 158
Volcanic activity, 36
Volcanic air pollution, 133–134
Volcanic eruptions, 37, 46
Volcanism, 30, 50

W

Warm fronts, 45
Waste. *See also* Toxic wastes
    composting, 24–25, 112
Waste disposal, 108–112
    methods of, 109–110
    reducing volume of waste and, 110–112
    waste stream and, 108–109
Waste stream, 108–109, 182
    input into, from building construction, 159
    reducing volume of waste in, 110–112
Waste treatment, 107–108
Wastewater treatment, 147–150
    Clean Water Act and, 148–149
    post-use, 147
    pre-use, 146–147
    toxic waste removal and, 148
Water
    erosion due to, 167
    polarity of, 14
    properties of, 33–34
    sea, desalinization of, 139
    as solvent, 34
    toxins in, treatment of, 107
    waste. *See* Wastewater treatment
Water currents, properties of water contributing to, 34
Water cycle, 22, 137
Water diversion projects, 141–142
Water erosion, 40
Waterlogging
    erosion due to, 168
    of soil, 41
Water pollution, 143–146
    in cities, 162–163
    control of, 149–150
    legislation on, 228
    pollutant types and, 143–146
Water Pollution Control Administration, 228
Water resources, 137–152
    availability of, 137–138
    compartments of, 137–138
    conservation of, 142–143
    groundwater as, 138
    management of, 140–142
    pollution of. *See* Water pollution
    precipitation and. *See* Precipitation; Rain
    use of, 139–140, 150
Watershed management, 142
Water table, 138
Water treatment, 24
Water vapor in atmosphere, 42, 44, 119
Waterways, 155
Watts, 189
Weather, 44–46. *See also* Climate
    air pollution and, 132–133
    clouds and, 44–45
    fronts and, 45
    hurricanes and, 46
    physical traits contributing to, 44
Weather fronts, 45
Weathering
    chemical, 37–38
    mechanical, 37
Wegener, Alfred, 34
Wells, 180
    artesian, 138

Wetlands, 67, 155, 178
    protection of, 73
Wet scrubbing, 122
Wilderness Act of 1964, 72, 156–157, 229
Wilderness areas, 72
Wilderness Society, 223
Wildlife, 175–178
    causes of extinction and, 176–177
    land management and, 175–178
    mitigation of problems affecting, 177–178
    traits of endangered species and, 175–176
Wildlife refuges, 73
Wind energy, 199
Wind erosion, 40, 167
Woburn, Massachusetts, 235–236
Wolves, restoring to Yellowstone, 74–75
World views, diverse, 220–222

Y

Yellowstone area succession, 61
Yellowstone hotspot, 49–50
Yellowstone hotsprings, 75
Yellowstone National Park, 74–75
Yosemite National Park, 222
Yucca Mountain, 207

Z

Zero population growth (zpg) (concept), 88
Zero Population Growth (ZPG) (organization), 223
Zooplankton, 64, 76
Zoos, 74

REA's Test Preps
The Best in Test Preparation

- REA "Test Preps" are **far more** comprehensive than any other test preparation series
- Each book contains full-length practice tests based on the most recent exams
- **Every** type of question likely to be given on the exams is included
- Answers are accompanied by **full** and **detailed** explanations

REA publishes hundreds of test prep books. Some of our titles include:

Advanced Placement Exams (APs)
Art History
Biology
Calculus AB & BC
Chemistry
Economics
English Language & Composition
English Literature & Composition
European History
French Language
Government & Politics
Latin
Physics B & C
Psychology
Spanish Language
Statistics
United States History
World History

College-Level Examination Program (CLEP)
Analyzing and Interpreting Literature
College Algebra
Freshman College Composition
General Examinations
History of the United States I
History of the United States II
Introduction to Educational Psychology
Human Growth and Development
Introductory Psychology
Introductory Sociology
Principles of Management
Principles of Marketing
Spanish
Western Civilization I
Western Civilization II

SAT Subject Tests
Biology E/M
Chemistry
French
German
Literature
Mathematics Level 1, 2
Physics
Spanish
United States History

Graduate Record Exams (GREs)
Biology
Chemistry
Computer Science
General
Literature in English
Mathematics
Physics
Psychology

ACT - ACT Assessment
ASVAB - Armed Services Vocational Aptitude Battery
CBEST - California Basic Educational Skills Test
CDL - Commercial Driver License Exam
CLAST - College Level Academic Skills Test
COOP, HSPT & TACHS - Catholic High School Admission Tests
FE (EIT) - Fundamentals of Engineering Exams
FTCE - Florida Teacher Certification Examinations

GED
GMAT - Graduate Management Admission Test
LSAT - Law School Admission Test
MAT - Miller Analogies Test
MCAT - Medical College Admission Test
MTEL - Massachusetts Tests for Educator Licensure
NJ HSPA - New Jersey High School Proficiency Assessment
NYSTCE - New York State Teacher Certification Examinations
PRAXIS PLT - Principles of Learning & Teaching Tests
PRAXIS PPST - Pre-Professional Skills Tests
PSAT/NMSQT
SAT
TExES - Texas Examinations of Educator Standards
THEA - Texas Higher Education Assessment
TOEFL - Test of English as a Foreign Language
USMLE Steps 1,2,3 - U.S. Medical Licensing Exams

For information about any of REA's books, visit www.rea.com

Research & Education Association
61 Ethel Road W., Piscataway, NJ 08854
Phone: (732) 819-8880

INSTALLING REA'S TESTWARE®

System Requirements

Pentium 75 MHz (300 MHz recommended), or a higher or compatible processor; Microsoft Windows 98 or later; 64 MB Available RAM; Internet Explorer 5.5 or higher.

Installation

1. Insert the AP Environmental Science TestWare® CD-ROM into the CD-ROM drive.
2. If the installation doesn't begin automatically, from the Start Menu, choose the Run command. When the Run dialog box appears, type d:\setup (where D is the letter of your CD-ROM drive) at the prompt and click ok.
3. The installation process will begin. A dialog box proposing the directory "Program Files\REA\AP_EnvironmentalScience" will appear. If the name and location are suitable, click ok. If you wish to specify a different name or location, type it in and click ok.
4. Start the AP Environmental Science TestWare® application by double-clicking on the icon.

REA's AP Environmental Science TestWare® is easy to learn and use. To achieve maximum benefits, we recommend that you take a few minutes to go through the on-screen tutorial on your computer. The "screen buttons" are also explained there to familiarize you with the program.

SSD accommodations for students with disabilities

Many students qualify for extra time to take the AP Environmental Science exam, and our TestWare® can be adapted to accommodate your time extension. This allows you to practice under the same extended time accommodations that you will receive on the actual test day. To customize your TestWare® to suit the most common extensions, visit our Website at *www.rea.com/ssd*.

Technical Support

REA's TestWare® is backed by customer and technical support. For questions about installation or operation of your software, contact us at:

Research & Education Association
Phone: (732) 819-8880 (9 a.m. to 5 p.m. ET, Monday–Friday)
Fax: (732) 819-8808
Website: *www.rea.com*
E-mail: info@rea.com

Note to Windows XP Users: In order for the TestWare® to function properly, please install and run the application under the same computer-administrator level user account. Installing the TestWare® as one user and running it as another could cause file access path conflicts.